A Modern Approach to Computational Biology

A Modern Approach to Computational Biology

Edited by Alfie Davies

MURPHY & MOORE
www.murphy-moorepublishing.com

Published by Murphy & Moore Publishing,
1 Rockefeller Plaza,
New York City, NY 10020, USA
www.murphy-moorepublishing.com

A Modern Approach to Computational Biology
Edited by Alfie Davies

International Standard Book Number: 978-1-63987-002-8 (Hardback)

Cataloging-in-Publication Data

A modern approach to computational biology / edited by Alfie Davies.
    p. cm.
Includes bibliographical references and index.
ISBN 978-1-63987-002-8
1. Computational biology. 2. Bioinformatics. 3. Systems biology. I. Davies, Alfie.
QH324.2 .M63 2022
570.285--dc23

TABLE OF CONTENTS

PREFACE

This book was inspired by the evolution of our times; to answer the curiosity of inquisitive minds. Many developments have occurred across the globe in the recent past which has transformed the progress in the field.

Computational biology refers to the science of using biological data to engineer algorithms or models for analyzing biological systems and relationships. It is one of the interdisciplinary approaches to the life sciences that draw from quantitative disciplines such as mathematics and information science. The introduction of a large amount of data in bioinformatics, molecular biology, and genomics makes computational biology a prominent discipline. Some of the subfields of computational biology are computational anatomy, biomodeling, genomics, neuroscience, pharmacology, evolutionary biology, cancer biology and neuropsychiatry. Some of the diverse topics covered in this book address the varied branches that fall under this category. Different approaches, evaluations, methodologies and advanced studies on computational biology have been included herein. This book includes contributions of experts and scientists which will provide innovative insights into this field.

This book was developed from a mere concept to drafts to chapters and finally compiled together as a complete text to benefit the readers across all nations. To ensure the quality of the content we instilled two significant steps in our procedure. The first was to appoint an editorial team that would verify the data and statistics provided in the book and also select the most appropriate and valuable contributions from the plentiful contributions we received from authors worldwide. The next step was to appoint an expert of the topic as the Editor-in-Chief, who would head the project and finally make the necessary amendments and modifications to make the text reader-friendly. I was then commissioned to examine all the material to present the topics in the most comprehensible and productive format.

I would like to take this opportunity to thank all the contributing authors who were supportive enough to contribute their time and knowledge to this project. I also wish to convey my regards to my family who have been extremely supportive during the entire project.

Editor

3D structure prediction of replication factor C subunits (RFC) and their interactome in *Arabidopsis thaliana*

Mohamed Ragab Abdel Gawwad[1], **Jasmin Šutković**[1], **Emina Zahirović**[1], **Faruk Berat Akcesme**[1], **Betul Akcesme**[1], **Lizhi Zhang**[2]

[1]Genetics and Bioengineering department, International University of Sarajevo, Ilidza, 71220 Bosnia and Herzegovina

[2]Department of Molecular Genetics, The Ohio State University, 484 West 12th Avenue, Columbus, OH 43210, USA

E-mail: mragab@ius.edu.ba

Abstract

DNA stress can causes potentially spontaneous genome damage during DNA replication process. Proteins involved in this process are DNA-dependent ATPases, required for replication and repair. In this study the 3-D structure of RFC protein subunits in *Arabidopsis thaliana*: RFC1, RFC2, RFC3, RFC4 and RFC5 are predicted and confirmed by Ramachadran plot. The amino acid sequences are highly similar to the sequences of the homologous human RFC 140-, 37-, 36-, 40-, and 38 kDa subunits, respectively, and also show amino acid sequence similarity to functionally homologous proteins from *E. coli*. All five subunits show conserved regions characteristic of ATP/GTP-binding proteins and have significant degree of similarity among each other. The segments of conserved amino acid sequences that define a family of related proteins have been identified. RFC1 is identical to CDC44, a gene identified as a cell division cycle gene encoding a protein involved in DNA metabolism. Subcellular localization and interactions of each protein RFC protein subunit is determined. It subsequently became clear that RFC proteins and their interactome have functions in cell cycle regulation and/or DNA replication and repair processes. In addition, AtRFC subunits are controlling the biosynthesis of salicylic and salicylic acid-mediated defense responses in Arabidopsis.

Keywords DNA; RFC protein; 3-D structure; structure prediction; interactome; *Arabidopsis thaliana*.

1 Introduction

Arabidopsis thaliana is widely used for studying plant sciences, including genetics, evolution, population genetics, and plant development. Although *A. thaliana* has little direct significance for agriculture, it has several traits that make it a useful model for understanding the genetic-, cellular-, and molecular biology of flowering plants. The small size of its genome makes *A. thaliana* useful for genetic mapping and sequencing with about 157 mega base pairs and five chromosomes. *A. thaliana* has one of the smallest genomes among plants. It was the first plant genome to be sequenced, completed in year 2000 by the Arabidopsis Genome

Initiative. The most up-to-date version of the *A. thaliana* genome is maintained by the Arabidopsis Information Resource (The Arabidopsis Genome Innovative, 2000).

Post-genomic research, such as metabolomics, has also provided useful insights to the metabolism of this species and how environmental perturbations can affect metabolic processes (Alberts et al., 2002). DNA replication depends on the coordinated action of numerous multiprotein complexes (Martinez-Antonio, 2011; Rahman et al., 2013). At the simplest level, it requires an initiator to establish the site of replication initiation, a helicase to unwind DNA, a polymerase to synthesize new DNA, and machinery to process the Okazaki fragments generated during discontinuous synthesis. Much is known about the DNA replication machinery in e.g. *S.cerevisiae* and animal model systems, but relatively little is known about the apparatus in plants. To understand plant DNA replication components it is necessary to combine already published experimental information and new bioinformatics analysis of genomic sequence data and therefore to examine the core of DNA replication machinery in the model plants of Arabidopsis species. In recent years there has been increased interest in plant DNA replication and repair machinery and in using plants as models for understanding the same processes in eukaryotes (Shultz et al., 2007). The DNA replication process represents a source of DNA stress that causes potentially spontaneous genome damage (Takahashi et al., 2010). DNA damage and replication errors might originate from DNA stress provoked either by exogenous and/or endogenous sources. The latter includes the replication process itself that necessitates the cooperation of many different proteins in a highly complex manner. Errors arisen during replication are preferentially repaired through homologous recombination between the replicated sister chromatids that lay in close proximity thanks to cohesion. The process of cohesion establishment is intimately connected with DNA replication. Cohesion depends on acetyltransferase (Eco1/Ctf7), which acetylates two lysine residues on the ATPase head domain of SMC3. Eco1/Ctf7 interacts physically and genetically with the PCNA and the RFC is observed to travel along the DNA with replication fork, suggesting that replication fork progression and sister chromatid cohesion are coupled events (Takahashi et al., 2010; Ashwin et al., 2011).

This model is supported by the observation that cohesion defects are caused by mutations in replisome components, such as DNA polymerase a-binding protein Ctf4, the Chl1 helicase and RFC components, which includes RFC2, RFC3, RFC4, and RFC5 protein subunits. Due to their sessile nature and need of sunlight, plants are particularly exposed to environmental genotoxins, which lead directly or indirectly, via generation of ROS, to DNA lesions including single and double strand breaks. Double strand breaks are particularly critical lesions, because if unrepaired they lead to major karyotypic instability and cell death (Kozak et al., 2009). RFC, a clamp-loader complex consisting of five different subunits, binds DNA at the template-primer junctions and displaces polymerase to terminate DNA primer synthesis. The binding of RFC to DNA creates a loading site for recruiting the DNA sliding clamp PCNA. RFC is an AAA+-type ATPase that requires ATP hydrolysis for opening and closing PCNA around DNA during DNA replication, repair, and recombination. Genetic studies in *A. thaliana* have identified many DNA replication- and repair related proteins are involved in the regulation of transcriptional gene silencing (Liu et al., 2010). Homotrimeric PCNA is considered an essential component in recruiting the general mediators for chromatin imprinting of epigenetic markers (Takashi et al., 2011). RFC can bind to DNA template-primer junction and load the PCNA clamp onto DNA with the assistance of ATP. PCNA loading recruits DNA polymerase to the site of DNA synthesis. Five subunits of RFC were identified as one large subunit RFC140/RFC1 and four small subunits RFC37/RFC2, RFC36/RFC3, RFC40/RFC4 and RFC38/RFC5, and have been found in all eukaryotes. RFC plays an essential role in DNA replication, DNA damage repair and check-point control during cell cycle progression. Recently, three RFC-like complexes (RLCs), namely, Rad24 RLC, Ctf18-RLC and Elg1-RLC, have been identified. Each RLC is made up of the four small subunits of the archetypal RFC, but the large subunit, RFC1, is replaced with an RFC related protein (Takashi et al., 2011).

Replication factor C (RFC), which is composed of five subunits, is an important factor involved in DNA replication and repair mechanisms. In DNA replication and repair processes, replication factor C (RFC) plays an important role as a clamp loader to load a clamp protein, PCNA, on the DNA strand in an ATP-dependent manner. After loading PCNA at a primer-template junction, DNA polymerase δ and ε bind to the protein-DNA complex and are tethered to the junction through protein-protein interaction. All five RFC subunits are found in all eukaryotes (Cullmann et al., 1995; Bunz et al., 1993; Li and Burgers, 1994; Noskov et al., 1994; Gary and Burgers, 1995; Gray and MacNeill, 2000).

RFC1 is the largest subunit of the RFC complex in Arabidopsis. Mutations in RFC1 lead to developmental defects and earlier flowering (Takashi et al., 2011). RFC1 is an important mediator of transcriptional gene silencing (TGS) DNA replication, DNA repair, homologues replication, and telomere length regulation. TGS controls the expression of transposable elements and of endogenous genes containing promoter repeats, and it is associated with increased DNA methylation.

RFC2 is a cellular component of replication factor C complex, found in *Arabidopsis thaliana* structure and twenty plants more. It is involved in biological ATP catabolic processes and in DNA replication. Molecular function of RFC2 is ATP-, DNA- and nucleotide binding mainly resulting in ATPase activity, DNA clamp loader activity, nucleoside-triphosphatase activity (Vladimir et al., 2008).

RFC3 is one of the small subunits of the RFC complex, originally purified from the HeLa cells, that is essential for the in vitro replication of Simian virus 40. RFC3 and other subunits, compromising the function of the protein complex, and leading to cell proliferation defects in the leaves and roots of Arabidopsis. Partial dysfunction in rfc3-1 leads to smaller plant size due to the reduced number of cells, suggesting defects in replication. Therefore, RFC3 plays an essential role in the process of cell proliferation. Also RFC3 is involved in ATP-, DNA- and nucleotide binding (Majka and Burgers, 2004).

RFC4 is located in DNA replication factor C complex, in nucleolus. It is involved in biological processes such as ATP catabolic process, embryo development ending in seed dormancy, and metabolic process. A molecular function of RFC4 is ATP-, DNA- and nucleotide binding, ATPase activity, DNA clamp loader activity, nucleoside-triphosphatase activity (Majka and Burgers, 2004).

RFC5 is a protein encoded by the RFC5 gene in humans and it is a part of the small subunits of the RFC complex originally. In *Arabidopsis thaliana* it encodes a protein with high homology to the RFC3 of yeast and other eukaryotes. RFC5 biological functions are mainly ATP catabolic process, metabolic process, and negative regulation of defense response. This protein is also involved in the processes of ATP-, DNA- and nucleotide binding, having the molecular functions of ATPase activity, DNA clamp loader activity and nucleoside-triphosphatase activity (Liu et al., 2010).

2 Materials and Methods
2.1 RFC family sequences modeling and multiple sequence alignment
The FASTA sequences of RFC family proteins (RFC1, RFC2, RFC3, RFC4 and RFC5) were obtained from NCBI databases. The Gene ID codes from The Arabidopsis Information Resource (TAIR) database are shown in Table 1.

The multiple sequence alignment searches for RFC proteins were performed with the ClustalW2 program. Default parameters were applied and aligned sequences were executed using this software.
2.2 Phylogenetic tree construction
Phylogenetic tree was constructed using Maximum likelihood method in Robust Phylogenetic Analysis software, where phylogenetic relationships between amino acids of AtRFC subunits were reconstructed and

analyzed (Dereeper et al., 2008).

Table 1 RFC subunits and their gene ID codes from TAIR

RFC proteins	Gene IDs
RFC1	AT5G22010
RFC2	AT1G63160
RFC3	AT1G77470
RFC4	AT1G21690
RFC5	AT5G27740

2.3 3-D structure prediction and conformation

Due to the missing information about the 3-D structure of AtRFC subunits in Protein Data Bank, Pdb files had been generated based on homology moduling using CPH models 3.2 servers (Nielsen et al., 2010). The 3-D structures for all AtRFCs were confirmed by Ramachadran plots using RAMPAGE program (Lovell et al., 2002).

2.4 Domain search and interaction prediction

The protein's genetically mobile domain annotations were obtained using *Pfam* database (Finn et al., 2010). Arabidopsis Interactions Viewer was used for the protein- protein interaction in order to find the partners of AtRFC subunits (Arabidopsis interaction viewer). Subcellular localization for each protein was predicted in WoLF PSORT program (Horton et al., 2007).

3 Result and Discussion

3.1 3-D structure prediction

The three-dimensional (3-D) structure details of proteins are of major importance in providing insights into their molecular functions. Replication factor C (RFC) is a five-subunit DNA polymerase accessory protein that functions as a structure-specific, DNA-dependent ATPase in *A.thaliana*. 3-D structure of factor C subunits; AtRFC1, AtRFC2, AtRFC3, AtRFC4and AtRFC5 were predicted. Their 3-D structure of the five subunits showed highly structural similarities. Additionally outputs have been confirmed by Ramachadran plots.

All five subunits show conserved regions characteristic of ATP/GTP-binding proteins and also have a significant degree of similarity among each other (Fig. 1).

3-D structure   **Ramachadran Plot**

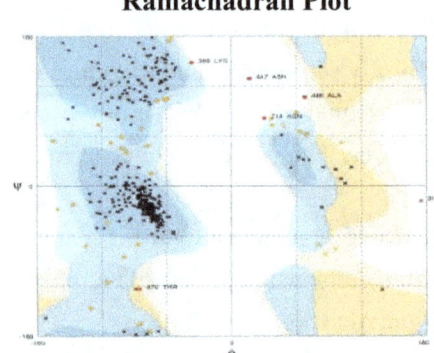

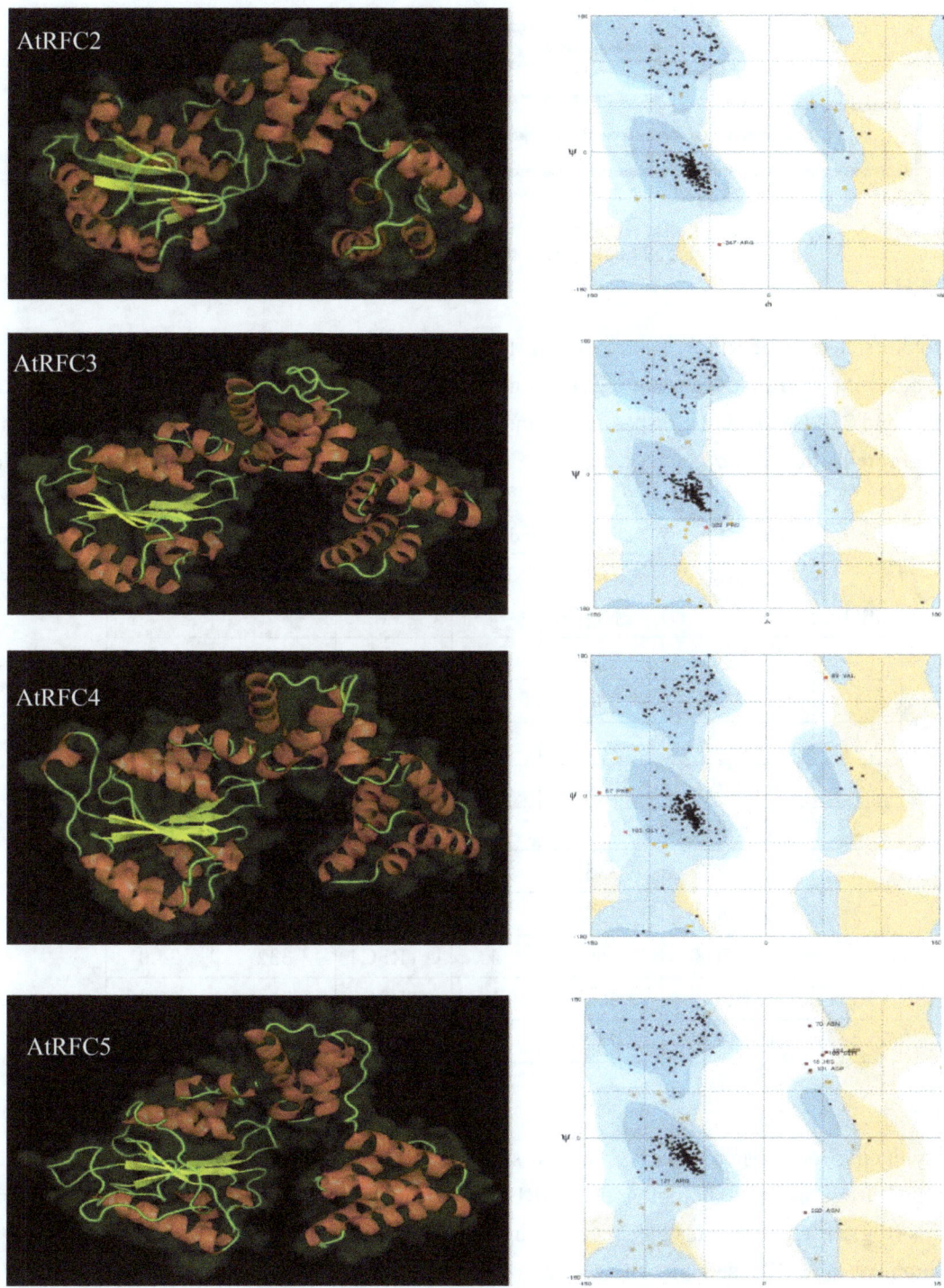

Fig. 1 Predicted 3-D structure of RFC homologous proteins.

3.2 Multiple sequence alignment and phylogenetic tree

Aligned sequences showed substantially significant homology, and constructed phylogenetic tree represents strong relationship between proteins in this family which play their major role as DNA-dependent ATPase in replication and repair systems (Fig. 2).

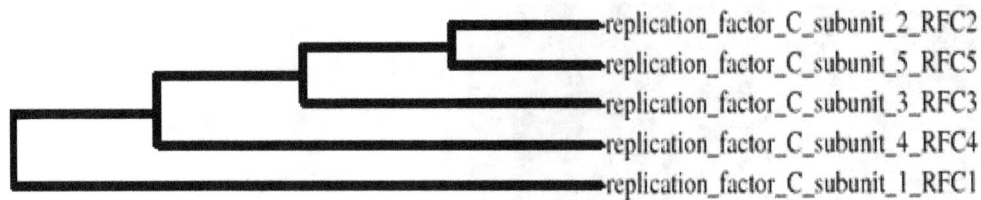

Fig. 2 Phylogenetic tree of AtRFC subunits in *A. Thaliana*.

The protein sequence of RFC1, RFC2, RFC3, RFC4, and RFC5 were obtained from the NCBI sequence database. The amino acid sequences are highly similar to the sequences of the homologous human RFC 140-, 37-, 36-, 40-, and 38 kDa subunits, respectively, and also show amino acid sequence similarity to functionally homologous proteins from *E. coli* and S. *cerevisiae*. The sequence alignment scores are shown in Table 2.

Table 2 ClustaW2 alignment and similarities among AtRFC subunits.

SEQA	NAME	LENGTH	SEQB	NAME	LENGTH	SCORE
1	ATRFC1	956	2	ATRFC2	333	19.0
1	ATRFC1	956	3	ATRFC3	369	21.0
1	ATRFC1	956	4	ATRFC4	332	22.0
1	ATRFC1	956	5	ATRFC5	354	16.0
2	ATRFC2	333	3	ATRFC3	369	35.0
2	ATRFC2	333	4	ATRFC4	332	37.0
2	ATRFC2	333	5	ATRFC5	354	23.0
3	ATRFC3	369	4	ATRFC4	332	34.0
3	ATRFC3	369	5	ATRFC5	354	23.0
4	ATRFC4	332	5	ATRFC5	354	23.0

The phylogenic relationships among the five subunits revealed that AtRFC4 and AtRFC1 subunits had been diverged from the two homologous (AtRFC2, AtRFC5) and AtRFC3 with reserved evolutionary relationships (Fig. 2). This confirms the similarity in function of AtRFC subunits.

The alignment of AtRFC subunits showed important conserved regions; RFC box I (ligase homology) and boxes II (ATP/GTP binding region) and III (Phosphate binding loop), which are conserved in all RFC subunits. The most conserved motif is RFC box III, forming a phosphate-binding loop (P loop, also known as Walker A) with the consensus sequence GxxxxGK(S/T). This loop usually contains additional glycines and prolines and has the consensus sequence PHUUUYGPPGTGKT(S/T), where U stands for a bulky aliphatic residue such as I, L, V or M. Substituting the third Gly for Asp in RFC3-1 mutant may affect the interaction between RFC3 and other subunits. The conserved regions are shown in Fig. 3 (Horton et al., 2007).

```
AtRFC1   MSDIRRWFMKAHERGNGSAPRSTSSRAGPVRNAAETAPIKSEQASEDLETADRRRTSKYF 60
AtRFC2   ------------------------------------------------------------
AtRFC3   ------------------------------------------------------------
AtRFC4   ------------------------------------------------------------
AtRFC5   ------------------------------------------------------------

AtRFC1   GRDKTKVRDEKEVEAIPARRKLKTESDDLVKPRPRKVTKVVDDDDDDFDVPISRKTRDTT 120
AtRFC2   ------------------------------------------------------------
AtRFC3   ------------------------------------------------------------
AtRFC4   ------------------------------------------------------------
AtRFC5   ------------------------------------------------------------

AtRFC1   PSKKLKSGSGRGIASRTVDNDDDDDGEDRETPLKSAGRGRGGRAAPGASTGGRGRGGGRG 180
AtRFC2   ------------------------------------------------------------
AtRFC3   ------------------------------------------------------------
AtRFC4   ------------------------------------------------------------
AtRFC5   ------------------------------------------------------------

AtRFC1   GFMNFGERRDPPHKGEREVPEGTPDCLAGLTFVISGTLDSLEREEAEDLIKRHGGRITGS 240
AtRFC2   ------------------------------------------------------------
AtRFC3   ------------------------------------------------------------
AtRFC4   ------------------------------------------------------------
AtRFC5   ------------------------------------------------------------

AtRFC1   VSKKTTYLLCDEDIGGRKSEKAKELGTKFLTEDGLFDIIRSSKPVKKSLPERSNKGTEKI 300
AtRFC2   ------------------------------------------------------------
AtRFC3   ------------------------------------------------------------
AtRFC4   ------------------------------------------------------------
AtRFC5   ------------------------------------------------------------

AtRFC1   CAPPRTSPQREETRGKPLAKSSPKKVPPAKGKNKIIETSLPWTEKYRPKVPNEIVGNQSL 360
AtRFC2   -MASSSSTSTGDGYNE------------------------PWVEKYRPSKVVDIVGNEDA 35
AtRFC3   -MTELTSAMDIDVDEIQPRKPINRGKDVVGFGPPPQSKATPWVEKYRPQSLDDVAAHRDI 59
AtRFC4   -MAPVLQSSQ------------------------------PWVEKYRPKQVKDVAHQEEV 29
AtRFC5   ------------------------------------MLWVDKYRPKSLDRVIVHEDI 21
                                             *.:*****.  .: ...
AtRFC1   VTQLHNWLSHWHDQFGGTGSKGKGKRLNDAGSKRAVLLSGTPGIGKTTSAKLVSQMLGFQ 420
AtRFC2   VSRLQ-----------------VIARDGNMPNLILSGPPGTGKTTSILALAHELLGT 75
AtRFC3   IDTID-----------------RLTNENKLPHLLLYGPPGTGKTSTILAVARKLYGP 99
AtRFC4   VRVLT-----------------NTLQTADCPHMLFYGPPGTGKTTTALAIAHQLFG- 68
AtRFC5   AQKLK-----------------KLVSEQDCPHLLFYGPSGSGKRTLIMALLKQIYGA 61
          :                   .         ::: *..*  **.:   :  :  :
AtRFC1   AVEVNASDSRGKANSNIAKGIGGSNANSVKELVNNEAMAANFD---------------- 463
AtRFC2   NYKEAVLELNAS----DDRGID-VVRNKIKMFAQKKVT-------------------- 108
AtRFC3   KYRNMILELNAS----DDRGID-VVRQQIQDFASTQSFS-------------------- 133
AtRFC4   -----VLELNAS----DDRGIN-VVRTKIKDFAAVAVGSNHR---------------- 100
AtRFC5   SAEKVKVENRAWKVDAGSRTID-LELTTLSSTNHVELTPSDAGFQDRYIVQEIIKEMAKN 120
           : .        :  *.       :.
AtRFC1   -----RSKHPKTVLIMDEVDGMSAGDRGGVADLIASIKISKIPIICICNDRYSQKLKSLV 518
AtRFC2   -----LPPGRHKVVILDEADSMTSGAQQALRRTIEIYSNSTR--FALACNTSAKIIEPIQ 161
AtRFC3   -----LGKSSVKLVLLDEADAMTKDAQFALRRVIEKYTKSTR--FALIGNHVNKIIPALQ 186
AtRFC4   --QSGYPCPSFKIIILDEADSMTEDAQNALRRTMETYSKVTR--FFFICNYISRIIEPLA 156
AtRFC5   RPIDTKGRKGYKVLVLNEVDKLSREAQHSLRRTMEKYSSSCR--LILCCNSSSKVTEAIK 178
          .:::::::*.*  ::      :  :     :  :  ::* *:  *:
AtRFC1   NYCLPLNYRKPTKQQMAKRLMHIAKAEGLEINEIALEELAERVNGDIRLAVNQLQYMSLS 578
AtRFC2   SRCALVRFSRLSDQQILGRLLVVVAAEKVPYVPEGLEAIIFTADGDMRQALNNLQAT--- 218
AtRFC3   SRCTRFRFAPLDGVHMSQRLKHVIEAERLVVSDCGLAALVRLSNGDMRKALNILQSTHM- 245
AtRFC4   SRCAKFRFKPLSEEVMSNRILHICNEEGLSLDGEALSTLSSISQGDLRRAITYLQSA--- 213
AtRFC5   SRCLNVRINAPSQEEIVKVLEFVAKKESLQLPQGFAARIAEKSNRSLRRAILSLETCR-- 236
          . *  ..         :      :   * :    :      :   :.:* *:  *:
AtRFC1   MSVIKYDDIRQRLLSSAKDEDISPFTAVDKLFGYNGGKLRMDERIDLSMSDPDLVPLLIQ 638
AtRFC2   ------------------------FSGF-SFVNQENVFKVCDQPHPLHVKNIV 246
AtRFC3   ------------------------ASKEITEEESKQITEEDVYLCTGNPLPKDIEQIS 279
AtRFC4   -----------------------TRLFGSTITSTDLLNVSGVVPLEVVNKLF 242
AtRFC5   ----------------------VQNYPFTGNQVISPMDWEEYVAEIATDMMK 266

AtRFC1   ENYLNYRPSGKDEAKRMDLLARAAESIADGDIINVQIRRYRQWQLSQSCCVASSILPASL 698
AtRFC2   RNVL--------ESKFDIACDGLRQLYD-------------------------- 266
AtRFC3   HWLL--------NKPFDECYKDVSEIKTR------------------------- 300
AtRFC4   TACK--------SGDFDIANKEVDNIVA-------------------------- 262
AtRFC5   EQSP--------KKLFQVRGKVYELLVN-------------------------- 286
                   ::                 . :
AtRFC1   LHGSREVLEQGERNFNRFGGWLGKNSTAGRNRRLMEDLHVHVLASRESSAGRETLRVDYL 758
AtRFC2   ------------------------------------------------------------
AtRFC3   ------------------------------------------------------------
AtRFC4   ------------------------------------------------------------
AtRFC5   ------------------------------------------------------------

AtRFC1   PLLLSRLTSPLQTLPKDEAVSEVVDFMNSYSISQEDFDTILELGKFKGRENPMEGVPPPV 818
AtRFC2   -----------LGYSPTDIITTLFRIIKN---------------------YDMAEYL 291
AtRFC3   -----------KGLAIVDIVKEITLFIFK-------------------IRMPSAV 325
AtRFC4   -----------EGYPASQIINQLFDIVAEA-----------------DSDITDMQ 289
AtRFC5   -----------CIPPEVILKRLLHELLK-------------------KLDSEL 309
                    . :          :: :          .         :
AtRFC1   KAALTKKYNEMNKTRMVRVADMVQLPGVKKAPKKRIAAMLEPTVDSLRDEDGEPLADNEE 878
AtRFC2   KLEFMKETGFAHMRICDGVGSYLQCLLLAK-LSIVRETARAP--------------- 333
AtRFC3   RVQLINDLADIEYRLSFGCNDKLQLGAIISTFTHARSIIVGAAK------------- 369
AtRFC4   KAKICKCLAETDKRLVDGADEYLQLLDVASSTICALSEMAQDF------------- 332
AtRFC5   KLEVCHWAAYYEHRMRLGQKAIFHIEAFVAKFMSIYKNFLISTFG------------- 354
          :  .  :    .               .  ::  .
```

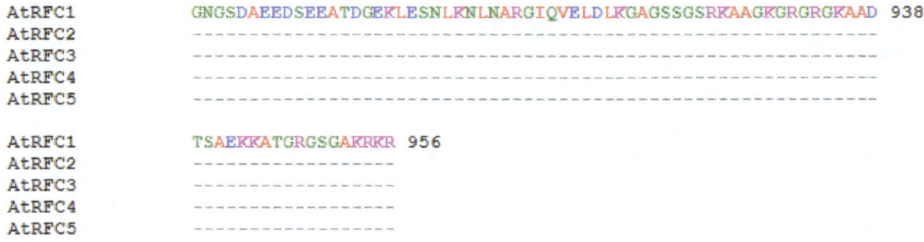

```
AtRFC1    GNGSDAEEDSEEATDGEKLESNLKNLNARGIQVELDLKGAGSSGSRKAAGKGRGRGKAAD  938
AtRFC2    -----------------------------------------------------------
AtRFC3    -----------------------------------------------------------
AtRFC4    -----------------------------------------------------------
AtRFC5    -----------------------------------------------------------

AtRFC1    TSAEKKATGRGSGAKRKR  956
AtRFC2    ------------------
AtRFC3    ------------------
AtRFC4    ------------------
AtRFC5    ------------------
```

Fig. 3 Alignments of AtRFC subunits protein by ClustalW2 software showing Box I, Box II and Box III respectively.

3.3 Domain search in *Pfam* database

Pfam domain search shows that RFC1 contains significant BRCT domain, which is found predominantly in proteins involved in cell cycle checkpoint functions, responsive to DNA damage reference and one AAA-ATPases domain which is associated with diverse cellular activities. BRCT domain extends from 204 to 282 amino acid residues, while AAA-ATPases domain extends from 392 to 530 residues.

Analysis showed that RFC2, RFC3a, RFC4, RFC5 contain significant AAA domains which extend from 47 to182 residues, 71 to 200 residues, 41 to 167 and 33 to 192 residues respectively.

3.4 Subcellular localization

The Subcellular localizations of RFCs were analyzed with using WoLF PSORT program. Results showed good and significant reliabilities that RFC1 is located in nucleus; RFC2, RFC4 and RFC5 are in cytoplasm whereas RFC3 is in cytoskeleton. TAIR software confirmed that RFC4 localization is in nucleus whereas others protein localization are unknown (Table 3).

Table 3 AtRFC family localization results

Protein	Subcellular localization
AtRFC1	Nucleus
AtRFC2	Cytoplasm
AtRFC3	Cytoskeleton
AtRFC3	Cytoplasm
AtRFC4	Cytoplasm
AtRFC4	Chloroplast
AtRFC4	Nucleus
AtRFC4	Vacuole
AtRFC4	Endoplasmic reticulum
AtRFC5	Cytoplasm
AtRFC5	Nucleus
AtRFC5	Mitochondria

3.5 Interactions

According to data obtained from the Arabidopsis Interactions Viewer we found many interactions belonging to all five subunits of AtRFC in different subcellular localizations, but mainly the partners were localized in nucleus. The interactome network of AtRFC subunits showed strongly interaction with AtPCNA1, AtPCNA2, CTF18 and RAD17 (Fig. 4). The common partner for RFC1/2/4 is AtPCNA whereas RFC2/3/4/5 interact strongly with CTF18 protein (Table 4). On the other hand we found that all AtRFC subunits interact with at least one of their AtRFC subunits.

Table 4 Interactome of AtRFC subunits

AtRFCs	Interactoms	Interolog Confidence	Interactome functions
RFC1	ATPCNA1	High	Proliferating cellular nuclear antigen 1
RFC1	ATRECQ4A	High	ATP-dependent DNA helicase
RFC1	ATRFC4	High	Rplication factroc C 4 ATPase family
RFC1	ATPCNA2	High	Proliferating cellilar nuclear antigen 2
RFC1	ATMSH2	High	Mismatch repair - MUTS homolog 2
RFC1	ATHDA19	High	Histone deacetylase 1
RFC1	ATRFC3	High	Rplication factroc C 3ATPase family
RFC1	ATRAD17	medium	Radiation sensitive 17
RFC2	ATCTF18	High	P-loop containing nucleoside triphosphate hydrolases
RFC2	ATRFC4	High	Rplication factroc C 4 ATPase family
RFC2	ATRFC5	High	Rplication factroc C 5 ATPase family
RFC2	ATRAD17	High	Radiation sensitive 17
RFC3	ATCTF18	High	P-loop containing nucleoside triphosphate hydrolases
RFC3	ATPCNA1	Medium	Proliferating cellular nuclear antigen 1
RFC3	ATRFC4	High	Rplication factroc C 4 ATPase family
RFC3	ATRFC2	High	Rplication factroc C 2 ATPase family
RFC3	ATPCNA2	High	proliferating cellular nuclear antigen 2
RFC3	AtRFC5	Medium	Rplication factroc C 5 ATPase family
RFC3	ATRAD17	High	Radiation sensitive 17
RFC4	ATCTF18	High	P-loop containing nucleoside triphosphate hydrolases
RFC4	ATPCNA1	High	Proliferating cellular nuclear antigen 1
RFC4	ATHDA19	Medium	Histone deacetylase 1
RFC4	ATPCNA2	High	Proliferating celltlar nuclear antigen 2
RFC4	ATRAD17	High	Radiation sensitive 17
RFC5	ATCTF18	High	P-loop containing nucleoside triphosphate hydrolases
RFC5	ATPCNA1	High	Proliferating cellular nuclear antigen 1
RFC5	ATRFC4	High	Rplication factroc C 4 ATPase family
RFC5	ATRFC1	High	Rplication factroc C 1 ATPase family
RFC5	ATRAD17	High	Radiation sensitive 17

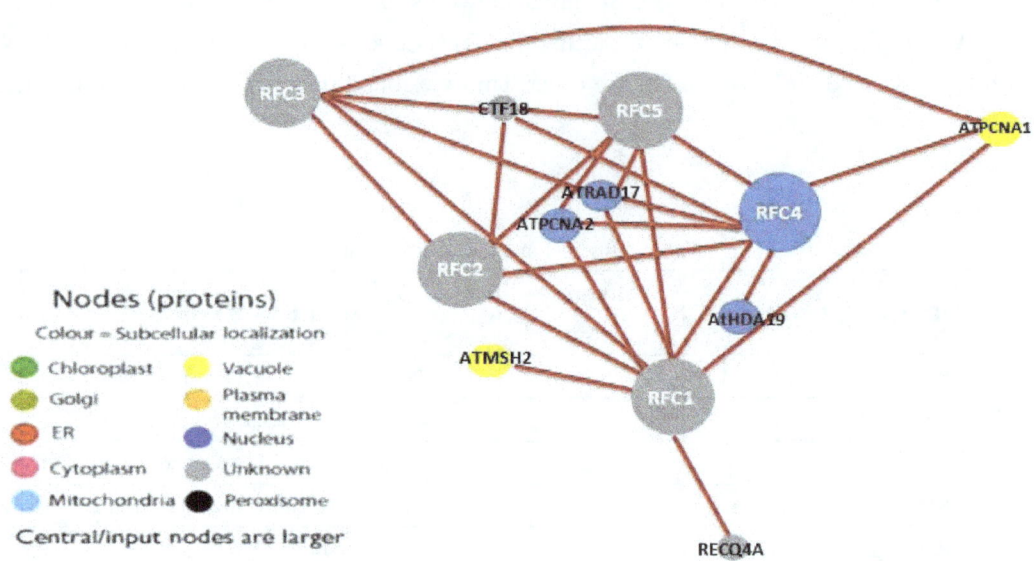

Fig. 4 Interactome of AtRFC subunits as obtained from Arabidopsis Interactions Viewer.

AtRFC1 is identical to CDC44 (Pai et al., 1989), identified as a cell division cycle gene encoding a protein involved in DNA metabolism. Due to their similarity, CDC44and RFC1 interact genetically with the gene encoding proliferating cell nuclear antigen, confirming previous biochemical evidence of their functional interaction in DNA replication. AtRFC1 strongly interacts with AtPCNA1 which is linked to a "sliding clamp" that functions as a mobile platform for the docking of an impressive array of enzymes responsible for the replication and repair of DNA (Yao et al., 2006). AtRFC1 also interacts with AtMSH2 which is involving in the repair of DNA replication errors (Leonard et al., 2003). The network of interactome for AtRFC1 showed an interesting dimerization with AtRFC3 and AtRFC4 respectively. The latter RFC subunits regulate cell proliferation and pathogen resistance in Arabidopsis (Depeiges et al., 2005). The unique property of AtRFC1 is the interaction with RECQ4 (ATP-dependent DNA helicase Q-like 4). The prominent functions of RECQ4 is the processing of double-holliday junctions (dHJs) that occur as intermediates during replication, DNA repair, or recombination and dissolve them in a manner which prevents deleterious crossover recombination (Johnson-Schlitz and Engels, 2006; Wu et al., 2005; Wu et al., 2006). Furthermore, RFC1 strongly correlates with HDA19, required for the repression of salicylic acid biosynthesis and salicylic acid-mediated defense responses in Arabidopsis. Loss of HDA19 activity increases the salicylic acid content , expressinf the genes required for accumulation of salicylic acid as well as *pathogenesis related* (*PR*) genes, resulting in enhanced resistance to *Pseudomonas syringae* (Choi et al., 2012).

AtRFC2 highly interacts with AtRFC4 and AtRFC5, which regulates cell proliferation and pathogen resistance in Arabidopsis (Noskov et al., 2008). AtRFC1, AtRFC2 and AtRFC4 interacts with CTF18, which is playing an important role in ATP binding, ATP catabolic process, ATPase activity, metabolic process, nucleoside-triphosphatase activity, nucleotide binding, replication fork and sister chromatid cohesion (Ghaemmaghami et al., 2003; Geisler-Lee et al., 2007).

The interactome partners of AtRFC3 are somehow similar to that of AtRFC1 and AtRFC2, with addition of AtRAD17, which encodes a homolog to yeast RAD17. AtRAD17 interacts with all RFC sububits , involved in the regulation of DNA damage repair and homologous recombination (Heitzeberg et al., 2004).

The possible proteins interaction with AtRFC4 are numerous containing three AtRFC subunits; AtRFC1,

AtRFC2 and AtRFC3. This conforms the strong   interaction of AtRFC4 with AtPCNA1 and AtPCNA2, playing regulatory roles in ATP catabolic process, ATPase activity and the machinery enzymes responsible for the replication and repair of DNA (Ghaemmaghami et al., 2003; Geisler-Lee et al, 2007).

AtRFC5 interaction with AtRFC1/2/3 and AtRFC4 confirms the connectivity   to CTF18, AtPCNA1 and AtRAD17. AtRFC5 interacts with with all RFC subunits, opening the door for the RFC coordination mechanism for regulation of DNA repair, DNA replication and DNA irradiation protection.

Measurements of the intracellular levels of the RFC subunits have been determined in a global analysis of protein expression in yeast (Geisler-Lee et al, 2007). The current Bioinformatics study confirms that subcomplexes of RFC, RFC2/3/4/5 and RFC2/5 (Yao et al., 2006) are capable of unloading PCNA clamps. Furthermore, due to the strong interaction of all RFC subunits interact with radiation sensitive 17 protein (RAD17), a notable new function could be assigned RFC family, which is the regulation of DNA damage repair. Due to the strong interaction of all RFC subunits interact with radiation sensitive 17 protein (RAD17), a notable new function could be assigned to our proteins family, which is the regulation of DNA damage repair.

4 Conclusion

Replication factor C (RFC) is a multi protein complex consisting of one large and four small subunits. RFC subunits have an associated ATPase activity that is stimulated by the binding of RFC to DNA and is further stimulated by proliferating cell nuclear antigen (PCNA). The interactome of all five AtRFC subunits assigned new functions to this subfamily, especially for DNA UV-protection, metabolic processes and controlling the biosynthesis of salicylic and salicylic acid-mediated defense responses in Arabidopsis.

Abbreviations

RFC: replication factor C
RFC1: Replication factor C subunit 1 protein
RFC2: Replication factor C subunit 2 protein
RFC3: Replication factor C subunit 3 protein
RFC4: Replication factor C subunit 4 protein
RFC5: Replication factor C subunit 5 protein
TAIR: Arabidopsis Information Resource
PCNA: the proliferating cell nuclear antigen
SMC3: structural maintenance of chromosome
ROS: reactive oxidative species
SSB: single strand breaks
DSB: double strand breaks
RLC: RFC-like complexes
TGS: Transcriptional gene silencing
RAD17:Radio Sensitive 17
CTF18:P loop containing nucleotide phosphate
HR: Homologous recombination
ClustalW2: multiple sequence alignment program, version 2
MSA: Multiple Sequence Alignment
Pymol: program to obtain 3D structure of the target proteins
SMART: Simple Modular Architecture Research Tool

WoLF PSORT: Subcellular localization prediction for each protein

PDB: Protein data bank

Acknowledgment

Authors thank the board of trustees, International University of Sarajevo for financial and moral support to realize this work.

References

Alberts B, Johnson A, Lewis J. et al. 2002. DNA replication, repair and recombination. In: Molecular Biology of the Cell (4th ed.). Garland Science, New York, USA

Arabidopsis interaction viewer. http://bar.utoronto.ca/interactions/cgi bin/arabidopsis_interactions_viewer.cgi

Bunz F, Kobayashiand R, Stillman B. 1993. cDNAs encoding the large subunit of human replication factor C. Proceedings of the National Academy of Sciences of USA, 90: 11014-11018

Choi SM, Song HR, Han SK, et al. 2012. HDA19 is required for the repression of salicylic acid biosynthesis and salicylic acid-mediated defense responses in Arabidopsis. Plant Journal, 71(1): 135-146

Cullmann G, Fien K, Kobayashi R, et al. 1995. Characterization of the five replication factor C genes of *Saccharomyces cerevisiae*. Molecular and Cellular Biology, 15(9): 4661-4671

Depeiges A, Farget S, Degroote F, et al. 2005. A new transgene assay to study microsatellite instability in wild-type and mismatch-repair defective plant progenies. Plant Science, 168: 939-947

Dereeper A, Guignon V, Blanc G, et al. 2008. Phylogeny.fr: Robust phylogenetic analysis for the non-specialist. Nucleic Acids Research, 36: 465-469

Elizabeth AG, Michael AA, David R, et al. 1994. CDC44: a putative nucleotide-binding protein required for cell cycle progression that has homology to subunits of replication factor C. Molecular and Cellular Biology, 14(1): 255-267

Finn RD, Mistry J, Tate J, et al. 2010. The Pfam protein families database: Bateman. Nucleic Acids Research, 38: 211-222

Ganpudi AL, Schroeder DF. 2011. UV Damaged DNA Repair & Tolerance in Plants. In: Selected Topics in DNA Repair (Chen CC, ed.). Intechopen, USA

Gary SL, Burgers MJ. 1995. Identification of the fifth subunit of Saccharomyces cerevisiae replication factor C. Nucleic Acids Research, 23(24): 4986-4991

Geisler-Lee J, O'Toole N, Ammar R, et al. 2007. A predicted interactome for Arabidopsis. Plant Physiology, 145: 317-329

Ghaemmaghami S, Huh WK, Bower K, et al. 2003. Reduced polymorphism in domains involved in protein-protein interactions. Nature, 425(6959): 737-741

Gray FC, MacNeill SA. 2000. The Schizosaccharomyces pombe rfc3_ gene encodes a homologue of the human hRFC36 and *S. cerevisiae* Rfc3 subunits of replication factor C. Current Genetics, 37: 159-167

Heitzeberg F, Chen IP, Hartung F, et al. 2004. The Rad17 homologue of Arabidopsis is involved in the regulation of DNA damage repair and homologous recombination. Plant Journal, 38(6): 954-968

Horton P, Park KJ, Obayashi T, et al. 2007. WoLF PSORT: protein localization predictor. Nucleic Acids Research, 35: 585-587

Johnson-Schlitz D, Engels WR. 2006. Template disruptions and failure of double Holliday junction dissolution during double-strand break repair in *Drosophila BLM* mutants. Proceedings of the National Academy of Sciences of USA, 103(45): 16840–16845

Kozak J, West CE, White C, et al. 2009. Rapid repair of DNA double strand breaks in Arabidopsis thaliana is dependent on proteins involved in chromosome structure maintenance. DNA Repair, 8(3): 413-419

Leonard JM, Bollmann SR, Hays JB. 2003. Reduction of stability of *Arabidopsis* genomic and transgenic DNA-repeat sequences (microsatellites) by inactivation of AtMSH2 mismatch-repair function. Plant Physiology, 133: 328-338

Li X, Burgers PMJ. 1994. Cloning and characterization of the essential Saccharornyces cerevisiue RFC4 gene encoding the 37 kDa subunit of replication factor C. Journal of Biological Chemistry, 269: 21880-21884

Liu Q, Wang J, Miki D, et al. 2010. DNA replication factor C1 mediates genomic Stability and transcriptional gene silencing in Arabidopsis. Plant Cell, 22(7): 2336-2352

Lovell SC, Davis IW, Arendall III WB, et al. 2002. Structure validation by Calpha geometry: phi,psi and Cbeta deviation. Proteins: Structure, Function & Genetics. 50: 437-450

Majka J, Burgers PM. 2004. The PCNA-RFC families of DNA clamps and clamp loaders. Progress in Nucleic Acid Research and Molecular Biology, 78: 227-260

Martinez-Antonio A. 2011. *Escherichia coli* transcriptional regulatory network. Network Biology, 1(1): 21-33

Mossi R, Hübscher U. 1998. Clamping down on clamps and clamp loaders – the eukaryotic replication factor C. Europe Journal of Biochemistry, 254: 209-216

Nielsen M, Lundegaard C, Lund O, et al. 2010. CPHmodels-3.0--Remote homology modeling using structure guided sequence profiles. Nucleic Acids Research, 38: 578-581

Noskov V, et al. 1994. The RFC2 gene encoding a subunit of replication factor C of *Saccharomyces cerevisiae*. Nucleic Acids Research, 22(9): 1527-1535

Noskov VN, Araki H, Sugino A. 2008. The RFC2 gene, encoding the third-largest subunit of the replication factor C complex, is required for an S-phase checkpoint in *Saccharomyces cerevisiae*. Molecular and Cellular Biology, 18(8): 4914-4923

Pai EF, Kabsch W, Krengel U, et al. 1989. Structure of the guanine-nucleotide-binding domain of the Ha-rasoncogene product p21 in the triphosphate conformation. Nature, 341: 209-214

Rahman KMT, Islam MdF, Banik RS, et al. 2013. Changes in protein interaction networks between normal and cancer conditions: Total chaos or ordered disorder? Network Biology, 3(1): 15-28

Shultz RW, Tatineni VM, Hanley-Bowdoin L, et al. 2007. Genome-wide analysis of the core DNA Replication machinery in the higher plants Arabidopsis and rice. Plant Physiology, 144(4): 1697-1714

Takahashi N, et al. 2010. The MCM-binding protein ETG1 aids sister chromatid cohesion required for postreplicative homologous recombination repair. PLoS Genetics, 6(1): 1-13

Takashi Y, Kobayashi Y, Tanaka K, et al. 2011. Arabidopsis replication protein A 70a is required for DNA damage response and telomere length homeostasis. Plant Cell Physiology, 50(11): 1965-1976

The Arabidopsis Genome Innovative. 2000. Analysis of the genome sequence of the flowering plant *Arabidopsis thaliana*, 408: 796-815

Wu L Bachrati CZ, Ou J, et al. 2006. BLAP75/RMI1 promotes the BLM-dependent dissolution of homologous recombination intermediates. Proceedings of the National Academy of Sciences of USA, 103(45): 4068-4073

Wu L, Chan KL, Ralf C, et al. 2005. The HRDC domain of BLM is required for the dissolution of double Holliday junctions. EMBO Journal, 24(14): 2679-2687

Xia ST, Xiao LT, Gannon P, et al. 2010. RFC3 regulates cell proliferation and pathogen resistance in Arabidopsis. Plant Signaling and Behavior, 5(2): 168-170

Yao NY, Johnson A, Bowman G, et al. 2006. Mechanism of PCNA clamp opening by RFC. Journal of Biological Chemistry, 281(25): 17528-17539

2

Bit by bit control of nonlinear ecological and biological networks using Evolutionary Network Control

Alessandro Ferrarini

Department of Evolutionary and Functional Biology, University of Parma, Via G. Saragat 4, I-43100 Parma, Italy

E-mail: sgtpm@libero.it, alessandro.ferrarini@unipr.it, a.ferrarini1972@libero.it

Abstract

Evolutionary Network Control (ENC) has been first introduced in 2013 to effectively subdue network-like systems. ENC opposes the idea, very common in the scientific literature, that controllability of networks should be based on the identification of the set of driver nodes that can guide the system's dynamics, in other words on the choice of a subset of nodes that should be selected to be permanently controlled. ENC has proven to be effective in the global control (i.e. the focus is on mastery of the final state of network dynamics) of linear and nonlinear networks, and in the local (i.e. the focus is on the step-by-step ascendancy of network dynamics) control of linear networks. In this work, ENC is applied to the local control of nonlinear networks. Using the Lotka-Volterra model as a case study, I show here that ENC is capable of locally driving nonlinear networks as well, so that also intermediate steps (not only the final state) are under our strict control. ENC can be readily applied to any kind of ecological, biological, economic and network-like system.

Keywords Evolutionary Network Control; genetic algorithms; intermediate control function; local dynamics; Lotka-Volterra model; network systems; nonlinear networks; predator-prey model.

1 Introduction

Evolutionary Network Control (ENC) has been recently developed to control any kind of ecological and biological networks from inside (Ferrarini, 2013) and from outside (Ferrarini, 2013b), by coupling network dynamics and evolutionary modelling (Holland, 1975). The endogenous control requires that the network is optimized at the beginning of its dynamics, by acting upon nodes, edges or both, so that it will inertially go to the desired state. The exogenous control requires that one or more exogenous controllers act upon the network at each time step.

ENC can be applied to both discrete-time (i.e., systems of difference equations) and continuous-time (i.e., systems of differential equations) networks. ENC opposes the common idea in the scientific literature that controllability of networks should be based on the identification of the set of driver nodes that can guide the system's dynamics, in other words on the choice of a subset of nodes that should be selected to be permanently controlled (Ferrarini, 2011). ENC makes use of an integrated solution (network dynamics - genetic optimization - stochastic simulations) to compute reliability of network control (Ferrarini, 2013c), and introduced the concepts of control success and feasibility (Ferrarini, 2013d). ENC makes use of intermediate control functions to locally (step-by-step) drive ecological and biological networks, so that also intermediate steps (not only the final state) are under its control (Ferrarini, 2014). ENC can also globally subdue nonlinear networks (Ferrarini, 2015), and impose early or late stability to any kind of ecological and biological network (Ferrarini, 2015b).

Table 1 Evolutionary network control and its developed variants.

Reference	Purpose
Ferrarini 2011	Theoretical bases of Evolutionary Network Control
Ferrarini 2013	Endogenous control of linear ecological and biological networks
Ferrarini 2013b	Exogenous control of linear ecological and biological networks
Ferrarini 2013c	Computing the uncertainty associated with network control
Ferrarini 2013d	Computing the degree of success and feasibility of network control
Ferrarini 2014	Local control of linear ecological and biological networks
Ferrarini 2015	Global control of nonlinear ecological and biological networks
Ferrarini 2015b	Imposing early/late stability to linear and nonlinear networks
This work	Local control of nonlinear ecological and biological networks

In this work, I show how ENC can locally control any kind of nonlinear networks, and I provide an applicative example based on the nonlinear, widely-used, Lotka-Volterra model (Lotka, 1925; Volterra, 1926). Of course, any other kind of ecological, biological, economic and network-like system is prone to be controlled by ENC.

2 Mathematical Formulation

2.1 Opening definitions

A generic ecological (or biological) dynamical system with n interacting actors is given as follows

$$\frac{d\mathbf{S}}{dt} = \varphi(\mathbf{S}(t)) \tag{1}$$

where S_i is the amount (e.g., number of individuals, total biomass, density, covered surface etc...) of the generic i-th actor (i.e., species, population, taxonomic group etc.) . If we also consider time-dependent inputs (e.g., species reintroductions) and outputs (e.g., hunting) from-to outside, we must write

$$\frac{d\mathbf{S}}{dt} = \varphi(\mathbf{S}(t)) + \mathbf{I}(t) + \mathbf{O}(t) \tag{2}$$

with initial values

$$\mathbf{S}_0 = <S_1(0), S_2(0)...S_n(0)> \tag{3}$$

and co-domain limits

$$\begin{cases} S_{1\min} \leq S_1(t) \leq S_{1\max} \\ \dots \\ S_{n\min} \leq S_n(t) \leq S_{n\max} \end{cases} \forall t \tag{4}$$

The nonlinear Lotka-Volterra model with logistic grow of the prey S_1 is a particular case of (1) and it reads as follows

$$\begin{cases} \dfrac{dS_1}{dt} = \alpha S_1(1 - \dfrac{S_1}{\kappa}) - \beta S_1 S_2 \\ \dfrac{dS_2}{dt} = \beta \gamma S_1 S_2 - \delta S_2 \end{cases} \tag{5}$$

with initial values

$$\mathbf{S}_0 = <S_1(0),\ S_2(0)> \tag{6}$$

and co-domain limits

$$\begin{cases} S_{1\min} \leq S_1(t) \leq S_{1\max} \\ S_{2\min} \leq S_2(t) \leq S_{2\max} \end{cases} \forall t \tag{7}$$

and equilibrium at

$$\begin{cases} \dfrac{dS_1}{dt} \leq \varphi \\ \dfrac{dS_2}{dt} \leq \varphi \end{cases} \text{with } \varphi \to 0 \tag{8}$$

2.2 Local control of nonlinear networks through the intermediate control function (*ICF*)

ENC has been developed so that it can also control intermediate dynamics of linear and nonlinear networks (Ferrarini, 2014) and not only their final state (Ferrarini, 2013; Ferrarini, 2013b).

To this purpose, ENC introduces the concept of intermediate control function (*ICF*; Ferrarini, 2014) and then controls each single step of network dynamics by subduing *ICF*. Using the *ICF*, ENC can effectively reach the following goals.

2.2.1 Freezing any network actor to a desired state

For instance, we want the prey $S_1(t)$ (or the predator $S_2(t)$) to stay as close as possible to a certain value during its dynamics. In this case, *ICF* must be formulated as follows

$$ICF = \int_{t=0}^{E} |k - S_i(t)|\, dt \tag{9}$$

where t is time (independent variable), E indicates the time at which the system goes to equilibrium, k is the value at which we want $S_i(t)$ to be tied to, vertical lines indicates the module of the difference.

ENC reaches its goal by optimizing the ecological/biological network in order to achieve the minimization of *ICF*. This goal can be achieved using genetic algorithms (Holland, 1975) for the endogenous optimization

(Ferrarini, 2013) of network edges (i.e., interaction among species) or nodes (i.e., species stocks), i.e. the modification of their values at the beginning of the network dynamics in order to minimize *ICF*.

Alternatively an exogenous control can be applied on exogenous node's edges (i.e., coefficients of interaction with the inner system) and exogenous node's stock (Ferrarini, 2013b).

As a result, ENC can constrain $S_i(t)$ as close as possible to k along its dynamics.

2.2.2 Binding two or more network actors together

For instance, we want to tie the prey $S_1(t)$ to the predator $S_2(t)$. In this case, ENC uses the following *ICF* to be minimized:

$$ICF = \int_{t=0}^{E} |S_1(t) - S_2(t)| \, dt$$

(10)

As a results, *S1* and *S2* are bound together (i.e. *S1≈S2*) along the whole network dynamics.

2.2.3 Freezing any actor to a desired function

We want here to tie the generic actor S_i to a function of the other actor S_j. For instance, we could force the predator S_2 to be always twice the prey S_1 during network dynamics.

In this case, ENC uses the following equation to be minimized

$$ICF = \int_{t=0}^{E} |f(S_j(t)) - S_i(t)| \, dt$$

(11)

As a results, S_i and $f(S_j)$ are tied together (i.e. $S_i \approx f(S_j)$) along the whole network dynamics.

2.2.4 Controlling each single change of any actor

We impose here that the generic actor S_i changes from step to step of a certain value u

$$ICF = \int_{t=0}^{E} \left| \frac{dS_i(t)}{dt} - u \right| \, dt$$

(12)

By minimizing *ICF*, we impose that S_i changes of exactly u at each time step.

2.2.5 Limiting the control to a certain time interval

In this case the previous equations must be changed to operate mathematical integration from time T_1 to time T_2 (and not from 0 to E), where T_1 and T_2 are generic points along the timeline.

3 An Applicative Example

Let's consider the Lotka-Volterra system of Eq. (5) with the following parameters and constants:

$S_1(0)=85$
$S_2(0)=20$
$\alpha=3$
$\beta=0.02$
$\gamma=1.2$
$\delta=2$
$\kappa=500$
$dt=0.01$

(13)

Fig. 1 shows its dynamical behavior. Fig. 2 depicts its phase plot.

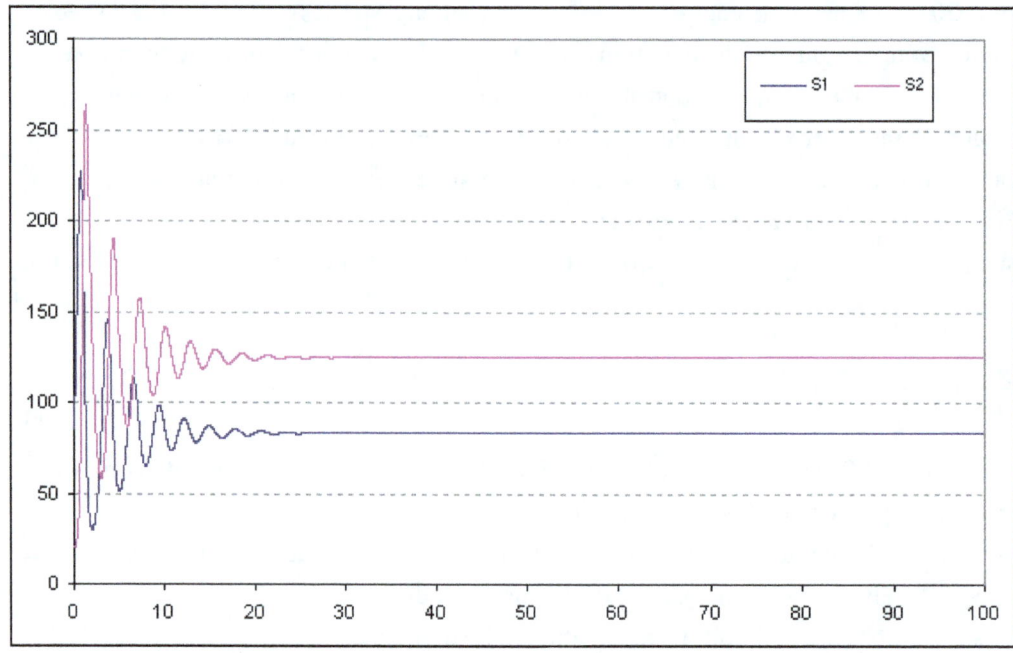

Fig. 1 Time plot of the nonlinear Lotka-Volterra dynamical system described in (13).

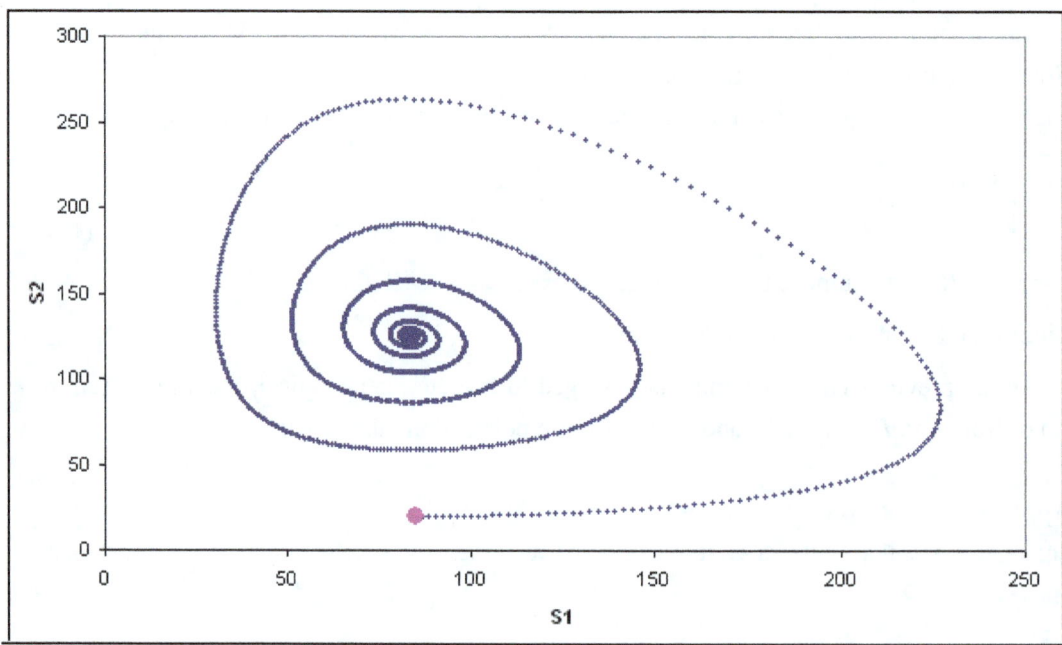

Fig. 2 Phase plot of the nonlinear Lotka-Volterra dynamical system described in (13).

The previous nonlinear system goes at equilibrium at E= 68.74 with S_1= 83.33 and S_2= 125.00. The average absolute distance between S_1 and S_2 in the time span [0 - 68.74] is equal to 44.86. Now let's suppose we want to halve this distance, i.e. to impose that it becomes equal to 22.43. By optimizing Eq. 10, ENC found the solution depicted in Fig. 3 with α= 2.4612, β= 0.0190, γ= 1.1886 while the other parameters were kept constant.

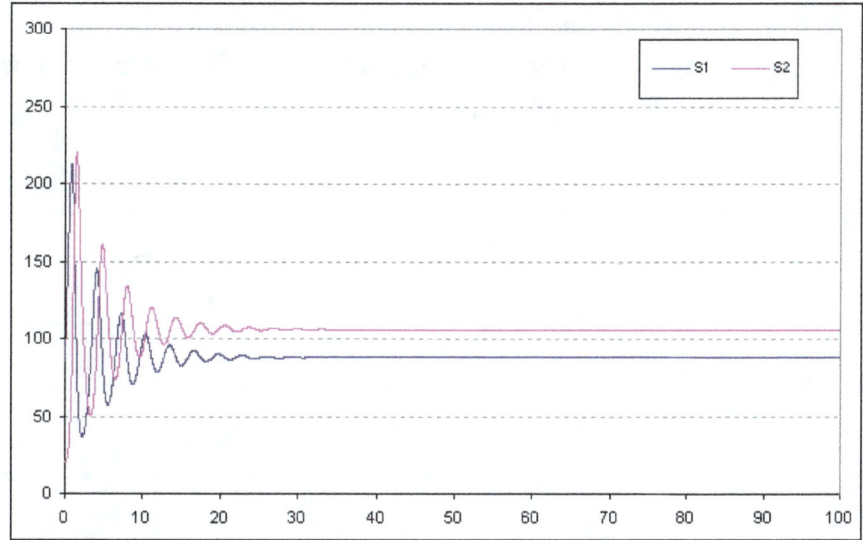

Fig. 3 ENC has found a solution to halve the average absolute distance between S_1 and S_2 (i.e., from 44.86 to 22.43) by acting upon α, β and γ while the other parameters were kept constant.

Now let's suppose we want to get the same result but only working on the carrying capacity κ. ENC found the solution that requires κ to be equal to 149.808. The system stabilizes at $E= 17.55$ and the average absolute distance between S_1 and S_2 results to be 22.44 (Fig. 4).

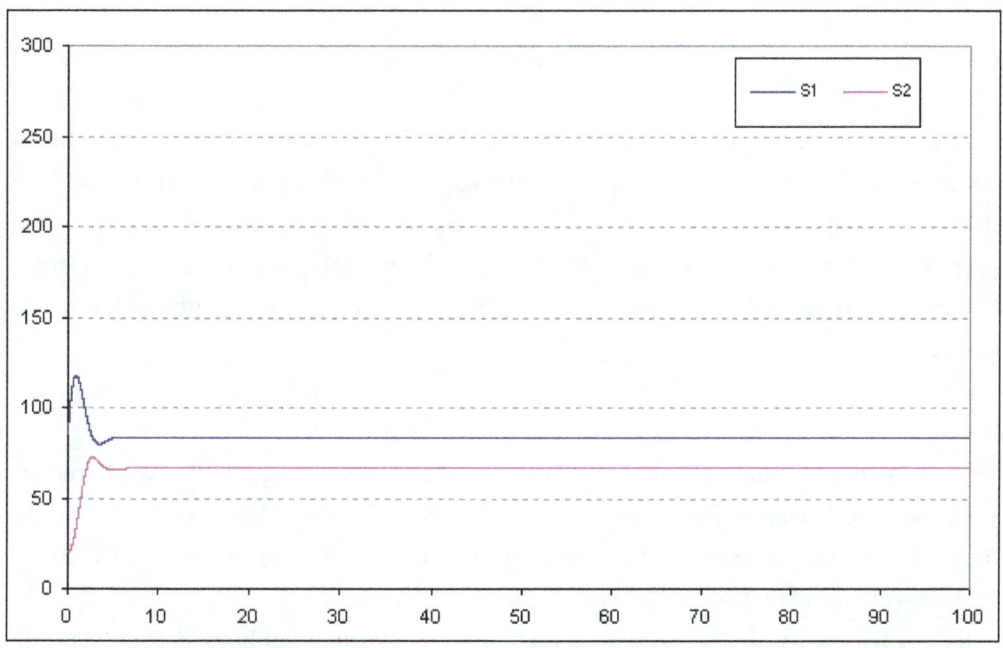

Fig. 4 ENC has found a solution to halve the average absolute distance between S_1 and S_2 (i.e., from 44.86 to 22.44) by acting upon the carrying capacity κ while the other parameters were kept constant.

With the parameters in (13), during the time span [0 – 68.74] the average absolute distance of two successive steps of S_1 was equal to 0.11. Now let's suppose we want to halve it by only working on the initial stocks of S_1 and S_2. By optimizing Eq. 12, ENC has found the solution depicted in Fig. 5, which stabilizes at E= 65.55 with initial stocks $S_1(0)$= 85 and $S_2(0)$= 62.67.

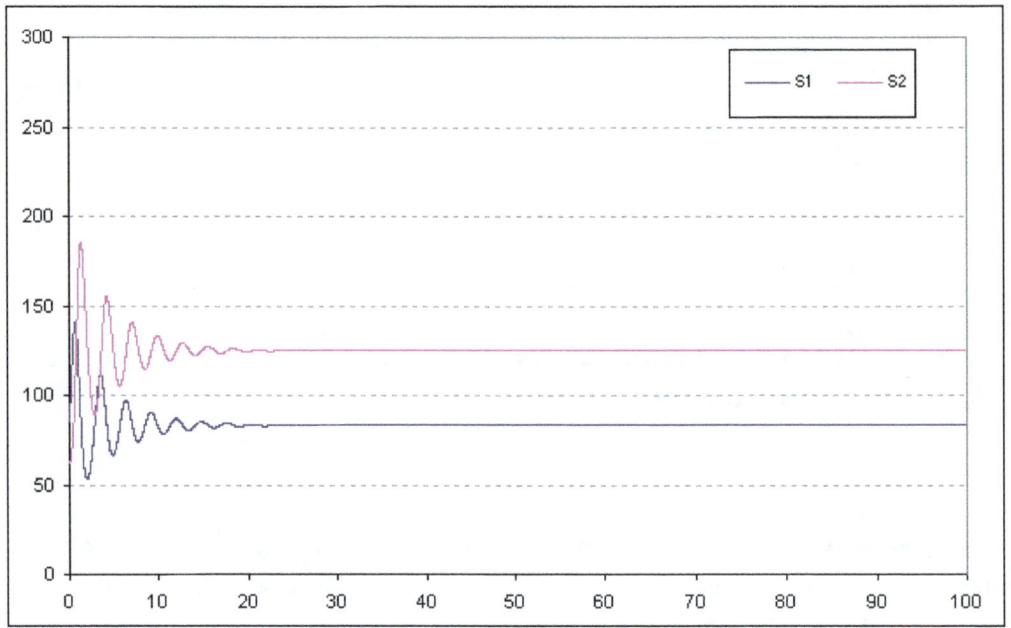

Fig. 5 ENC has found a solution to halve the average absolute distance between two successive steps of S_1 (i.e., from 0.11 to 0.055) by acting upon the initial stocks of S_1 and S_2 while the other parameters were kept constant.

Of course, by optimizing equations from (9) to (12) ENC is able to realize any other kind of control on nonlinear networks, as already showed for linear networks as well (Ferrarini, 2014). In addition, global and local controls can be coupled in ENC in order to achieve a complete control of network's dynamics. The framework proposed here might also be applied to semi-quantitative networks (Ferrarini, 2011b).

ENC has been applied using the software Control-Lab 5 (Ferrarini, 2015c) written in Visual Basic (Balena, 2001; Pattison, 1998).

4 Conclusions

The control of ecological and biological networks is a pivotal and trendy topic. In this work, the theoretical and methodological framework named Evolutionary Network Control (ENC) has showed to be able to locally control any kind of nonlinear network, while in previous works it showed to be on top of globally and locally subduing linear networks, and of globally taming nonlinear networks.

The potential applications of ENC in ecology and biology are virtually unlimited, for instance: a) neutralize damages to ecological and biological networks, b) safeguard rare and endangered species, c) manage ecological systems at the least possible cost, d) counteract the impacts of climate change, and e) balance the negative pressure due to human activities. ENC has been developed exactly with these purposes in mind.

References

Balena F. 2001. Programming Microsoft Visual Basic 6.0. Microsoft Press, Redmond, WA, USA

Ferrarini A. 2011. Some thoughts on the controllability of network systems. Network Biology, 1(3-4): 186-188

Ferrarini A., 2011b. Some steps forward in semi-quantitative network modelling. Network Biology, 1(1): 72-78

Ferrarini A. 2013. Controlling ecological and biological networks via evolutionary modelling. Network Biology, 3(3): 97-105

Ferrarini A. 2013b. Exogenous control of biological and ecological systems through evolutionary modelling. Proceedings of the International Academy of Ecology and Environmental Sciences, 3(3): 257-265

Ferrarini A., 2013c. Computing the uncertainty associated with the control of ecological and biological systems. Computational Ecology and Software, 3(3): 74-80

Ferrarini A., 2013d. Networks control: introducing the degree of success and feasibility. Network Biology, 3(4): 115-120

Ferrarini A., 2014. Local and global control of ecological and biological networks. Network Biology, 4(1): 21-30

Ferrarini A., 2015. Evolutionary network control also holds for nonlinear networks: Ruling the Lotka-Volterra model. Network Biology, 5(1): 34-42

Ferrarini A. 2015b. Imposing early stability to ecological and biological networks through Evolutionary Network Control. Proceedings of the International Academy of Ecology and Environmental Sciences, 5(1): 49-56

Ferrarini A. 2015c Control-Lab 5: a software for the application of Ecological Network Control. Manual, 108 pages (in Italian)

Holland J.H. 1975. Adaptation in Natural and Artificial Systems: An Introductory Analysis with Applications to Biology, Control and Artificial Intelligence. University of Michigan Press, Ann Arbor, USA

Lotka, A. J. 1925. Elements of Physical Biology. Baltimore: Williams & Wilkins Co., USA

Pattison T. 1998. Programming Distributed Applications with COM and Microsoft Visual Basic 6.0. Microsoft Press, Redmond, WA, USA

Volterra, V. 1926. Variazioni e fluttuazioni del numero d'individui in specie animali conviventi. Mem. R. Accad. Naz. dei Lincei. Ser. VI, vol. 2.

3

Commonality in structure among food web networks

Carrie J. Byron[1], **Craig Tennenhouse**[2]

[1]Department of Marine Sciences, University of New England, 11 Hills Beach Road, Biddeford, ME 04005, USA

[2]Department of Mathematical Sciences, University of New England, 11 Hills Beach Road, Biddeford, ME 04005, USA

E-mail: cbyron@une.edu, ctennenhouse@une.edu

Abstract

A goal of this study was to determine similarities in structure among food webs that are otherwise disparate with regard to species, population, and size. Food webs were examined as directed, unweighted graphs in order to normalize food webs with regard to biomass and population/species distinctions. The graphs were further normalized with regard to topological size and existence of circuits through the reduction of each strongly connected component to a single node. This had the added benefit of resulting in networks with more clear delineation between trophic levels. Finally, common induced subgraphs were considered for their obvious value in characterizing network structure. Through this study we determined not only that there are pairs of systems that are highly similar in structure once appropriately normalized for size, makeup, and geographical location, but also that a majority of food webs have similar structural components when compared with random food webs.

Keywords network ecology; graph theory; food webs; directed graphs.

1 Introduction

Ecosystems are complex. Management of ecosystems for environmental and societal goods and services can be even more complex. Finding commonalities and patterns in this complexity will help streamline management efforts. The complexity in ecosystems can be captured in its food web which describes predator prey relationships and flow of energy between species.

Food webs are one type of network that has been subject to examination through the founding disciplines of network ecology and graph theory. Dunne (2002, 2004, 2009) has done much work characterizing network structure of food webs demonstrating the role of connectedness, size, and robustness. Food webs, like most real world networks, are not random (Williams and Martinez, 2000). Food webs from different types of ecosystems share fundamental structural and ordering characteristics (Dunne et al., 2004).

Finding ways to describe and quantify these fundamental structural and ordering characteristics can be challenging due to the complexity of the food webs. Several metrics and techniques have already been developed in the field of network ecology. Here we attempt to advance this field by proposing a methodology for examining commonality among food web models that is strongly rooted in graph theory and builds from prior work of Allesina et al. (2005).

We focus on "wet" food webs, including marine, brackish, and freshwater systems from around the world and propose a set of metrics derived from graph theory that can be used to evaluate similarities and differences across these complex systems. Using a combined graph theory and network ecology approach, we first establish that, like other real-world networks, food webs have more organization than randomly generated webs (Watts and Strogatz, 1998; Williams and Martinez, 2000). We then test for similarities and differences across all food web types to examine whether there are mathematical properties of food webs that are ubiquitous around the world.

Our approach is advancement over other studies because of our large sample size (21) and ecosystem diversity (i.e. fresh-water, estuarine, marine). Many other comparative studies of food webs only look at 3-5 networks and are typically all a similar type of ecosystem (Allesina et al., 2005; Bascompte and Melián, 2005; Dunne et al., 2004). We also examined larger subgraph structures (up to 7 nodes) compared to other studies (3 nodes) (Borrelli, 2015).

2 Method

We will use the following terminology throughout this manuscript.

Definition: A <u>directed graph</u> (henceforth referred to as a *graph*) is an ordered pair of sets *(V, E)* of vertices (or *nodes*) *V* and edges *E*. Each edge *e* in *E* is itself an ordered pair *(u, v)* of distinct elements from *V*, (i.e. we do not allow loops or multiple edges, but circuits of length *2* are permitted). The <u>in-degree</u> (alternatively <u>out-degree</u>) of a node *v* in a graph *G* is the number of edges in *G* of the form *(x, v)* (alternatively *(v, x)*) for any vertex *x* in *G*.

Definition: A graph *G* is <u>connected</u> if, for every pair *u,v* of vertices in *G*, there is a set of edges in *G* (e_1, e_2,…,e_n) such that *u* is in e_1, *v* is in e_n, and for each *i* the edges e_i and e_{i+1} have a non-empty intersection.

Definition: An <u>*n*-path</u> is a connected graph on *n* vertices in which one vertex has out-degree *1* and in-degree *0* (the *source*), one vertex has out-degree *0* and in-degree *1* (the *sink*), and the remaining vertices each have in-degree and out-degree *1*. An <u>*n*-circuit</u> is a connected graph on *n* vertices in which each vertex has both in-degree and out-degree *1*. A graph that contains no circuit is <u>acyclic</u>.

Definition: A graph *G* = (V_G, E_G) is an <u>induced subgraph</u> of a graph *H* = (V_H, E_H) if there is an injective function *f* from V_G to V_H such that for any edge *(u, v)* in E_G the corresponding edge *(f(u), f(v))* is in E_H, and if *(u, v)* is not an edge in E_G then *(f(u), f(v))* is not an edge in E_H.

As an example, the *4*-path is not an induced subgraph of the *4*-circuit, but it is an induced subgraph of the *n*-circuit for all *n>* 4.

Definition: The <u>order</u> of a graph *G* is the number of vertices in *G*, denoted *n(G)*. Its <u>size</u> *e(G)* is the number of edges in *G*.

2.1 Food web networks

Our goal with this project was to determine commonality among food web structures, in particular simplified, acyclic models of food webs irrespective of geographic location, size, and species particulars. We began with 21 food webs (*FW*) imported from Ecopath (www.ecopath.org) to Sagemath (www.sagemath.org) as graphs based on the diet matrix (Table 1). The Ecopath diet matrix captures the percent of diet of prey items for every predator in the food web. Each Ecopath food web was created by different researchers for different purposes and therefore has unique classifications and groupings of species (Table 2). To standardize these diverse models, each food web was given two treatments prior to any analysis: (1) removal of detritus and (2) removal of circuits.

Table 1 Networks used in analysis. Networks published prior to 2010 were downloaded from the Ecopath website (www.ecopath. org). All networks were assigned both a three letter abbreviation (Abbr.) and a number (Num.) that is used in subsequent tables and figures.

Name	Abbr.	Num.	Reference
Aleutian Islands, Alaska, USA	ALE	1	(Guenette and Christensen, 2005; Guenette et al., 2007)
Weddell Sea, Antarctica	WED	2	(Jarre-teichmann et al., 1997)
Lake Tanganyika, Burundi	TAN	3	(Moreau et al., 1993b)
Lake Ontario, Canada	ONT	4	(Halfon and Schito, 1993)
Northern Gulf of St. Lawrence, Canada	GSL	5	(Morissette et al., 2003)
Lake Aydat, France	AYD	6	(Reyes-Marchant et al., 1993)
Great Barrier Reef, Australia	GBR	7	(Gribble, 2005)
Gironde Estuary, France	GIR	8	(Lobry, 2004)
Iceland	ICE	9	(Buchary, 2001)
Lake Kinneret, Israel	KIN	10	(Walline et al., 1993)
Lake Victoria, Kenya	VIC	11	(Moreau et al., 1993a)
Narragansett Bay, Rhode Island, USA	NAR	12	(Byron et al., 2011a)
Laguna de Bay, Philippines	LDB	13	(Delos Reyes, 1995)
Lagoons, Rhode Island, USA	LRI	14	(Byron et al., 2011b)
Saco River Marsh, Maine, USA	SRM	15	Byron, unpublished
Southeast Alaska, USA	SAK	16	(Guenette et al., 2007)
PrakramaSamudra Reservoir, Sri Lanka	PRA	17	(Moreau et al., 2001)
Prince William Sound, Alaska, USA (pre oil spill)	PW1	18	(Dalsgaard and Pauly, 1997)
Prince William Sound, Alaska, USA (post oil spill)	PW2	19	(Okey and Pauly, 1999)
West Florida Shelf, USA	WFS	20	(Okey et al., 2004)
Lake Kariba, Zimbabwe	KAR	21	(Machena et al., 1992)

Table 2 Metadata describing the motivation for creating the food web model and the different emphasis on species groupings in each study system. See Table 1 for full ecosystem names associated with the first column, ' Abbr.'. 'Percent of FW nodes aggregated' is the number of species groups that contain multiple species divided by the total number of species groups in the original food web model, as defined by the author, 'FW nodes'. 'RAM nodes' is the number of nodes in the RAM after removing cyclicity and detritus.

Abbr.	Study Goal	Species Focus	FW nodes	Percent of FW nodes aggregated	RAM nodes
ALE	Evaluate whether predation by killer whales might explain the decline of Steller sea lions in the central and western Aleutian Islands.	Steller sea lion and their principal prey species	40	60%	34
WED	Integrate the results of the various research efforts directed towards the shelf communities into a coherent whole.	dominant groups of benthic shelf community	20	100%	15
TAN	Quantify the food web and the production of pelagic fish and invertebrates.	pelagic fish and invertebrates	7	100%	4
ONT	Characterize the food web.	Phytoplankton, zooplankton, benthos were aggregated. Fish, mysid, amphipod species were left independent.	14	29%	13
GSL	Impact of groundfish collapse.	phytoplankton and detritus to marine mammals and seabirds, including harvested species of pelagic, demersal,	32	59%	30

		and benthic domains			
AYD	Understand functioning of eutorophic ecosystem.	emphasis on two dominat fish species, perch (*Percafluviatilis*) and roach (*Rutilusrutilus*)	11	82%	10
GBR	Identify the effects of the major fisheries in (1) mangrove, (2) lagoon-seagrass, and (3) coral reef systems, and the possible confounding effects of independently developed fisheries management plans.	fish and prawn	32	75%	8
GIR	Improve understanding of the complexity of estuarine ecosystems and response to various pressures.	estuarine fish	18	89%	17
ICE	Describe North Atlantic marine ecosystem with fisheries prior to expansion of large-scale commercial fisheries.	two primary producer groups, five invertebrate groups, twelve fish groups (including one juvenile group for cod), one seabirds group, three marine mammals groups and one detritus group	24	63%	20
KIN	Characterize the food web.	Good data for biomass and production of phytoplankton and zooplankton and diet and catches of main fish species.	14	71%	13
VIC	Evaluate change in dynamics of fish community after introduction of Nile perch.	fish-centric model	16	75%	12
NAR	Calculate carrying capacity for shellfish aquaculture.	Includes all trophic levels with emphasis on filter feeders.	15	93%	14
LDB	*Text not available, only data tables*	*Text not available, only data tables*	17	71%	15
LRI	Calculate carrying capacity for shellfish aquaculture.	Includes all trophic levels with emphasis on filter feeders.	16	88%	15
SRM	Characterize the food web.	fish and birds	29	62%	27
SAK	Understand why sea lions increased in the presence of killer whales in Southeast Alaska.	Steller sea lion and their principal prey species	40	68%	19
PRA	Describe trophic relationships and importance of unexploited fish stocks.	commercial fisheries and introduced tilapiine fish	17	71%	16
PW1	Characterize trophic interactions prior to oil spill.	not fish-centric, plankton to mammals	19	95%	18
PW2	Understand structure and functional characteristics of food web after oil spill.	primary producers, zooplankton, benthic invertebrates, planktivorous 'forage fishes', larger fishes, birds, mammals, and detritus	48	73%	13
WFS	Community effects of seafloor shading by plankton blooms.	primary producers	59	92%	6
KAR	Assess trophic interrelationships and community structure.	Trophic groups selected based on known importance and availability of data from the literature. Some groups were left out because of perceived minor importance for overall trophic flows. Some fish species were grouped both because commercial landing statistics do not separate individuals species and also because their biology is similar.	10	70%	9

2.2 *RAM* generation process

Despite the unique species groupings in each web, there is one exception - detritus. Every Ecopath model must include a detritus (decaying organic matter) component. Detritus is used to capture any unused energy in the ecosystem and in that way is common across all webs. The vertex with this detritus label shares at least one edge with every other vertex in each food web. Because we are interested in examining similarities and differences across food webs, the detritus vertex was removed from every food web. What remains from *FW* is a graph *F*.

It is common that food webs contain circuits whereby energy flows cyclically between specific predator and prey groups, typically across 2 adjacent trophic levels. Because we are interested in how energy moves across multiple trophic levels, smaller circuits such as these become less consequential. Cycles or hereafter, circuits, in F are located and reduced to a single vertex, resulting in a reduced acyclic model (RAM) of the food web (see Allesina et al. 2005 for another treatment of reduced acyclic graphs). The resulting 21 $RAMS$ are more uniform in size than the graphs in FW and contain fewer vertices, leading to simpler computational analysis.

2.3 Random RAM generation process

Because one of our goals is to find properties of networks that are unique to food webs, random graphs were created for comparison. Each graph F in FW has a density $d(F)$, the ratio of the number of edges to $n(F)(n(F)+1)$. Note that $n(F)(n(F)+1)$ is the maximum number of edges in a graph on $n(F)+1$ vertices. A random graph on $n(F)+1$ vertices was generated through inclusion of each potential edge among each pair of vertices with probability equal to $d(F)$. The resulting graph is then stripped of a vertex of highest total degree (i.e. the sum of in-degree and out-degree). What results is a graph R_F with the same order as and similar density to F. The set of all graphs of the form R_F is denoted RFW. Each graph in RFW undergoes the RAM generation process, described above, to result in a random equivalent of each RAM, or an $RRAM$.

2.4 Induced subgraphs

There exist 243,262 graphs on up to 7 nodes that are acyclic and connected, which were placed into array A. Each graph in A was examined for inclusion in each $RAMR$ as an induced subgraph, resulting in a binary array A_R. These binary arrays were subsequently added component-wise and the resulting array A_t, with integer components between zero and 21 inclusive, was generated. The array A_t is an indicator of occurrence for each graph in A among the $RAMS$. Graphs from A with at least five vertices were examined for high occurrence. Those that appeared often were included in a set which we will refer to as S, consisting of induced subgraphs of approximately two-thirds of the $RAMS$ examined (Fig. 1). It was discovered that no graph in A with greater than 6 nodes is an induced subgraph of a significant number of $RAMS$. Connected, acyclic graphs with more than 6 nodes did not appear as induced subgraphs with high enough frequency to be studied. Next, each graph in S was examined for its inclusion in each $RRAM$.

2.5 Network metrics

To examine similarities and differences across food webs, individual metrics of each FW and RAM were calculated and standardized against the metrics of order or size. The eight metrics listed in Table 3 were used for cluster and principle component analyses. Each metric captures a unique quality of the networks useful for examining similarities and differences across ecosystems.

We define some relevant graph theoretic terms below

Definition: The <u>clique number</u> of a graph is the order of the largest complete subgraph. An undirected graph is <u>complete</u> if each vertex is adjacent to every other vertex.

Definition: The <u>connectivity</u> (alternatively <u>edge-connectivity</u>) of a graph is the fewest number of vertices (edges) the removal of which results in a disconnected graph.

Definition: A graph's <u>density</u> is the ratio of its size to the maximum possible number of edges among its vertices.

Definition: The <u>distance</u> between vertices u, v in a graph G is the length of the shortest path from u to v. The <u>diameter</u> of G is the length of the greatest distance. If there are vertices with no path between them then the <u>finite diameter</u> is defined to be the length of the greatest distance between vertices that do have a connecting path.

Definition: An <u>independent set</u> is a collection of vertices with no adjacencies among one other.

Definition: An <u>induced path</u> is simply an induced graph in the form of a path. That is, no vertex along the path is adjacent to any other besides its neighbors along the path.

Definition: The <u>eccentricity</u> of a vertex v is the greatest distance from v to any other vertex in a graph. The <u>radius</u> of a graph is the lowest eccentricity.

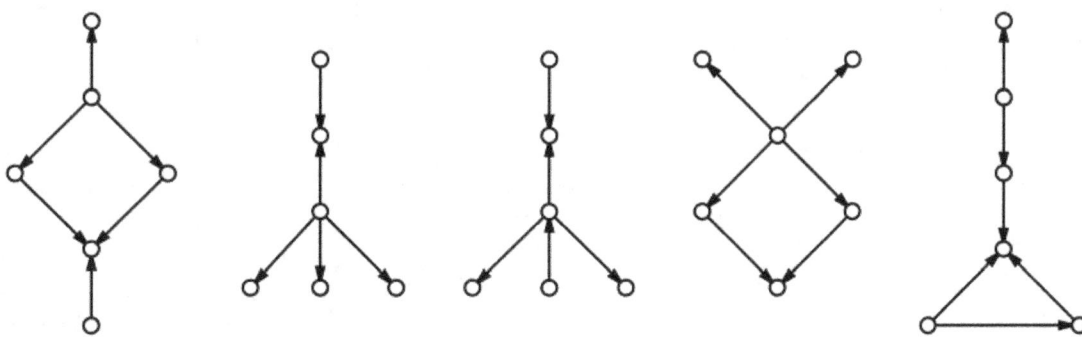

Fig. 1 The five six-node graphs below are those that appear in *14 RAMS* (Fig. 1a) and 13 *RAMS* (Figs. 1b-1e). None of the graphs in Fig. 1 appear as induced subgraphs in any of the *RRAMS*.

2.6 Cluster analysis

A cluster analysis was performed to examine relatedness among food webs. Clusters are based on the shortest Euclidean distances between computed metrics. The resulting dendrogram plot is of the hierarchical binary cluster tree where the height of the U-shaped bars are the distances (y-axis) between networks (x-axis) being connected.

2.7 PCA

A Principle Component Analysis (PCA) was performed to examine relationship between metrics. Since variances among metrics were similar due to standardizing against order or size, raw data was used to perform the PCA. The first principal component and second principal component were plotted for each network for both food webs and *RAMS*. A biplot of the principal component coefficients showing variables represented as vectors. This biplot allows visualization of the magnitude and sign of each variable's contribution to the first two principal components, and how each observation is represented in terms of those components.

3 Results

3.1 Induced subgraphs

No graph in *S* was found as an induced subgraph in any *RRAM*. It was determined that each *RRAM* is much smaller than the *RAMS*. Ecologically, the naturally-occurring *RAMS* have a more complex structure than randomized ones, and naturally-occurring food webs appear to have much lower cyclicity than randomized food webs.

Large graphs that were found to be induced subgraphs of a majority of *RAMS*, i.e. the graphs in *S*, were not represented in all *RAMS*. There are five systems in which no graph from *S* appeared as an induced subgraph (Lake Tanganyika in Burunidi, Lake Aydat in France, the Great Barrier Reef in Australia, West Florida Shelf in USA, Lake Kariba in Zimbabwe) (Table 3). The remaining 16 *RAMS* did contain at least one induced subgraph from the set *S*, and 9 of those *RAMS* contained all 5 common induced subgraphs (Aleutians in USA, Northern Gulf of St. Lawrence in Canada, Narragansett Bay in USA, Laguna de Bay in Philippines,

Rhode Island coastal lagoons in USA, Saco River estuary marsh in USA, Southeast Alaska, Lake Prakrama Samudra in Sri Lanka, and Prince William Sound Alaska prior to the oil spill) (Table 3). There appears to be no geographic or environmental pattern to the number of these common induced subgraphs a network contains (Fig. 2).

Table 3 Metrics computed for each Food Web (FW) and *RAM*. Metrics were standardized by order or edges making all values a relative proportion. The number of induced subgraphs for each network is specified in the first column, *Ind. Sub.* Networks either contained all 5 of the listed induced subgraphs (5), at least one induced subgraph (1+), or no induced subgraphs. Network names are listed in Table 1.

Ind. Sub.	Network Abbr. Num.	clique number/ order (undirected)		connectivity/ order (undirected)		density (edges/ max possible edges)		edge connectivity/ edges (undirected)		finite diameter/ order		independent set/ order (undirected)		induced path/ order (undirected)		radius/ order (undirected)	
		FW	RAM	FW	RAM	FW	RAM	FW	RAM	FW	RAM	FW	RAM	FW	RAM	FW	RAM
5	ALE 1	0.28	0.29	0.00	0.06	0.44	0.44	0.00	0.01	0.10	0.09	0.33	0.35	0.31	0.32	0.00	0.06
5	GSL 5	0.29	0.30	0.16	0.17	0.60	0.60	0.02	0.02	0.10	0.10	0.29	0.30	0.29	0.27	0.06	0.07
5	LDB 13	0.25	0.27	0.00	0.07	0.35	0.40	0.00	0.02	0.19	0.20	0.50	0.47	0.44	0.47	0.00	0.13
5	LRI 14	0.27	0.27	0.13	0.13	0.37	0.37	0.05	0.05	0.13	0.13	0.40	0.40	0.47	0.47	0.13	0.13
5	NAR 12	0.36	0.36	0.21	0.21	0.46	0.46	0.07	0.07	0.14	0.14	0.36	0.36	0.57	0.57	0.14	0.14
5	PRA 17	0.19	0.19	0.13	0.13	0.40	0.40	0.04	0.04	0.13	0.13	0.44	0.44	0.50	0.50	0.13	0.13
5	PW2 19	0.15	0.38	0.06	0.08	0.32	0.47	0.01	0.03	0.11	0.23	0.32	0.54	0.32	0.38	0.04	0.08
5	SAK 16	0.33	0.32	0.00	0.11	0.60	0.47	0.00	0.02	0.10	0.16	0.26	0.37	0.23	0.42	0.00	0.11
5	SRM 15	0.18	0.19	0.04	0.04	0.24	0.24	0.01	0.01	0.11	0.11	0.50	0.48	0.32	0.33	0.07	0.07
1+	GIR 8	0.29	0.29	0.06	0.06	0.38	0.38	0.02	0.02	0.18	0.18	0.41	0.41	0.47	0.47	0.12	0.12
1+	ICE 9	0.48	0.55	0.04	0.05	0.62	0.63	0.01	0.01	0.09	0.10	0.22	0.25	0.35	0.30	0.09	0.10
1+	KIN 10	0.31	0.31	0.08	0.08	0.42	0.42	0.03	0.03	0.23	0.23	0.46	0.46	0.54	0.54	0.15	0.15
1+	ONT 4	0.23	0.23	0.23	0.23	0.38	0.38	0.10	0.10	0.15	0.15	0.54	0.54	0.46	0.46	0.15	0.15
1+	PW1 18	0.28	0.28	0.11	0.11	0.35	0.35	0.04	0.04	0.17	0.17	0.44	0.44	0.50	0.50	0.11	0.11
1+	VIC 11	0.60	0.50	0.33	0.42	0.75	0.68	0.06	0.11	0.13	0.17	0.33	0.42	0.33	0.42	0.07	0.08
1+	WED 2	0.26	0.27	0.05	0.07	0.29	0.28	0.02	0.03	0.21	0.20	0.47	0.53	0.37	0.40	0.11	0.13
0	AYD 6	0.60	0.60	0.40	0.40	0.71	0.71	0.13	0.13	0.20	0.20	0.40	0.40	0.40	0.40	0.10	0.10
0	GBR 7	0.29	0.50	0.13	0.25	0.49	0.61	0.02	0.18	0.26	0.25	0.29	0.50	0.32	0.38	0.06	0.13
0	KAR 21	0.33	0.33	0.11	0.11	0.39	0.39	0.07	0.07	0.22	0.22	0.44	0.44	0.56	0.56	0.22	0.22
0	TAN 3	0.83	1.00	0.67	0.75	1.07	1.00	0.25	0.50	0.33	0.25	0.33	0.25	0.50	0.50	0.17	0.25
0	WFS 20	0.24	0.50	0.07	0.17	0.49	0.40	0.00	0.17	0.07	0.17	0.22	0.67	0.29	0.50	0.03	0.17

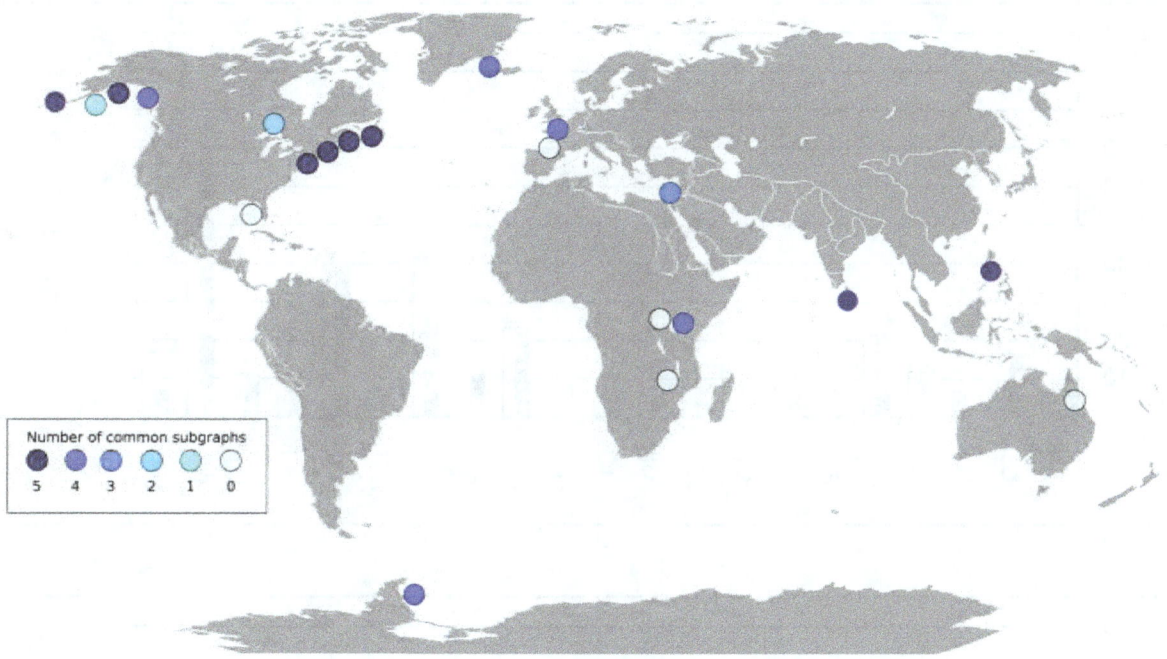

Fig. 2 Map of the world showing geographic locations of all the networks used in the study. Colored dots are shaded according to the number of the five common induced subgraphs in each network.

3.2 Network metrics

Metric values were similar for both food webs and *RAMS* across most networks (Fig. 3). Because metrics were standardized against order or size, there was low variance across the metrics.

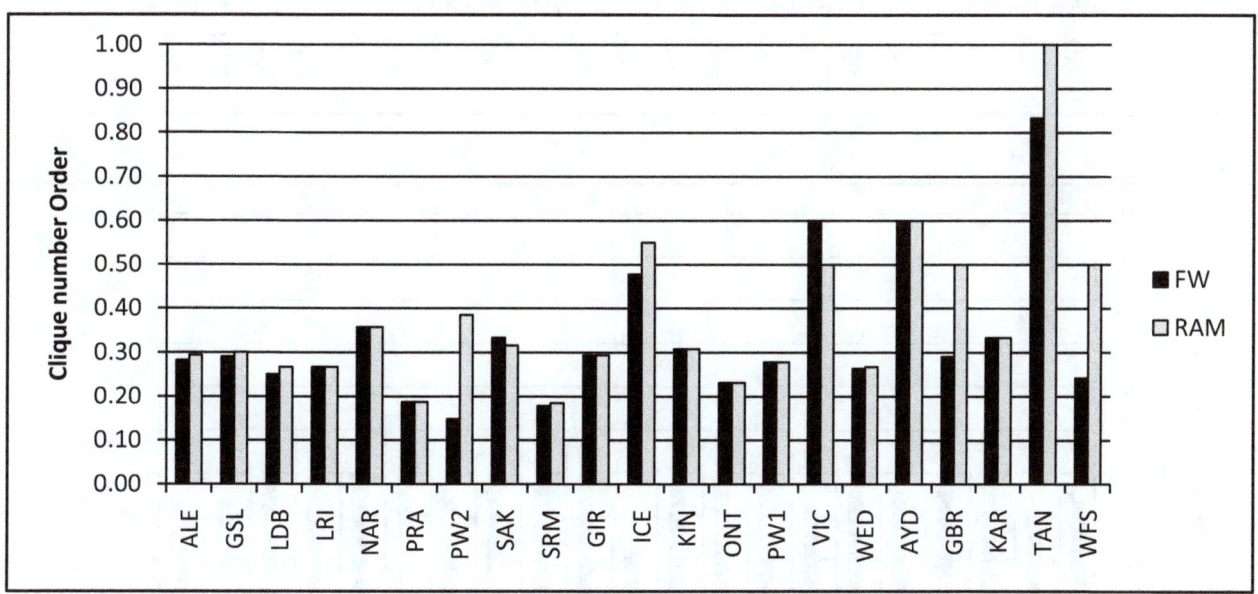

Fig. 3 a

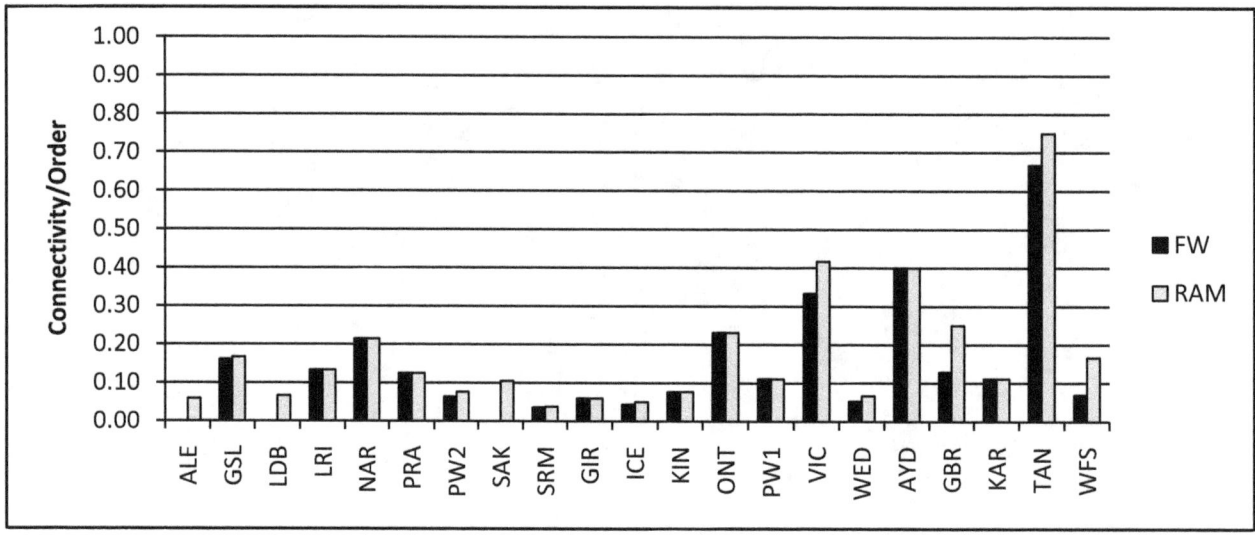

Fig. 3 b

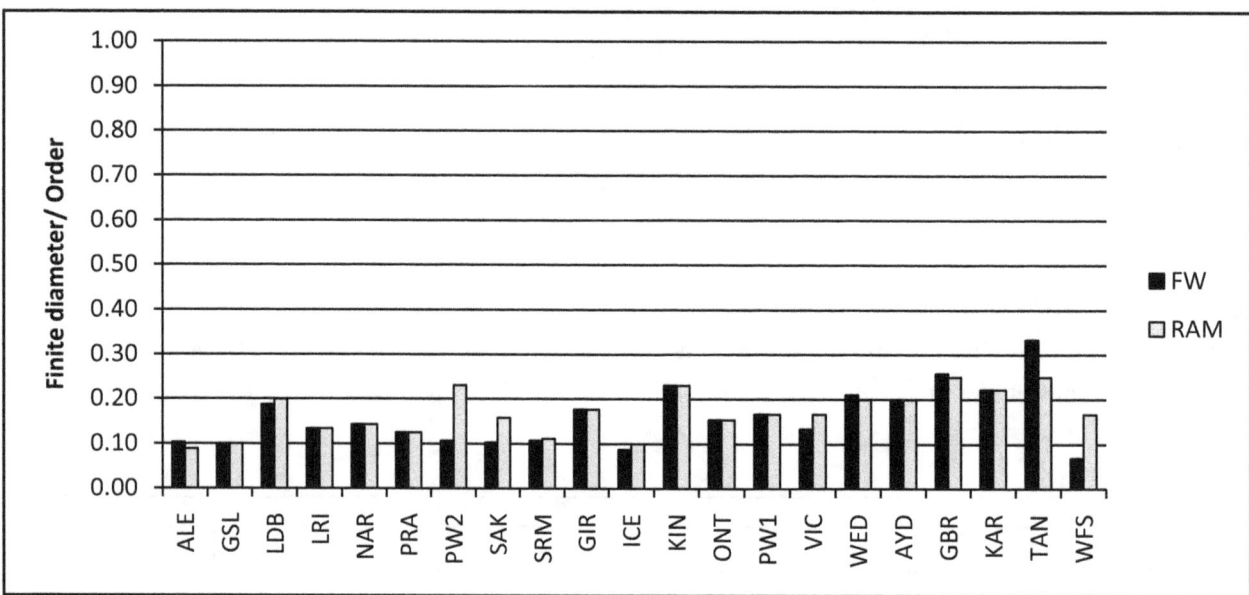

Fig. 3 c

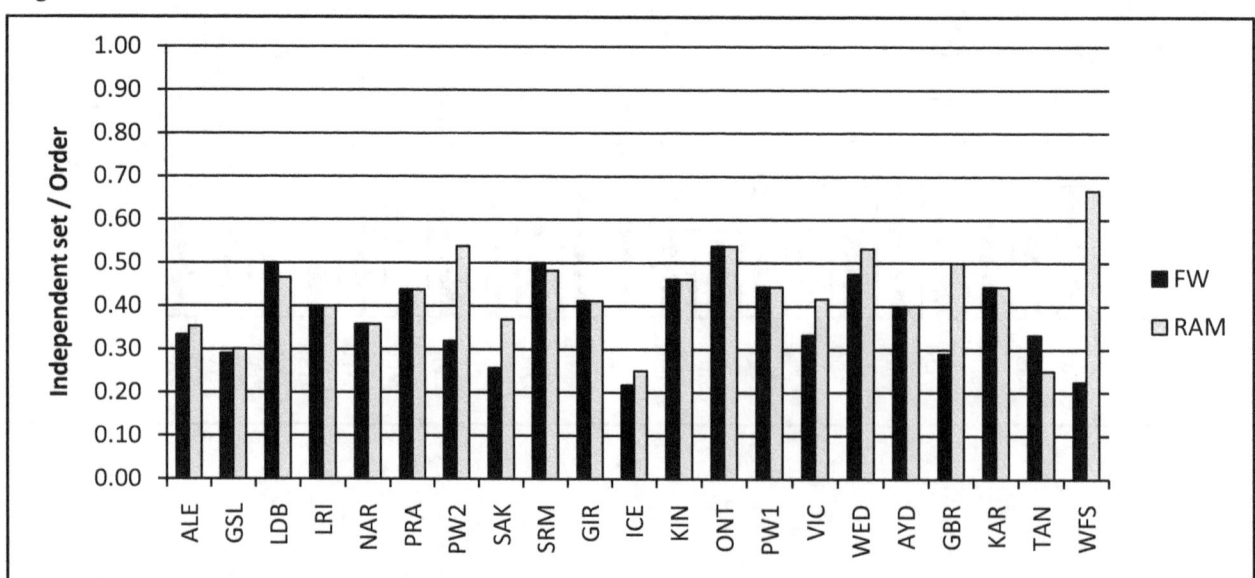

Fig. 3 d

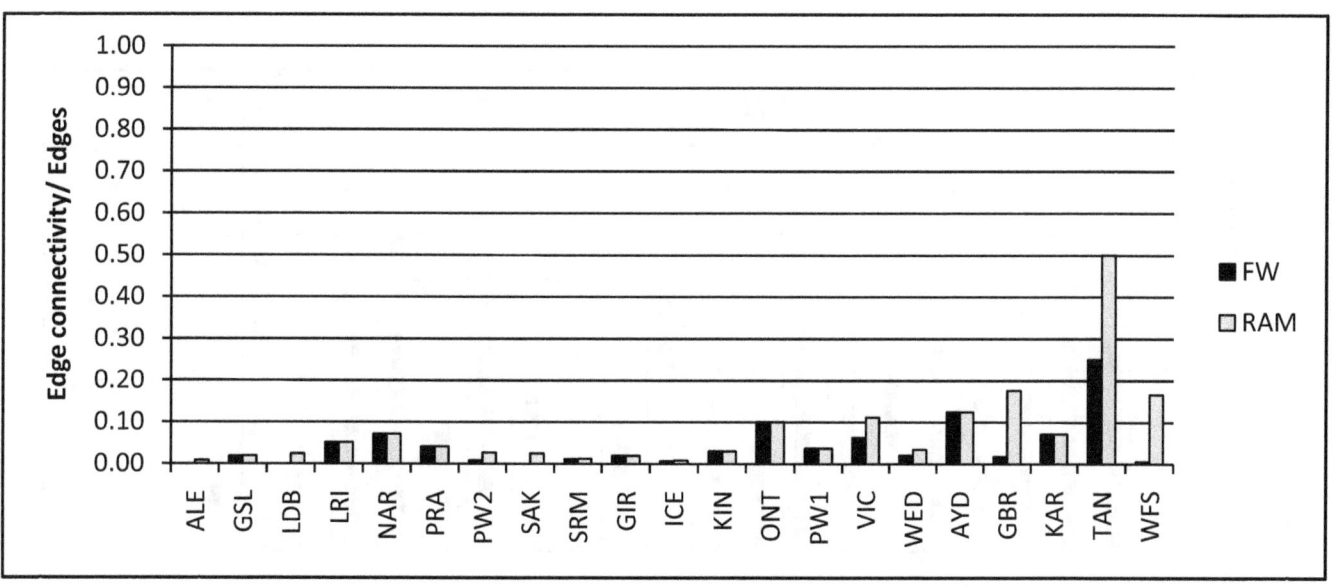

Fig. 3 e

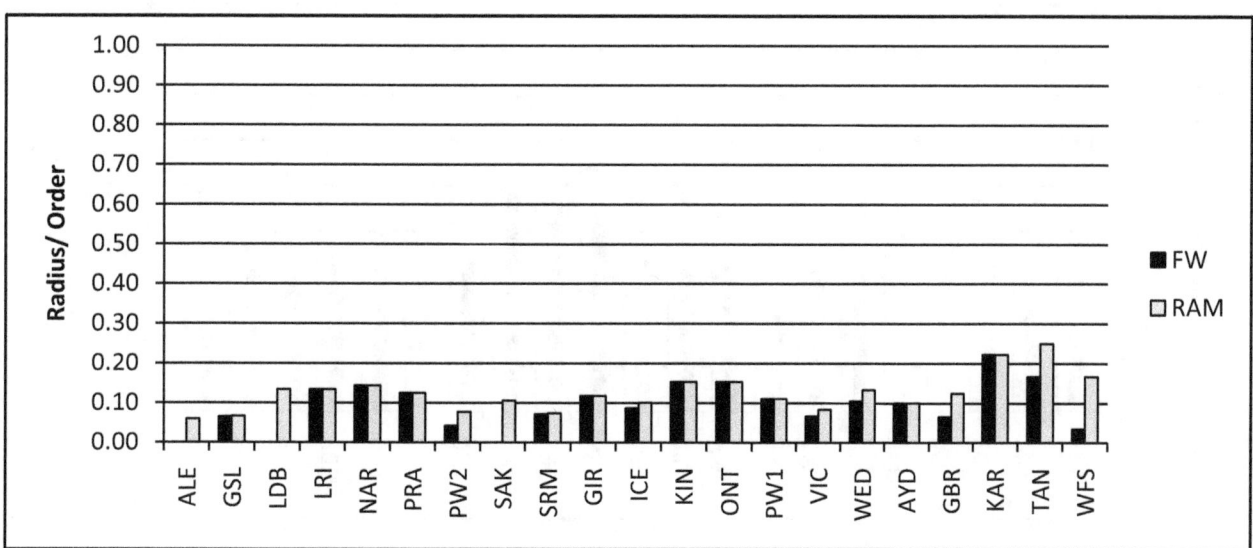

Fig. 3 f

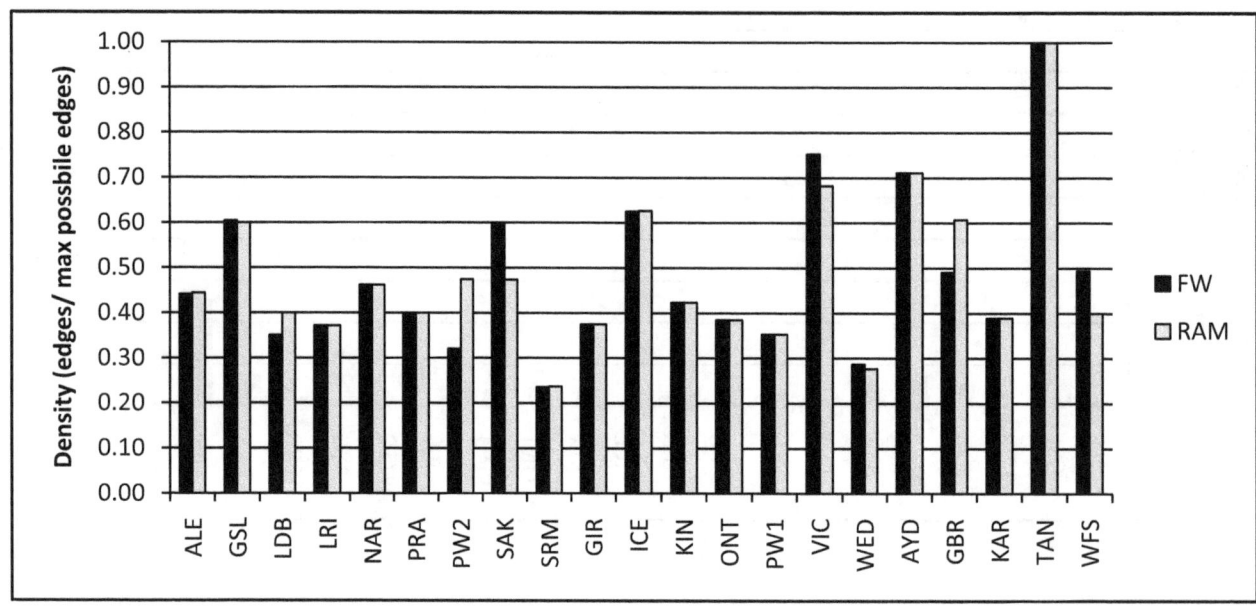

Fig. 3 g

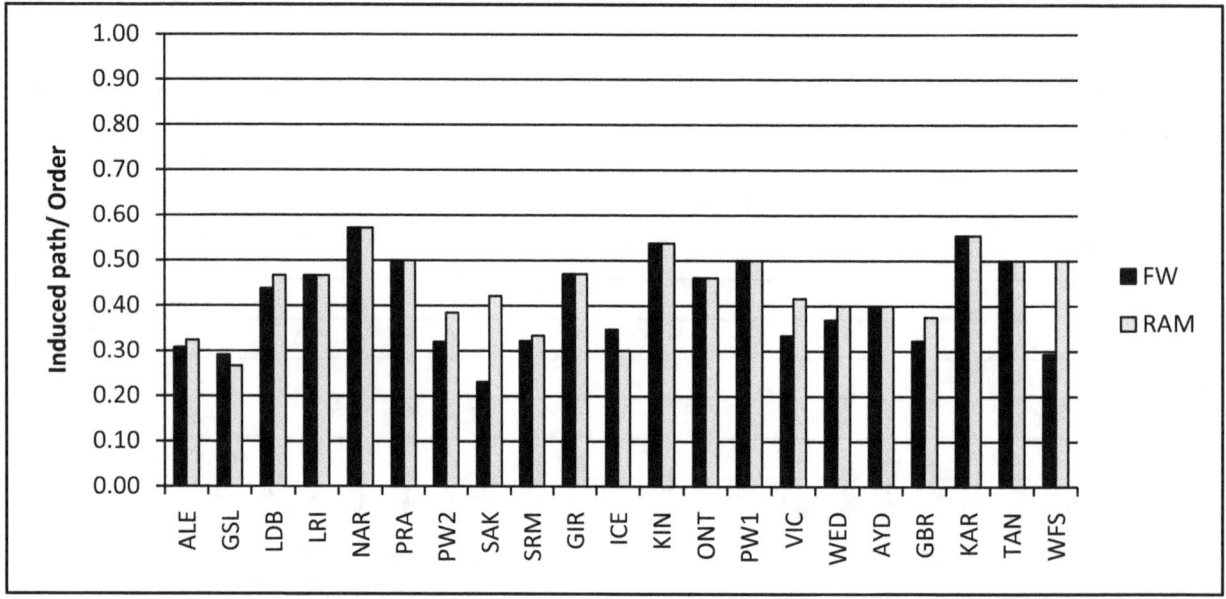

Fig. 3 h

Fig. 3 a-h Depicts data shown in Table 3. Each panel, a-h, is a different metric calculated on the Food Web (FW in dark gray) and associated *RAM* (light gray) for each network system. Network name abbreviations are listed in Table 1.

3.3 Cluster analysis

The food web network of the West Florida Shelf in the USA was unique to all other food webs (Figs. 4a, 5a). Conversely, the *RAMS* of Weddell Sea in Antarctica, Antarctica and Lake Aydat in France were similar to each other and different from all other *RAMS* (Figs. 4b, 5b).

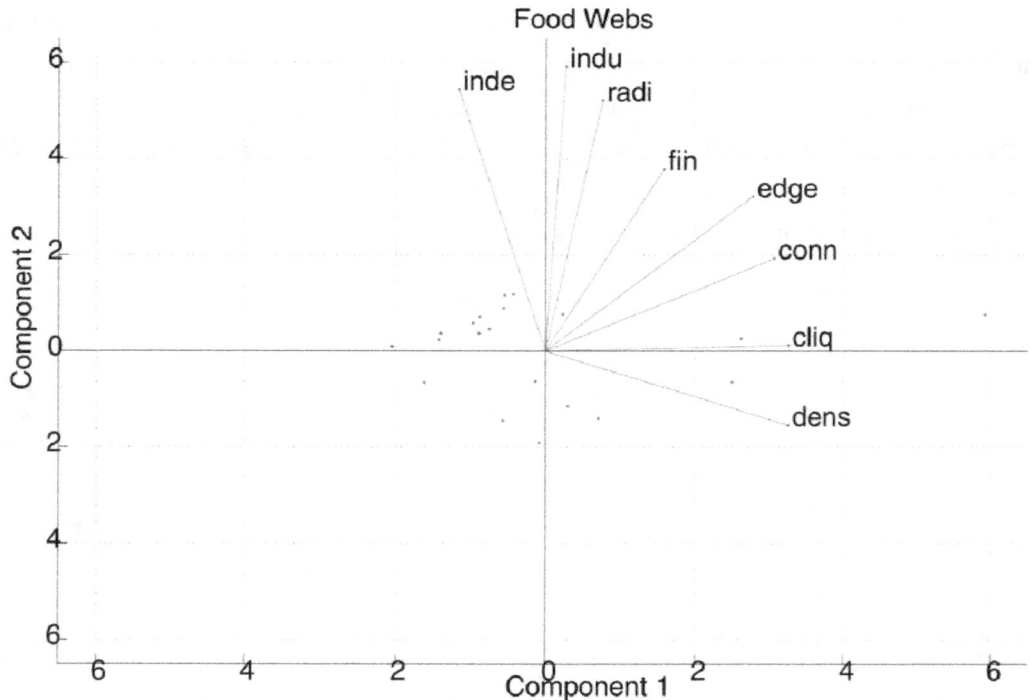

Fig. 4 a

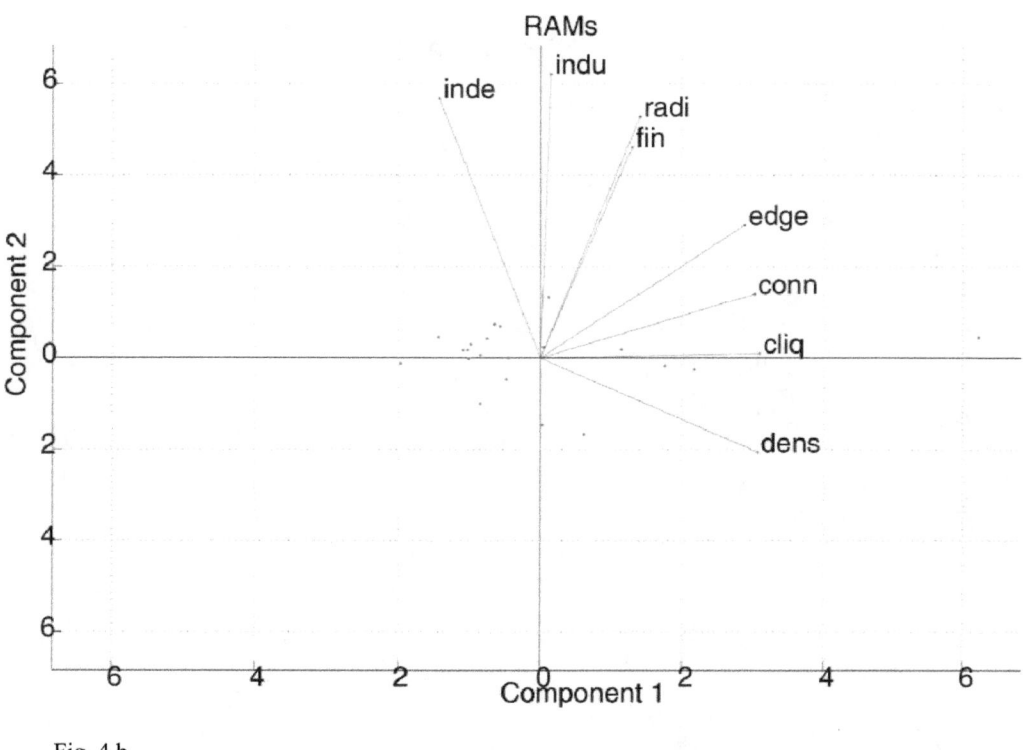

Fig. 4 b

Fig. 4 a-b PCA biplot of (a) food web and (b) *RAM* networks. The names of all the metrics are abbreviated by the first four letters as appear in bold face type in Table 3.

3.4 PCA

Because of the strong influence of size, all metrics were then standardized by size, or order. Preliminary analysis performed on raw data showed that the first PC was largely explained by size in both food webs (>90%) and *RAMS* (>55%). Using standardized data (Table 3), clique number became the most influential metric. The first principal component explained 70% of the variance (Fig. 5).

There were a few outlier networks, primarily Burundi Lake in Tanganyika. Lake Aydat in France and Lake Victoria in Kenya are strongly influenced by this first principal component. Marine food webs tend to group more closely together than freshwater systems (Fig. 6).

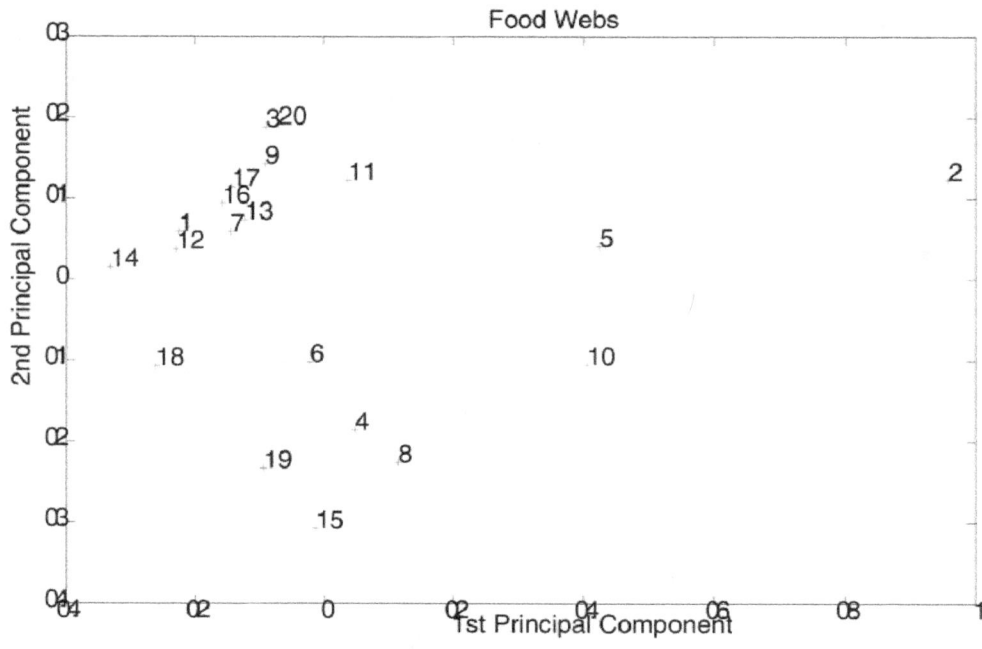

Fig. 5 a

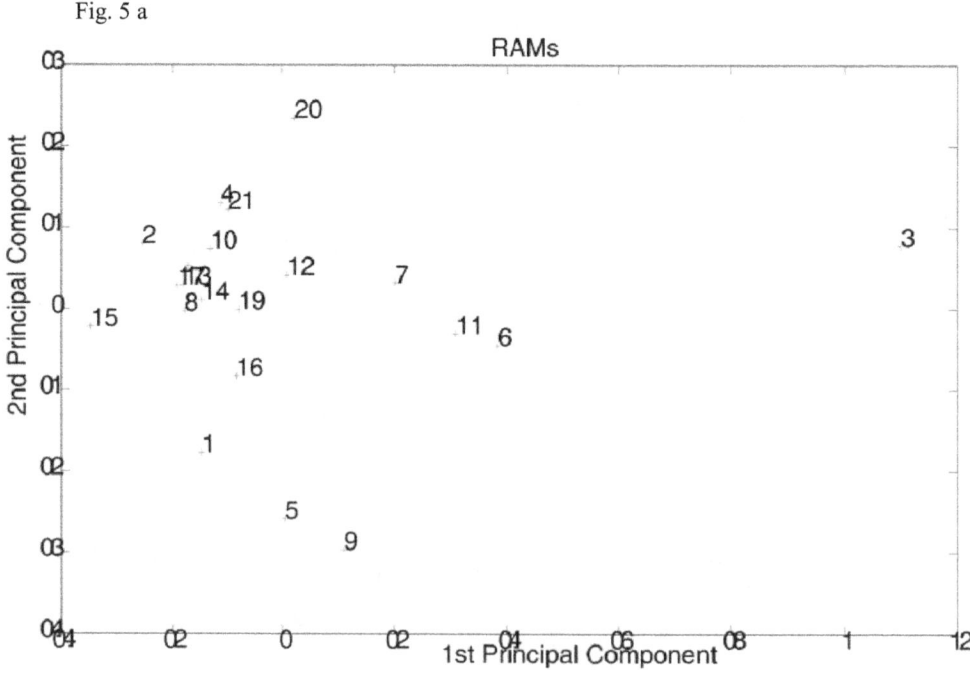

Fig. 5 b

Fig. 5 a-b Principal components of (a) food web and (b) *RAM* networks. See Table 1 for key of numeric labels of networks.

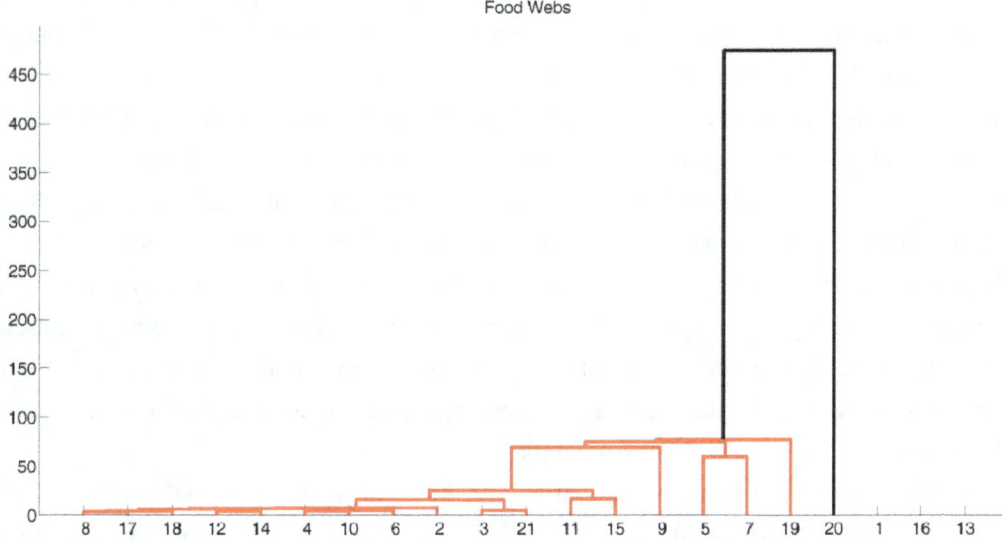

Fig. 6a

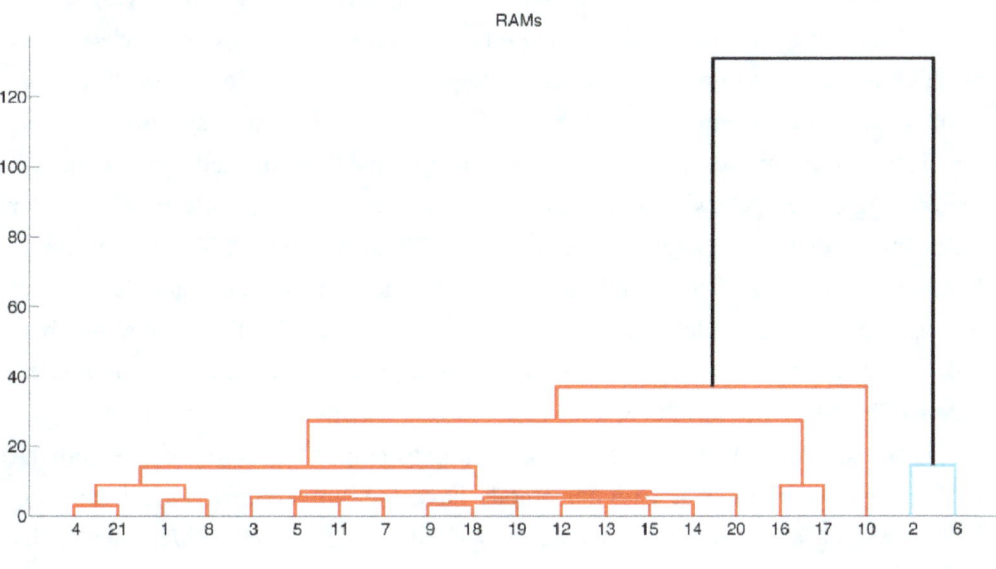

Fig. 6 b

Fig. 6 a-b Dendrogram resulting from cluster analysis of (a) food web and (b) *RAM* networks. A unique color is assigned to each group of nodes within the dendrogram whose linkage is less than 70% of the maximum linkage. X-axis: the numeric network abbreviations specified in Table 1. Y-axis: distance between two networks being connected.

4 Discussion

A goal of this study was to determine similarities in structure among food webs that are otherwise disparate with regard to species, population, and size. Through this study, we determined not only that there were pairs of network systems that were highly similar in structure, once appropriately normalized for size and makeup, but also that a majority of food webs have similar structural components when compared with random food webs. Structural similarities have been identified in sub components within food webs (Stouffer et al., 2007).

Other studies have also demonstrated that food webs from different types of ecosystems (i.e. marine, estuarine, fresh-water, terrestrial) share fundamental structural and ordering characteristics, despite variable diversity and complexity inherent in the web (Camacho et al., 2002; Dunne et al., 2004).

Based on the cluster analysis, environmental and geographical characteristics have little to do with how food webs are related to each other. There are no apparent environmental or geographical distinctions that easily explain why the West Florida Shelf, USA system was unique from all other systems or why the Weddell Sea, Antarctica and Lake Aydat, France cluster separately from all other systems. Most likely, West Florida Shelf system stands out from other systems because of the original motivation for creating the model. The modelers wanted to investigate the effect of phytoplankton shading on benthic primary production. This research question is quite different than that motivating any of the other study system (Table 2). Therefore, the uniqueness of ecosystems may be attributed to the research question structuring the model, rather than the inherent organization or structure of the ecosystem itself.

We attempted to control for some of the variability, inherent model construction for different research purposes and goals, by removing circuits and reducing full food webs into *RAMS*. Several other studies that examined food webs for similar network structures only considered substructures on full food webs (Borrelli, 2015; Stouffer et al., 2007). Despite our attempt to normalize models against initial construction biases, it is possible that *RAMS* still capture some of these model construction biases. For example, if food web A has one group called 'planktivorous fish' preying on zooplankton compared to food web B having three groups called 'herring' and 'mackerel' and 'sandlance' all preying on zooplankton, then those two webs could look different after reducing them to *RAMS*. How modelers aggregate groups of species does impact network ecology and graph theory metrics. Modelers often make their decision on how to group species based on a particular research focus (Table 2). In this study, we used all Ecopath derived models and it is a common practice that Ecopath modelers aggregate species into functional groups, based on available data, in order to limit the number of nodes and produce a manageable sized network (Christensen et al., 2008). We found a high degree of aggregation, where most models contained a high percentage of nodes that included several functionally similar species, as opposed to a single species (Table 2). It is suggested that the optimal number of nodes, 12-24, above and below which mass balance food web models become less helpful for understanding these complex systems (Christensen et al., 2008; Plagányi, 2004; Plagányi, 2007). In this study, we then further aggregate species across trophic levels by removing cycling to reduce SCCs to a single node (Allesina et al., 2005) (Table 2). These types of aggregation methods have the similar objective of reducing complexity. Furthermore, PCA and cluster analyses results are very similar for FW and *RAMS* which is justification for creating *RAMS* in the first place.

Alesina et al. (2005) only found four out of 17 food webs that did not reduce to a single node after a process similar to our *RAM* construction. Allesina et al. (2005) uses the term DAGs – Directed Acyclic Graphs to describe the same thing as our *RAMS*. The term DAG describes a particular type of mathematical object that implies an abstract mathematical structure whereas we feel the term *RAM* more clearly represents the object itself. We discovered that the food webs in our data set also had high cyclicity before removing detritus. Once the nodes associated with detritus were removed from our food webs, we found many more "interesting" reduced graphs.

This study further emphasizes that food webs are not random and that they are more similar across a range of ecosystem types than may be predicted. The similarities we were able to identify were embedded in the graphical structure of the network and are not necessarily connected to obvious environmental or geographical properties of the ecosystems themselves. These findings emphasize the importance of using innovative techniques for investigating and quantifying complex networks for the purpose of ecosystem management.

Acknowledgment

The authors would like to thank Dr. Nick Record for his editorial comments and the anonymous reviewers that helped advance this manuscript.

References

Allesina S, Bodini A, Bondavalli C. 2005. Ecological subsystems via graph theory: the role of strongly connected components. Oikos, 110: 164-176

Bascompte J, Melián CJ. 2005. Simple trophic modules for complex food webs. Ecology, 86: 2868-2873

Borrelli JJ. 2015. Selection against instability: stable subgraphs are most frequent in empirical food webs. Oikos, 1-6

Buchary EA. 2001. Preliminary reconstruction of the Icelandic marine ecosystem in 1950 and some predictions with time series data, Fisheries impacts on the North Atlantic ecosystems: Models and analysis, vol. 4. 296-306, FCRR

Byron C, Link J, Bengtson D, Costa-Pierce B. 2011a.Calculating Carrying Capacity of Shellfish Aquaculture Using Mass-Balance Modeling: Narragansett Bay, Rhode Island. Ecological Modelling, 222: 1743-1755

Byron C, Link J, Costa-Pierce B, Bengtson D. 2011b. Modeling Ecological Carrying Capacity of Shellfish Aquaculture in Highly Flushed Temperate Lagoons. Aquaculture, 314: 87-99

Camacho, J., Guimera, R., Amaral, L., 2002. Robust patterns in food web structure. Physical Review Letters, 88: 228102

Christensen V. 2008. Ecopath with Ecosim version 6 User Guide.

Dalsgaard J, Pauly D. 1997. Preliminary mass-balance model of Prince William Sound, Alaska, for the pre-spill period 1980-1989. Fisheries Centre, University of British Columbia, Canada

Delos Reyes MR. 1995. Geology of Laguna de Bay, Philippines: Long-term alterations of a tropical-aquatic ecosystems 1980-1992. Universitaet Hamburg, Germany

Dunne JA. 2009. Ecological Networks: Linking Structure to Dynamics in Food Webs. In: Workshop on Theoretical Ecology and Global Change (Pascual M, Dunne JA, eds). Oxford University Press, UK

Dunne JA, Williams RJ, Martinez ND. 2002. Food-web structure and network theory: The role of connectance and size. Proceedings of the National Academy of Sciences of USA, 99: 12917-12922

Dunne J.A., Williams, R.J., Martinez, N.D., 2004.Network structure and robustness of marine food webs. Marine Ecology Progress Series 273: 291-302

Gribble NA. 2005. CD Ecosystem modeling of the Great Barrier Reef: A balanced trophic biomass approach., in: Zerger, A., Argent, R.M. (eds.), MODSIM 2005 International Congress on Modelling and Simulaitons. 170,176, Modelling and Simulation Society of Australia and New Zealand

Guenette S, Christensen V. 2005. Foodweb models and data for studying fisheries and environmental impact on Eastern Pacific ecosystems. Fisheries Center, University of British Columbia, Vancourver, BC, Canada

Guenette S, Heymans SJJ, Christensen V, Trites AW. 2007. Ecosystem models of the Aleutian Islands and southeast Alaska show that Steller sea lions are impacted by Killer Whale predation when sea lion numbers are low. In: Proceedings of the Fourth Glacier Bay Science Symposium Vol. 2007-5047 (Piatt JF, Gende SM, eds). 150-154, US Geological Survey Scientific Investigations Report, USA

Halfon E, Schito N. 1993. Lake Ontario food web, an energetic mass balance. In: Trophic models of Aquatic ecosystems (Christensen V, Pauly D, eds). 29-39, ICLARM, Manila, Phillipines

Jarre-teichmann A, Brey T, Bathmann UV, Dahm C, et al. 1997. Trophic flows in the benthic shelf community of the eastern Weddell Sea, Antarctica. In: Antarctic Communities: Species, Structure and Survival (Bettaglin B, Valencia J, Walton DWH, eds). 118-134, Cambridge University Press, Cambridge, UK

Lobry J. 2004. Which reference pattern of functioning for estuarine ecosystems? The case of fish succession in the Gironde estuary. PhD Thesis. University of Bordeaux, France

Machena C, Kolding J, Sanyanga RA. 1992. Preliminary assessment of the trophic structure of Lake Kariba, Africa. In: Trophic models of aquatic ecosystems (Christensen V, Pauly D, eds). 130-137, ICLARM, Manila, Phillipines

Moreau J, Ligtvoet W, Palomares MLD. 1993a. Trophic relationship in the fish community of Lake   Victoria, Kenya, with emphasis on teh impact of Nile perch (Latesniloticus). In: Trophic models of aquatic ecosystems (Christensen V, Pauly D, eds). 144-152, ICLARM, Phillipines

Moreau J, Nyakageni,B., Pearce M, Petit P. 1993b. Trophic relationships in the pelagic zone of Lake Tanganyika (Burundi Sector). In: Trophic models of aquatic ecosystems (Christensen V, Pauly D, eds). 138-143, ICLARM, Manila, Phillipines

Moreau J, Villaneauva MC, Amarasinghe US, Schiemer F. 2001. Trophic relationships and possible evolution of the production under various fisheries management strategies in a Sri Lankan reservoir, In: Reservoir and Culture-based Fisheries: Biology and Management Vol. 98 (DeSilva, ed). Australian Centre for International Agricultural Research, Canberra, Australia

Morissette L, Despatie SP, Savenkoff C, Hammill MO, Bourdages H, Chabot D., 2003. Data gathering and input parameters to construct ecosystem models for the northern Gulf of St. Lawrence (mid-1980s). Canadian Technical Report of Fisheries and Aquatic Sciences 2497 Vol. 2497. Department of Fisheries and Oceans, Quebec, Canada

Okey TA, Pauly D. 1999. A trohpic mass balance model of Alaska's Prince William Sound Ecosystem for the post-spill period 1994-1996 (2nd edition). Fisheries Centre Research Reports Vol. 7(4). University of British Columbia, Vancouver, Canada

Okey TA, Vargo GA, Mackinson S, Vasconcellos M, Mahmoudie B, Meyer CA. 2004. Simulating community effects of sea floor shading by plankton blooms over the West Florida Shelf. Ecological Modelling, 172: 339-359

Plangányi EE. 2007. Models for an ecosystem approach to fisheries. FAO Fisheries Technical Paper Series 477. FAO, Italy

Plagányi ÉE, Butterworth DS. 2004. A critical look at the potential of Ecopath with Ecosim to assist in practical fisheries management. South African Journal of Marine Science, 26: 261-287

Reyes-Marchant P, Jamet JL, Lair N, Taleb H, Palomares MLD. 1993. A preliminary model of a eutrophic lake (Lake Aydat, France). In: Trophic models of aquatic ecosystems (Christensen V, Pauly D, eds). ICLARM, Manila, Phillipines

Stouffer DB, Camacho J, Jiang W, Amaral LAN. 2007. Evidence for the existence of a robust pattern of prey selection in food webs. Proceedings of the Royal Society B: Biological Sciences, 274: 1931-1940

Walline PD, Pisanty S, Gophen M, Berman T. 1993. The ecosystem of Lake Kinneret, Israel. In: Trophic models of aquatic ecosystems (Christensen V, Pauly D, eds). 103-109, ICLARM, Manila, Phillipines

Watts DJ, Strogatz SH. 1998. Collective dynamics of 'small-world' networks. Nature, 393: 440-442

Williams RJ, Martinez ND. 2000. Simple rules yield complex food webs. Nature, 404: 180-183

4

Comparative structural analysis of HAC1 in *Arabidopsis thaliana*

Amar Ćemanović, Jasmin Šutković, Mohamed Ragab Abdel Gawwad

Genetics and Bioengineering department, International University of Sarajevo, Ilidza, 71220 Bosnia and Herzegovina

E-mail: mragab@ius.edu.ba,mrgawad@hotmail.com

Abstract

Histone acetylation is an important posttranslational modification correlated with gene activation. In *Arabidopsis thaliana*, the histone acetyltransferase 1 (AtHAC1) is homologous to animal p300/CREB (cAMP-responsive element-binding protein)-binding proteins, which are the main histone acetyltransferases participating in many physiological processes, including proliferation, differentiation, and apoptosis. In this study the 3-D structure of the HAC1 protein in *Arabidopsis thaliana* was predictedusing 4 homology-based prediction servers: ESyPred3D, 3D-JIGSAW, SWISS-MODEL and PHYRE2. The homology modeled structureswere evaluated and stereochemical analysis done by Ramachadran plot analysis. The amino acid sequences of *Arabidopsis thaliana* HAC1protein are highly similar to the sequence of the homologous human p300/CREB. SWISS MODEL and Phyre2 servers computed the identical 3D structures. Validation and verification methods, using Z-score and 3D-1D score, showed that these 3D models are of good and acceptable quality.

Keywords *Arabidopsis thaliana*; HAC1 protein; 3-D structure; structure prediction, homology.

1 Introduction

Arabidopsis thaliana, also known as thale-cress or mouse-ear cress, or simply Arabidopsis,is a small flowering plant belonging to the *Brassicaceae* family, a plant family comprising species such as radish and cabbage. Although not of agronomic importance, it is a widely known plant due to its use as a model organism in various genetic and plant biology studies. In addition, it is the first plant that had its genome sequenced. The prominent features of Arabidopsis are its small genome (135 Mb), rapid life cycle and ease of cultivation in laboratory conditions (The Arabidopsis Genome Innovative, 2000).

In cells of eukaryotic organisms, including plants, the DNA is associated with histone proteins to form structures called nucleosomes. In nucleosomes DNA is wound around a group of eight histones, which makes the region inaccessible to transcription factors and therefore transcriptionally inactive. Histone acetyltransferases (HATs) are a group of enzymes which function in transcriptional regulation by means of histone modification. Specifically, they transfer acetyl groups from acetyl-CoA to specific residues in histone tails, thereby lowering the affinity of histones for DNA and, accordingly, making the DNA more accessible to the transcription machinery. The process of acetylation is reversible, with the deacetylation being performed by deacetylase enzymes (Sterner and Berger, 2000).

Based on sequence similarity, histone acetyltransferases have been organized into four families: GNAT, MYST, p300/CBP and TAF$_{II}$250. The p300/CBP family is especially important since it acetylates not only histones, but also various other proteins, including CREB, p53, Stat3 and HIV-1 Tat protein (Sterner and Berger, 2000).This protein family in *Arabidopsis thaliana*includes the following members: HAC1, HAC2, HAC4, HAC5 and HAC12 (Pandey et al., 2002).

From the members of the p300/CBP family, HAC1 has the highest homology with the human and animal p300/CREB protein (Deng et al., 2007). Although the human p300/CREB is well characterized and has an experimentally determined crystal structure, little is known about its homolog in *Arabidopsisthaliana*. It has been found that it is located in the nucleus and exhibits, besides its acetyltransferase activity, also transcription cofactor activity. Furthermore, it is found to acetylate primarily histones H3 and H4 (Earley et al., 2007).

The aim of this paper is to predict the best 3D model for the HAC1 protein in *Arabidopsis thaliana*, using homology modeling techniques by model comparison and evaluation of the predicted structures.

2 Materials and Methods

The protein sequence and information (Name and Origin, Protein attributes, General annotation and Entry information) of HAC1 was retrieved from the National centre of Biotechnology information (NCBI), in FASTA format, with the accession number NP_565197. Additional information were taken from the Arabidopsis Information Resource (TAIR), with the identification number AT1G79000. Secondary Structure Prediction of HAC1 protein was done by GOR IV server prediction server, based on the information theory (Garnier et al., 1996). There is no defined decision constant. GOR IV uses all possible pair frequencies within the window of 17 amino acid residues.

The protein sequence was subjected for comparative homology modelling via SWIS-MODEL Server (Schwede et al., 2003) and PHYRE2 server (Kelley LA and Sternberg MJE, 2009). SWISMODEL software assured a better control of the homology modelling process and promoted a deeper understanding of various features of the protein structure meant to be modeled.Phyre2 is a major update to the original Phyre server with a range of new features, accuracy is improved, using the alignment of hidden Markov models via HH search to significantly improve accuracy of alignment and detection rate (Jefferys et al., 2010).For additional structural and comparative analysis, ESyPred3D (Cristophe et al., 2002) and 3DJIGSAW (Bates et al., 2001) online servers predicted the 3D structure of HAC1.

Finally, once the 3D structures were generated, structural evaluation and stereo-chemical analysis was performed using different evaluation and validation tools. Backbone conformation of all models was evaluated by analysis of Psi/Phi Ramachandran plot using RAMPAGE program (Lovel et al., 2002). Energy calculations on a protein chain were performed by atomic empirical mean force potential ANOLEA (Melo et al., 1997). The program performs energy calculations on a protein chain, evaluating the "Non- Local Environment" (NLE) of each heavy atom in the molecule. A plot obtained from the Verify3D (Eisenberg, D., 1997) structure Evaluation Server that represented the average3D-1D profile score. Verify3D analyzes the compatibility of an atomic model (3D) with its own amino acid sequence (1D).

3 Results and Discussion

The most successful general approach for predicting the structure of proteins involves the detection of homolog's of known three-dimensional (3D) structure—the so-called template-based homology modeling or fold-recognition. Protein 3D structure is very important in understanding the protein functions, interactions and localizations (Denise Chen, 2009). In this study, for *Arabidopsis thaliana* histone acetyltransferase 1 protein , with 1697 amino acids, the best homolog found in all modeling programs, is the human p300 histone acetyltransferase (HAT), with protein data bank ID '3BIY'. The best homolog found in both modelers, Phyre2and SWISS modeler, with 45.8% sequence similarity, is a human p300 histone acetyltransferase (HAT). HAT1 regulates gene expression by acetylating histones and other transcription factors (Xu et al., 2001).

The secondary structural analysis of the protein was done and random coil was found to be most frequent (54.63%), followed by alpha helix (28.87%). Extended strand (Ee) was found to be least frequent (16.50%) (Data not shown).The dominance of the coiled regions indicates the high level of conservation and stability of the protein structure (Neelamathi et al., 2009).The predicted models were visualized by PyMol software, in the mode of cartoon view, with the lines removed.

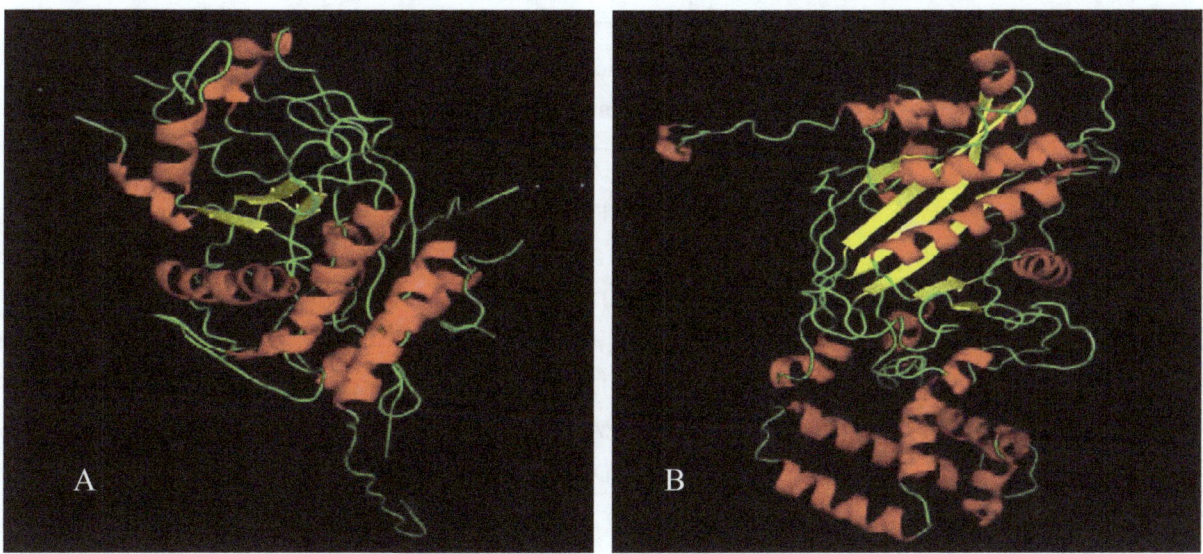

Fig. 1 Predicted models of HAC1 protein by (A)ESyPred3D and (B)3D-JIGSAW servers.

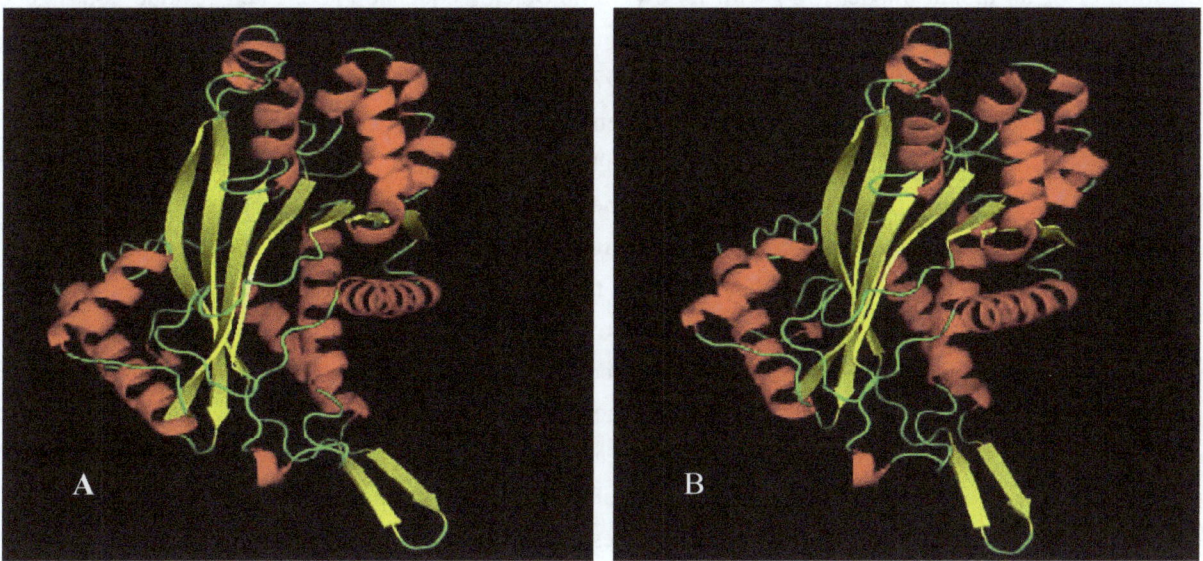

Fig. 2 Predicted models of HAC1 protein by (A) Phyre2and (B) SWISS-MODEL servers.

Benchmark of 4 different homology modeling programs: 3DJIGSAW, Swiss-Model, EasyPred3D and Phyre2, has been performed used to transform the alignment to a 3D model. The difference between the programs is how the information contained in the alignment is used to build a 3D model.

One of the most frequently used modelling programs; SWISS-MODEL generates the 3D models from a small number of rigid bodies, obtained from the core of the aligned regions (Blundell et al., 1987). The final three-dimensional (3D) structure was then confirmed by the Phyre2 server, an automatic fold recognition server for predicting the structure and the function of proteins based on homology modeling (Lawrence, 2011).

Errors of the model are usually estimated either from the energy of the model or from theresemblance of a given characteristic of the model to real structures (Sippl, 1995). Predicted models of HAC1 (Figs. 1-4) are evaluated by ANOLEA and Verify3D tools (Table 1).

Table 1 Model evaluation of proteins by using ANOLEA and Verfy3D tools.

Protein	Modeling program	Evaluation program	High energyzone (%)	Z-score	3D-1D profile
AtHAC1	ESyPred3D	ANOLEA	88.27%	18.04	-
		Verify3D	-	-	0.68
	3DJIGSAW	ANOLEA	51.67%	6.15	-
		Verify3D	-	-	0.72
	SWISS MODEL	ANOLEA	36.95%	3.51	-
		Verify3D	-	-	0.71
	PHYRE2	ANOLEA	36.95%	3.51	-
		Verfy3D	-	-	0.71

Verify3D method assesses protein structures using three-dimensional profiles. This program analyzes the compatibility of an atomic model (3D) with its own amino acid sequence (1D). Each residue is assigned a structural class based on its location and environment (alpha, beta, loop, polar, apolar etc).The scores ranges from -1 (bad score) to +1 (good score) (Eisenberg, 1997). ANOLEA program performs energy calculations on a protein chain, evaluating the "Non- Local Environment" (NLE) of each heavy atom in the molecule (Melo et al., 1997).ANOLEA results (Tab.1) are shown as High energy (%) and pseudo Z-score. High energy zone (HEz) in the protein profile correlates with errors in the structure. Z-score is represents as pseudo energies of target protein sequences, where lower Z-score means higher reliability and vice versa.

Energy calculations performed by ANOLEA, gave the Z-score of 3.51 in 3D models generated by Phyre 2 and SWISS-MODEL server, indicating same quality of the models. While the HAC1 models obtained from EasyPred3D and 3DJIGSAW gave 18.04 and 6.15 Z-scores, two to six fold higher scores compared to SWISS Models and Phyre2 servers. The high Z-score obtained from EasyPred3D server is caused by the specific alignment method, obtained by combining, weighting and screening the results of several multiple alignment programs. Therefore the final three-dimensional structures is built using the modeling package MODELLER, which is not done in this study. 3D-1D profile scores, obtained from Verify3D, of HAC1 models is 0.71 obtained from Phyre 2 and SWISS-MODEL servers. EasyPred3D and 3DJIGSAW homology modeler resulted in 0.72 and 0.68 respectively.

The 3D structures generated in SWISS-MODEL and Phyre2 servers are identical, having the same number of α helixes and β sheets. Z-score of SWISS-MODEL and Phyre2 modeler is much smaller compared to EasyPred3D and 3DJIGSAW server. Furthermore, the 3D-1D score of 0.72 (maximum is 1) indicates a good quality 3D model, compared to the human homolog model. Ramachandran plots analysis, generated by the RAMPAGE program (Lovel et al., 2002) for all four 3D models, shows all residues in the most favorable regions and disallowed regions (Fig. 3).

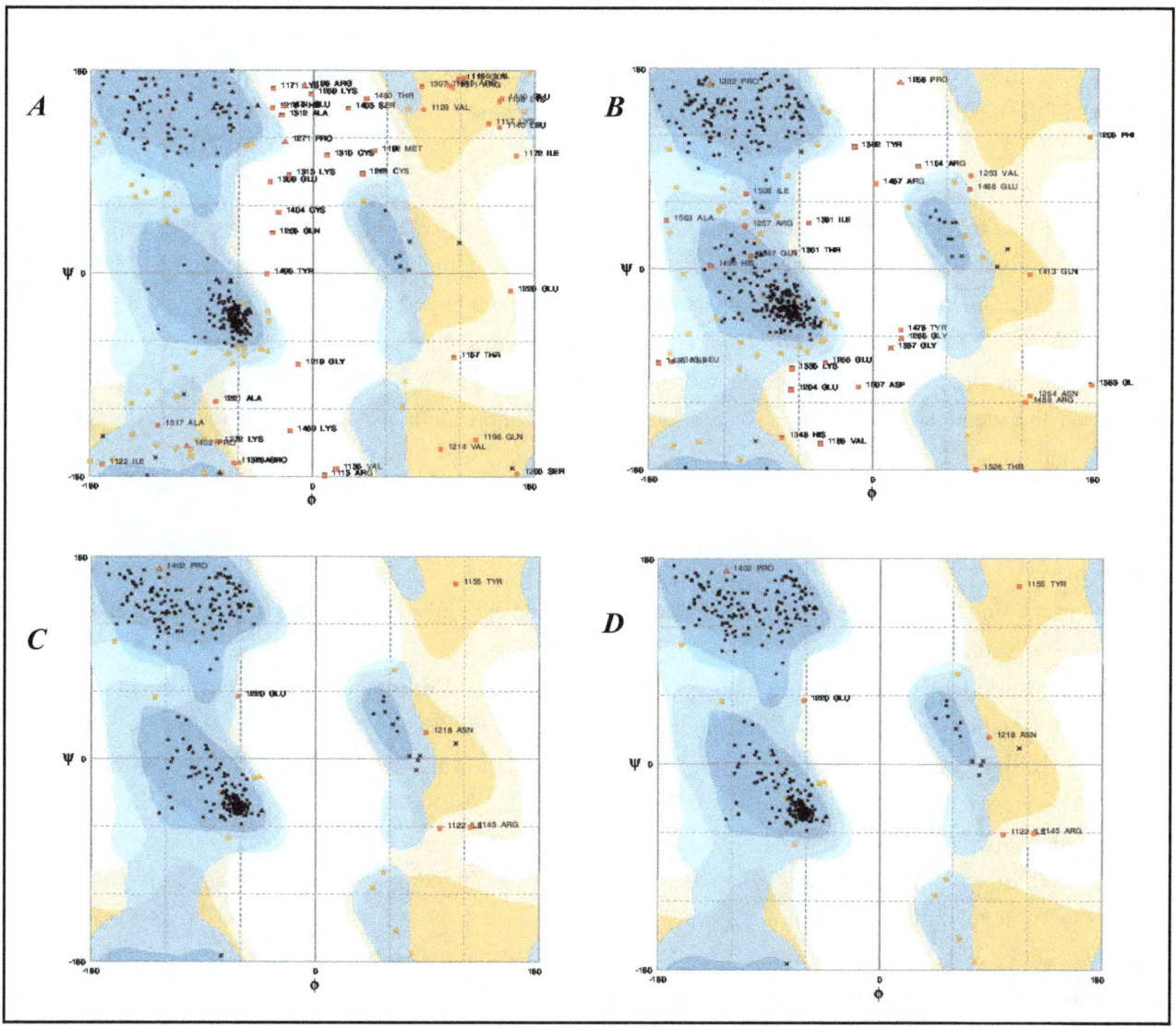

Fig. 3 *A*. Ramachandran plot of the ESyPred3D HAC1 model; ***B*.** Ramachandran plot obtained from the 3D-JIGSAW server; ***C*.** Ramachandran plot from the SWISS-MODEL server; ***D*.** Ramachadran plot of the PHYRE2 HAC1 model.

The number of residues in favored region in the ESyPred3D 3D model of HAC1 is 73.0% (Fig. 3A). Ramachandran plot analysis for 3DJIGSAW HAC1 model (Fig. 3B), resulted in 83.9% of all residues placed in favored regions. These two results confirm the model check analysis of Verify3D and ANOLEA, indicating that 3DJIGSAW and ESyPred3D servers predicted the models with insufficient residues in the expected favored region. However, SWISS-MODEL and PHYRE2 servers were capable to produce a good 3D structure of *Arabidopsis thaliana* HAC1 protein, having 95% of all residues located in expected regions in Ramachandran plot analysis (Fig. 3 C and D).

Although the high similarity of 45.8 with thep300/CBP HAT homolog from humans , with a well predicted X-ray crystallography structure and known function, it is difficult to attribute some the functional characteristics to our predicted AtHAC1(Arabidopsis histone acetyltransferase of the CBP family) just according the structural similarities, that are stated in this paper.

The p300/CBP HAT protein from humans, with the sequence length of 2414 amino acids, is a bigger protein than HAC1 in Arabidopsis thaliana, with 1697 amino acids.

For future research, structural prediction and conformations of other *Arabidopsis thaliana* members p300/CBP histone acetyltransferase proteins, this HAC1 structural analysis could be used to find out the other proteins that not yet have been identified in this pathway. Furthermore, interactome analysis of all HAC member of the CBP family is recommended in order to assign new functions to this family.

4 Conclusion

An attempt was made to predict the three-dimensional structure of Arabidopsis histone acetyltransferase 1 of the CBP family. The structural analysis approach was based on homology modeling, using SWISS-MODEL, Phyre2, ESyPred3D and 3DJIGSAW model servers. For 3D model validations, Verify3D and ANOLEA tools have been used. Additionally, Ramachandran plot analysis was performed in order to verify whether all residues of our predicted models lye in the most favorable regions. SWISS MODEL and Phyre2 servers computed the identical 3D structure. Validation and verification methods, using Z-score and 3D-1D score, showed that these 3D models are of good and acceptable quality.

References

Arabidopsis Genome Initiative. 2000. Analysis of the genome sequence of the flowering plant *Arabidopsis thaliana*. Nature, 408(6814): 796-815

Arnold K, Bordoli L, Kopp J, et al. 2006. The SWISS-MODEL Workspace: A web-based environment for protein structure homology modelling. Bioinformatics, 22: 195-201

Bates PA, Kelley LA, MacCallum RM, et al. 2001. Enhancement of Protein Modelling by Human Intervention in Applying the Automatic Programs 3D-JIGSAW and 3D-PSSM. Proteins: Structure, Function and Genetics, Suppl(5): 39-46

Blundell TL, Sibanda BL, Sternberg MJ, et al. 1987. Knowledge-based prediction of protein structures and the design of novel molecules. Nature, 326: 347-352

Chen D. 2009. Structural GENOMICS: Exploring the 3D Protein Landscape. Simbios, the NIH National Center for Physics-Based Simulation of Biological Structures,11-18

Deng W, Liu C, Pei Y, et al. 2007. Involvement of the histone acetyltransferase AtHAC1 in the regulation of flowering time via repression of FLOWERING LOCUS C in Arabidopsis. Plant Physiology, 143: 1660-1668

Earley KW, Shook MS, Brower-Toland B, et al. 2007. In vitro specificities of Arabidopsis co-activator histone acetyltransferases: implications for histone hyperacetylation in gene activation. The Plant Journal, 52: 615-626

Eisenberg D, Luthy R, et al. 1997. VERIFY3D: assessment of protein models with three-dimensional profiles. Methods in Enzymology, 277: 396-404

Garnier J, Gibrat JF, Robson B. 1996. GOR secondary structure prediction method version IV. Methods in Enzymology, 266: 540-553

Greer J. 1990. Comparative Modeling Methods: Application to the Family of the Mammalian Serine Proteases.

Proteins, 7: 317-334

Kelley LA, Sternberg MJE. 2009. Protein structure prediction on the web: a case study using the Phyre server. Nature Protocols, 4: 363-371

Kelley LA, Bennett-Lovsey R, Herbert A, et al. 2011. Phyre: Protein Homology/analogY Recognition Engine. Structural Bioinformatics Group, Imperial College, London, UK

Lovell SC, Davis IW, Arendall III WB, et al. 2002. Structure validation by Calpha geometry: phi,psi and Cbeta deviation. Proteins: Structure, Function and Genetics, 50: 437-450

Melo F, Feytmans E. 1998. Assessing protein structures with a non-local atomic interaction energy. Journal of Molecular Biology, 2775: 1141-1152

Neelamathi E, Vasumathi E, Bagyalakshmi S, et al. 2009. Insilico prediction of structure and functional aspects of a hypothetical protein of Neurosporacrassa. Journal of Cell and Tissue Research, 93: 1889-1894

Pandey R, Müller A, Napoli CA. 2002. Analysis of histone acetyltransferase and histone deacetylase families of *Arabidopsis thaliana* suggests functional diversification of chromatin modification among multicellular eukaryotes. Nucleic Acids Research, 30(23): 5036-5055

Raghava GPS. 2002. APSSP2 : A combination method for protein secondary structure prediction based on neural network and example based learning. CASP5. A-132

Sippl MJ. 1995. Knowledge based potentials for proteins. Current Opinion in Structural Biology, 5: 229-235

Söding J, Biegert, A, Lupas AN. 2005 TheHHpred interactive server for protein homology detection and structure prediction. Nucleic Acids Research, 33(Web Server issue): W244-8

Sterner DE, Berger SL. 2000. Acetylation of Histones and Transcription-Related Factors. Microbiology and Molecular Biology Reviews, 64(2): 435

The PyMOL Molecular Graphics System, Version 1.3. Schrödinger, LLC

Xu W, Chen H, Du K, et al. 2001. A transcriptional switch mediated by cofactor methylation. Science, 294: 2507-2511

5

Determination of keystone species in CSM food web: A topological analysis of network structure

LiQin Jiang[1], **WenJun Zhang**[1,2]

[1]School of Life Sciences, Sun Yat-sen University, Guangzhou 510275, China

[2]International Academy of Ecology and Environmental Sciences, Hong Kong

E-mail: zhwj@mail.sysu.edu.cn, wjzhang@iaees.org

Abstract

The importance of a species is correlated with its topological properties in a food web. Studies of keystone species provide the valuable theory and evidence for conservation ecology, biodiversity, habitat management, as well as the dynamics and stability of the ecosystem. Comparing with biological experiments, network methods based on topological structure possess particular advantage in the identification of keystone species. In present study, we quantified the relative importance of species in Carpinteria Salt Marsh food web by analyzing five centrality indices. The results showed that there were large differences in rankings species in terms of different centrality indices. Moreover, the correlation analysis of those centralities was studied in order to enhance the identifying ability of keystone species. The results showed that the combination of degree centrality and closeness centrality could better identify keystone species, and the keystone species in the CSM food web were identified as, *Stictodora hancocki*, small cyathocotylid, *Pygidiopsoides spindalis*, *Phocitremoides ovale* and *Parorchis acanthus*.

Key words keystone species; topological parameters; centrality indices; biological networks.

1 Introduction

Food webs are complex ecological networks describing trophic relationships between species in a certain area (Pimm, 1982; Belgrano et al., 2005; Arii et al., 2007). If the entire food web is treated as a graph, the nodes in the graph represent different species (individuals) in the ecosystem and the edges denote the interactions between species (individuals). As a kind of network, food webs provide a new way to study communities (Albert and Barabasi, 2002; Newman, 2003). To some extent, such a network is a formalized description for complex relationships between species within the system.

The concept of keystone species originated from the thought that species diversity of an ecosystem was controlled by the predators in the food chains, and they affected many other creatures in the ecosystem.

Keystone species refer to those that biomass is disproportionate with its impact on the environment, and the extinction of keystone species may lead to the collapse of communities (Paine, 1969; Mills et al., 1993; Springer et al., 2003). The concept of keystone species means that an ecological community is not just a simple collection of species (Mouquet, 2013). As a result, the ecologically important species might not necessarily be the rare species conservation biologists always believed (Simberloff, 1998), because rare species are associated with the little biomass and abundance of species, and the importance of species is a kind of functional properties of the network. Therefore, the traditional protection pattern for rare species should be gradually transformed into the maintenance of keystone species (Wilson, 1987).

Keystone species strongly affect species richness and ecosystem dynamics (Piraino et al., 2002), so the research of keystone species is an important area for predicting and maintaining the stability of ecosystem (Naeem and Li, 1997; Tilman, 2000). Definition of keystone species emphasizes the functional advantages of species in the ecosystem, and whether a species is a keystone species depends upon if it has a consistent effect in ecological function (Power et al., 1996), namely its sensitivity to environmental changes, such as competition, drought, floods and other ecological processes. In the past, researchers used many field experimental methods to study keystone species, but they mainly focused on the impact of changes in the abundance of a species on the other species (Paine, 1992; Wootton, 1994; Berlow, 1999). The main identification methods include control simulation method (Paine, 1995; Bai, 2011), equivalent advantage method (Khanina, 1998; Ji, 2002), competitive advantage method (Yeaton, 1988; Bond, 1989), the relative importance of species interactions method (Tanner and Hughes, 1994), community importance index method (Power et al., 1996), keystone index method (Jordán et al., 1999) and functional importance index method (Hurlbert, 1997). However, these methods mainly concentrated on a few species. Thus researchers need to do an assessment of the interactions between species in the community before the experiment, in order to determine species not important or interesting. So these methods are obvious subjective and produce certain mistake on identifying keystone species (Wootton 1994; Bustamante et al., 1995). Furthermore, monitoring species reaction to changes in the external environment through the above experimental methods requires that experimenters have a high professional quality. And because of the longer experimental time span, greater cost (Ernest and Brown, 2001), as well as other factors during the experiment, they are only suitable for semi-artificial or simple controllable ecosystems. It is more difficult to judge whether a species is a keystone species based on certain characteristics of species (Menge et al., 1994). So far, we don't have a perfect and universally applicable method to identify keystone species.

Research of keystone species has evolved from the initial direct experimental methods to network/software analysis. For example, Libralato et al. (2006) analyzed keystone indicators of functional groups of a species or a group of species in food web model through the ecosystem modeling (the Ecopath with Ecosim, EwE), and then ranked the level of the key indicators to obtain the keystone species. Jordán et al. (2008) pointed out that there were at least two methods to quantitatively assess the importance of species in communities. One was the structural importance of network analysis and another for the functional importance of network analysis. So they calculated the structural importance and the functional importance of species in the food web in Prince William Sound by CosBiLaB Graph software, and evaluated the advantages and disadvantages of the two methods. They believed that the combination of these two methods in the future would be the most important way to research dynamic mechanism. Kuang and Zhang (2011) analyzed the topological properties of the food web in Carpinteria Salt Marsh and found that parasites played a very important role in the food web, and the addition of parasites in the food web would change some properties and greatly increase the complexity of the food web. Therefore, the relationship between keystone species and topological characteristics can provide an effective method to understand and describe the topological structures, dynamic characteristics and the complexity of functions between species within the food web. And it also can provide valuable theory and evidence for conservation ecology, biodiversity, habitat management, as well as the dynamics and stability of the ecosystem.

Nevertheless, so far we lack of effective methods to identify keystone species and quantify their relative importance, so the quantitative assessment of species importance in the food web is becoming increasingly important and urgent (Paine, 1966; Power et al, 1996; Jordán, 2008). In recent years, there have been some major discoveries about the topological properties of complex systems (Strogatz, 2001; Albert and Barabási, 2002; Newman, 2003), and these also affect the definition and identification of keystone species. For example, the highly connected species were found to have more important influence on sustainability of food webs (Soulé and Simberloff, 1986), which promoted the generation of the concept of degree. Degree of nodes thus become the most widely used topological parameter to measure the keystone species (Dunne et al., 2002a). Degree refers to the direct impacts between species (Callaway et al., 2000; West, 2001; Zhang, 2011, 2012a, 2012b, 2012c, 2012d). However, indirect impacts between species are also important (Wooton, 1994; Huang, et al., 2008). For example, Darwin (1859) described the influence of cats on the clovers. Although indirect effects of chemical and behavioral studies may be difficult to quantify, some indirect impacts of network links have been proposed (Ulanowicz and Puccia, 1990; Patten, 1991). Thus the concept of centrality is proposed to address this problem. Centrality focuses on the indirect effects between species. The impacts of food webs are generally spread through indirect ways, so it may require detailed research and quantitative description on the effective range of indirect interactions from a specific point to the entire network (Jordán, 2001). In other words, it is necessary to determine how relevant these species are in the food web (Yodzis, 2000; Williams et al., 2002). The concept of centrality stemmed from the social network analysis (Wasserman and Faust, 1994), namely the ability of a node communicates with other nodes or the intimacy of a node with the others (Go'mez et al., 2003). These have resulted in a series of topological parameters relating to the relative importance of a node, such as degree centrality, betweenness centrality, closeness centrality, clustering coefficient centrality, eigenvector centrality and information centrality, etc. In present paper, we used various methods to detect and quantify relative importance of species in a famous food web, CSM (Carpinteria Salt Marsh) food web, reported by Lafferty et al. (2006a, 2006b, 2008), and further studied the correlation between topological parameters of the food web, aiming to evaluate the effectiveness of various methods in quantifying relative importance of species and detecting the keystone species in the food webs.

2 Materials and Methods

2.1 Data source

Data were collected from the food web, Carpinteria Salt Marsh, California, reported by Lafferty et al. (2006a, 2006b, 2008) (http: //www.nceas.ucsb.edu/interactionweb/html/carpinteria.html). CSM food web includes four sub-webs, predator-prey sub-web, predator-parasite sub-web, parasite-host sub-web, and parasite-parasite sub-web.

2.2 Methods

2.2.1 Pajek software

Pajek is a software platform for network analysis, which contains various methods/algorithms/models on analysis of topological properties.

2.2.2 Centrality measures

Centrality indices are used to measure impact and importance of nodes in a network. The most commonly used centrality indices are degree centrality, betweenness centrality, closeness centrality, clustering coefficient centrality and eigenvector centrality (Navia et al., 2010; Zhang, 2012a, b).

(1) Degree centrality (DC)

DC is the simplest measure which considers the degree of a node (species) only. The degree of species i is: $D_i = D_{in,i} + D_{out,i}$, where $D_{in,i}$: number of prey species of species i, and $D_{out,i}$: number of predator species of species i. The degree of species was calculated by Net/Partitions/DC/All in Pajek.

(2) Betweenness centrality (BC)

BC is calculated by the following formula

$$BC_i = 2\sum_{j \leq k} g_{jk}(i)/g_{jk} / [(N-1)(N-2)]$$

where $i \neq j \neq k$, g_{jk}: the shortest path between species j and k, $g_{jk}(i)$: number of the shortest paths containing species i, N: total number of species in the food web. A greater BC_i means that the effect of losing species i will promptly disperse across the food web (Zhang, 2012a, b).

(3) Closeness centrality (CC)

CC_i refers to the mean shortest path of species i

$$CC_i = (N-1)/\sum_{j=1}^{N} d_{ij}$$

where $i \neq j$, d_{ij} is the length of the shortest path between species i and j. A greater CC_i means a more importance of species i.

In Pajek, we use Net/Vector/Centrality/Closeness/All and Net/Vector/Centrality/Betweenness to calculate BC (Wasserman and Faust, 1994).

(4) Clustering coefficient centrality (CU)

Clustering coefficient centrality denotes the ratio of the actual edges E_i of node i connected with its neighbors divided by the most possible edges $D_i(D_i-1)/2$ between them (Watts and Strogatz, 1998). In other words, it refers to the ratio of the directly connected neighboring pairs divided by all the neighboring pairs in the neighboring points of the node, that is

$$CU_i = 2E_i / [D_i(D_i-1)]$$

It measures how close the current node is to its neighboring nodes. The averag clustering coefficient of all nodes is the clustering coefficient of the entire network. Obviously, the clustering coefficient of a network is weighted by the clustering coefficient of all nodes whose degree must be at least 2. $0 \leq CU \leq 1$; if $CU=0$, all nodes in the network are isolated, and if $CU=1$, the network is fully connected. Furthermore, studies have shown that clustering coefficient is related to network modularity. Clustering coefficient of the entire network reflects the overall trend of all the nodes gathering into a module (Eisenberg and Levanon, 2003; Ravasz et al, 2002).

(5) Eigenvector centrality (EC)

Eigenvector centrality is the dominant eigenvector of the adjacency matrix A of the network (Bonacich, 1987), i.e., the extent of a node connected to the node with the highest eigenvector centrality. In the word of social networks, a person tends to occupy the central place more likely if he (she) has contacted more people in the center position. Eigenvector centrality reflects the prestige and status of nodes. This measure tries to find the keystone node in the entire network rather than in the local structure. Here, eigenvector is e, and $\lambda e = Ae$, where A is the adjacency matrix of a food web. Therefore, the EC of node i is

$$EC_i = e_1 (i)$$

where e_1 is the eigenvector corresponding to the maximum eigenvalue λ_1. A greater value of EC_i means a greater number of the neighboring nodes connected with node i, and it indicates that the node is in the core position.

3 Results

3.1 Degree centrality

As shown in Fig. 1, 2 and Table 1, the species with the greater DC values in the full CSM food web are largely consistent with that in the predator-parasite sub-web, parasite-parasite sub-web and parasite-host sub-web. And these species are substantially parasites. The species with the maximum DC value in the predator-prey sub-web is *Pachygrapsus crassipes*, and the species with the forth DC value is *Willet*. Although DC values of the two species are larger, they are slightly lower than nine parasite species, such as *Mesostephanus appendiculatoides*, etc. In addition, the basal species, Marine detritus, is of greater importance also.

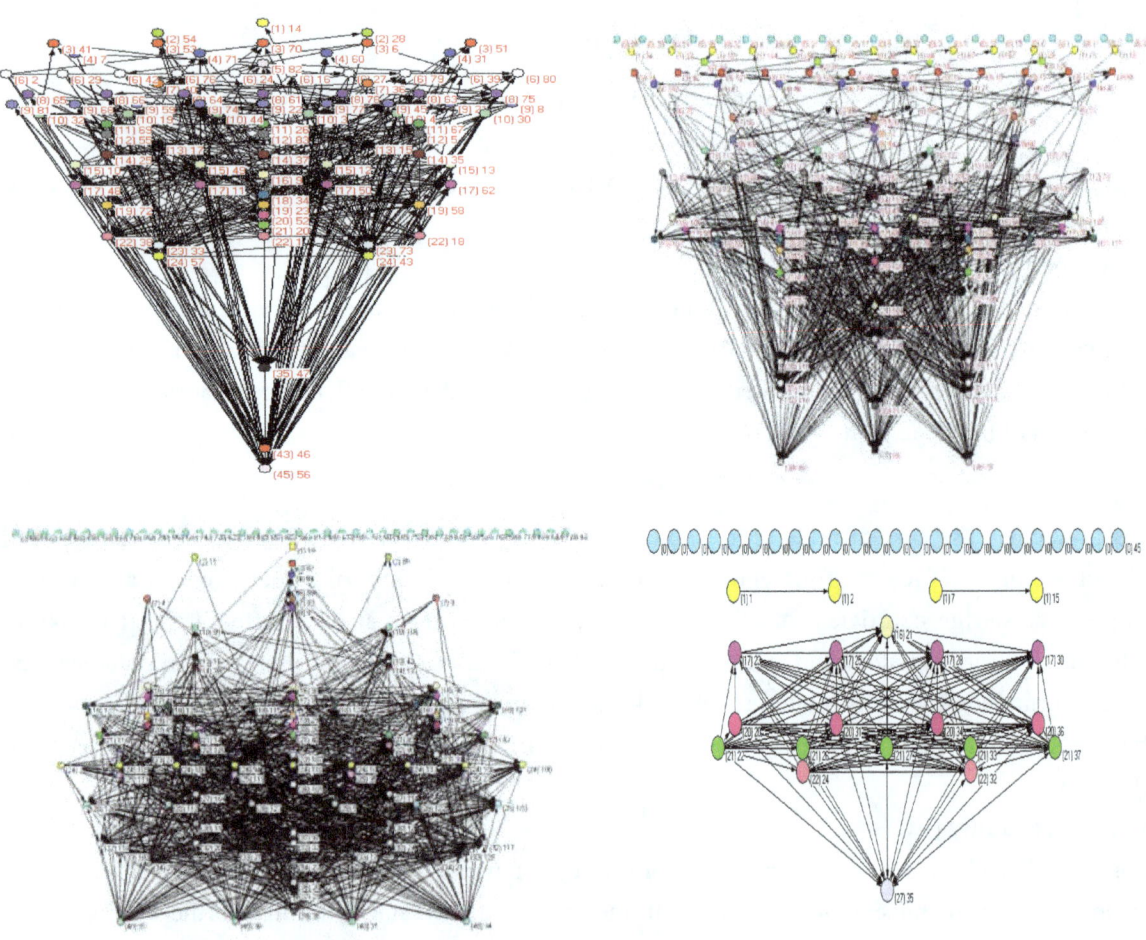

Fig. 1 Results of degree centrality for the four sub-webs of CSM food web (upper left: predator-prey sub-web; upper right: predator-parasite sub-web; bottom right: parasite-parasite sub-web; bottom left: parasite-host sub-web). The numbers in parentheses are total links (degree, or incoming degree + outgoing degree) and the numbers outside parentheses are species ID codes. The ID codes of different sub-webs are different from the original species.

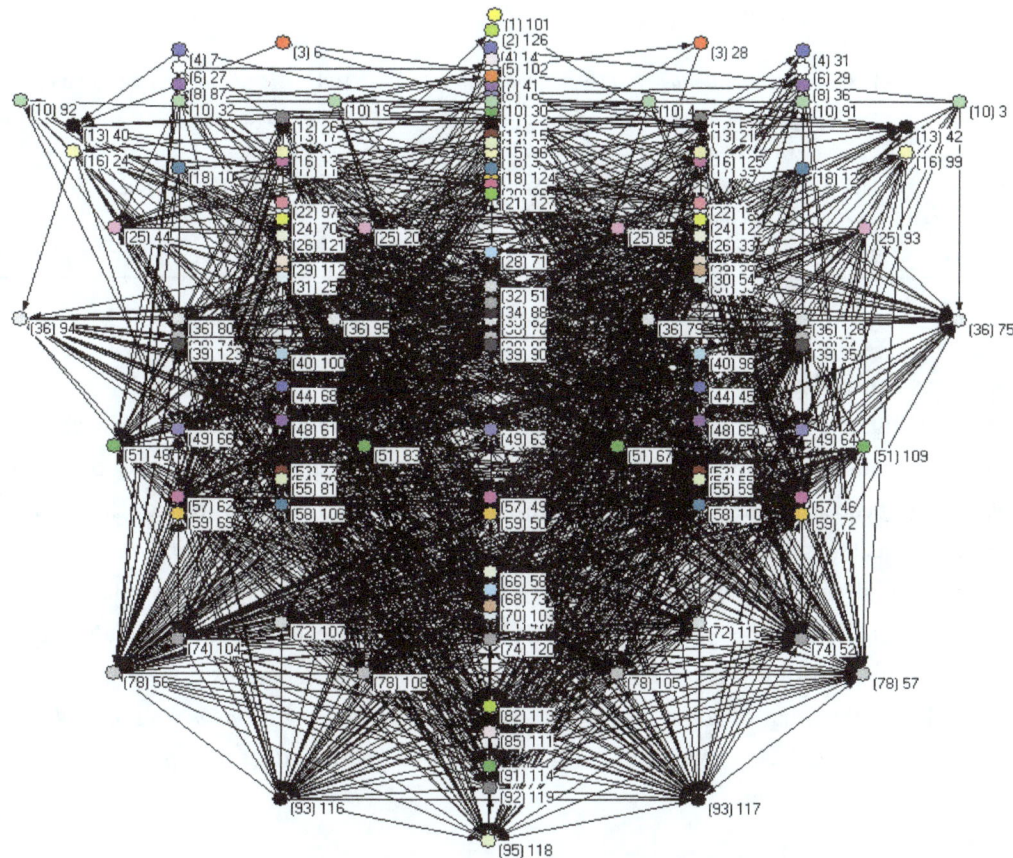

Fig. 2 Results of degree centrality for the full CSM food web. The numbers in parentheses are total links (degree, or incoming degree + outgoing degree) and the numbers outside parentheses are species ID codes.

Table 1 Species with greater DC values in the full CSM food web and four sub-webs.

Predator-prey sub-web		Predator-parasite sub-web		Parasite-parasite sub-web		Parasite-host sub-web		Full CSM food web	
ID	Species	ID	Species	ID	Species	ID	Species	ID	Species
56	*Pachygrapsus crassipes*	90	*Culex tarsalis*	118	*Mesostephanus appendiculatoides*	117	*Stictodora hancocki*	118	*Mesostephanus appendiculatoides*
46	*Hemigrapsus oregonensis*	89	*Aedes taeniorhynchus*	115	*Renicola cerithidicola*	114	*Phocitremoides ovale*	117	*Stictodora hancocki*
47	*Fundulus parvipinnis*	98	Plasmodium	107	*Renicola buchanani*	119	*Pygidiopsoides spindalis*	116	Small cyathocotylid
57	Willet	117	*Stictodora hancocki*	120	Microphallid 1	116	Small cyathocotylid	119	*Pygidiopsoides spindalis*
43	*Cleavlandia ios*	119	*Pygidiopsoides spindalis*	116	Small cyathocotylid	118	*Mesostephanus appendiculatoides*	114	*Phocitremoides ovale*
73	*Gillycthys mirabilis*	116	Small cyathocotylid	110	Large xiphideocercaria	111	*Parorchis acanthus*	111	*Parorchis acanthus*
33	*Macoma nasuta*	114	*Phocitremoides ovale*	109	*Catatropis johnstoni*	113	*Cloacitrema michiganensis*	113	*Cloacitrema michiganensis*
18	*Anisogammar*	111	*Parorchis*	105	*Probolocoryphe*	104	*Himasthla*	105	*Probolocoryphe*

us confervicolus		*acanthus*		*uca*		*rhigedana*		*uca*	
1	Marine detritus	118	*Mesostephanus appendiculatoides*	119	*Pygidiopsoides spindalis*	105	*Probolocoryphe uca*	108	*Acanthoparyphium sp.*
38	*Geonemertes*	113	*Cloacitrema michiganensis*	117	*Stictodora hancocki*	108	*Acanthoparyphium sp.*	56	*Pachygrapsus crassipes*
		31	Mosquito larva					57	Willet

3.2 Betweenness centrality

As illustrated in Fig. 3 and 4, the BC values of all nodes in the predator-parasite sub-web and parasite-host sub-web are 0, because these species do not locate between other species in the network. But the radius of *Mesostephanus appendiculatoides* in the parasite-parasite sub-web is very obvious, indicating that some species in the parasite-parasite sub-web need to go through *Mesostephanus appendiculatoides*. Once this species is removed, all the interaction chains will collapse and largely destruct the whole sub-web. From Table 2, the BC values of the top four species in the CSM food web are identical with that in the predator-prey sub-web, while some parasites with larger DC values, such as *Mesostephanus appendiculatoides*, etc., whose BC values are lower than that of some free-living species, such as *Hemigrapsus oregonensis*. It indicates that the nutritional flow of free-living species in the food web has a greater effect than parasites.

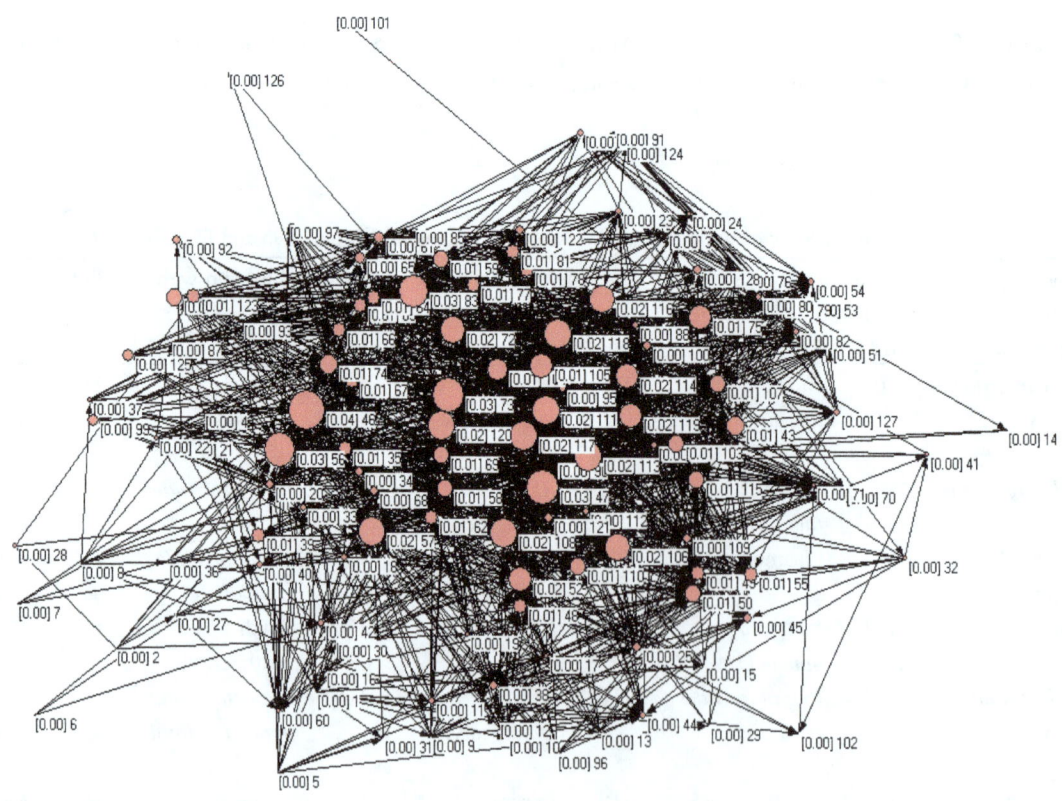

Fig. 3 Results of betweenness centrality for the full CSM food web. The numbers in parentheses are betweenness centralities and the numbers outside parentheses are species ID codes.

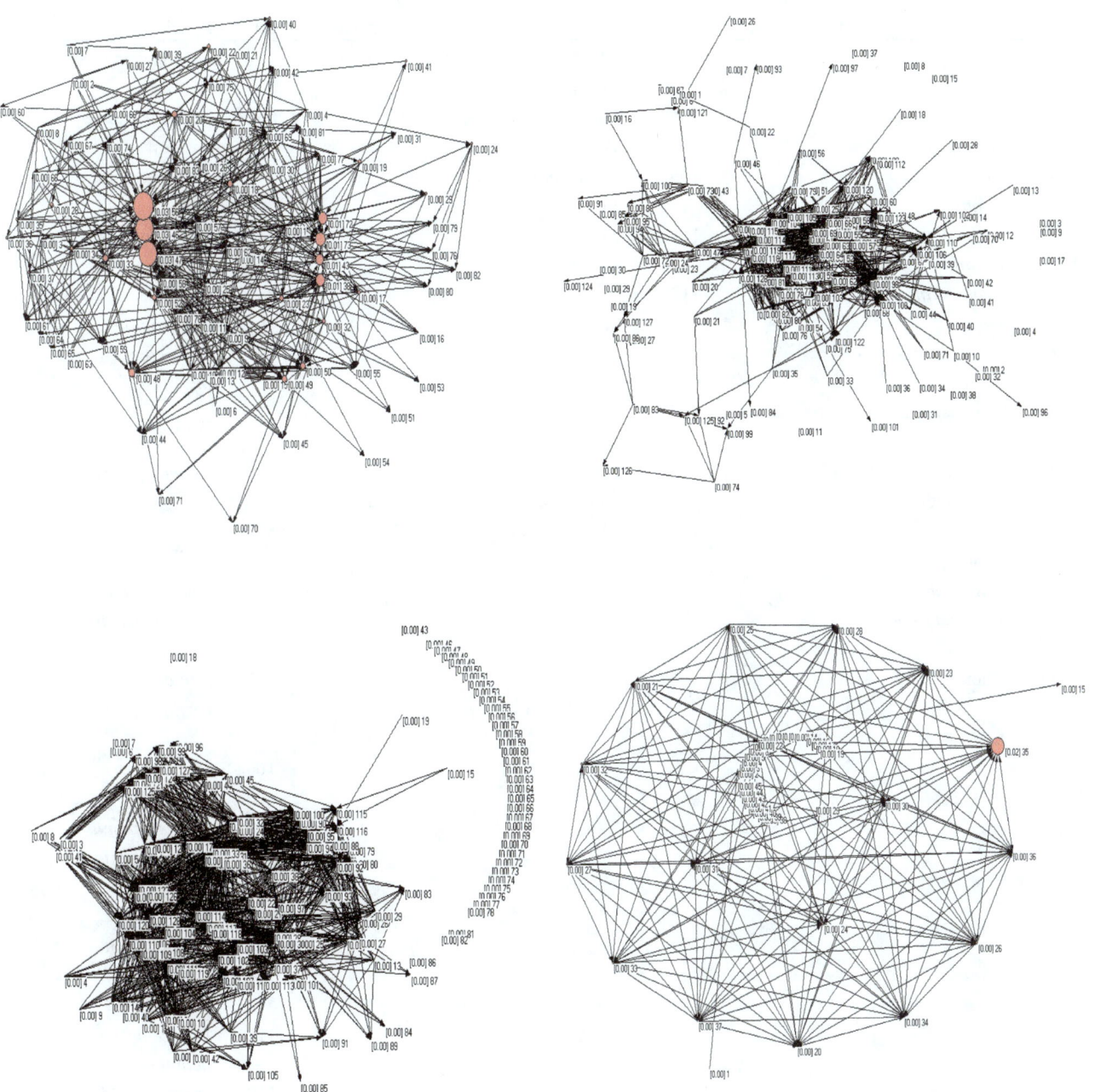

Fig. 4 Results of betweenness centrality for the four sub-webs of CSM food web (upper left: predator-prey sub-web; upper right: predator-parasite sub-web; bottom right: parasite-parasite sub-web; bottom left: parasite-host sub-web). The numbers in parentheses are betweenness centralities and the numbers outside parentheses are species ID codes. The ID codes of different sub-webs are different from the original species. The size of the node relates to the value of BC; the greater BC is, the bigger the node radius is. The species ID codes of different sub-webs are different from the original species, and the magnification of each figure is different.

Table 2 Species with greater BC values in the full CSM food web and four sub-webs.

Predator-prey sub-web		Predator-parasite sub-web		Parasite-parasite sub-web		Parasite-host sub-web		Full CSM food web	
ID	Species	ID	Species	ID	Species	ID	Species	ID	Species
46	*Hemigrapsus oregonensis*			118	*Mesostephanusap pendiculatoides*			46	*Hemigrapsusore gonensis*
56	*Pachygrapsus crassipes*			106	Himasthla species B			56	*Pachygrapsuscr assipes*
47	*Fundulusparv ipinnis*			109	*Catatropisjohnsto ni*			47	*Fundulusparvipi nnis*
73	*Gillycthys mirabilis*			111	*Parorchis acanthus*			73	*Gillycthys mirabilis*
72	*Leptocottusar matus*			115	*Renicola cerithidicola*			83	*Triakis semifasciata*
38	Geonemertes			105	*Probolocoryphe uca*			72	*Leptocottus armatus*
43	Cleavlandiaio s			110	Large xiphideocercaria			57	Willet
48	Western Sandpiper			116	Small cyathocotylid			108	*Acanthoparyphi um sp.*
50	Least Sandpiper			120	Microphallid 1			52	Dowitcher
18	*Anisogammar usconfervicol us*							113	*Cloacitrema michiganensis*
11	Phoronid							115	*Renicola cerithidicola*
								106	Himasthla species B
								118	*Mesostephanus appendiculatoid es*
								116	Small cyathocotylid
								117	*Stictodora hancocki*
								111	*Parorchis acanthus*
								119	*Pygidiopsoides spindalis*
								120	Microphallid 1

3.3 Closeness centrality

CC values of species in food webs increases with the increase of species richness and completeness of food web. Connection between species in the full CSM food web is closer than the other four sub-webs (Fig. 5 and 6, Table 3). Combined with Table 2, the species with the maximum CC value is *Pachygrapsus crassipes* (species 56) in the full CSM food web, and it is also the greatest in the predator-prey sub-web, indicating it is closer than other species in food web. The species with the tenth CC value is *Fundulus parvipinnis* (species 47) in the full CSM food web, but it is the third in the predator-prey sub-web, just following behind *Pachygrapsus crassipes* and *Hemigrapsus oregonensis*.

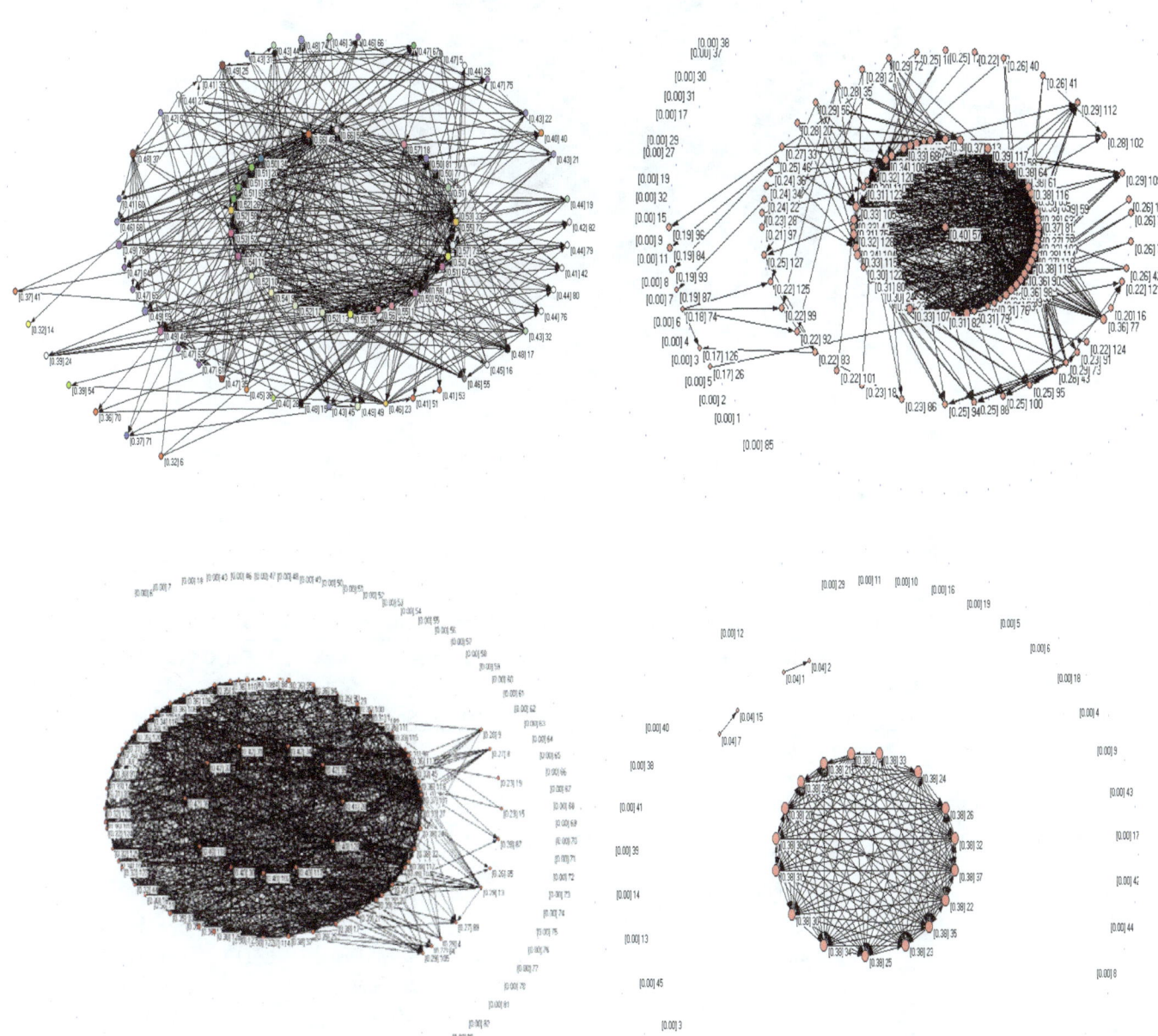

Fig. 5 Results of closeness centrality for the four sub-webs of CSM food web (upper left: predator-prey sub-web; upper right: predator-parasite sub-web; bottom right: parasite-parasite sub-web; bottom left: parasite-host sub-web). The numbers in parentheses are closeness centralities and the numbers outside parentheses are species ID codes. Species ID codes of different sub-webs are different from the original species.

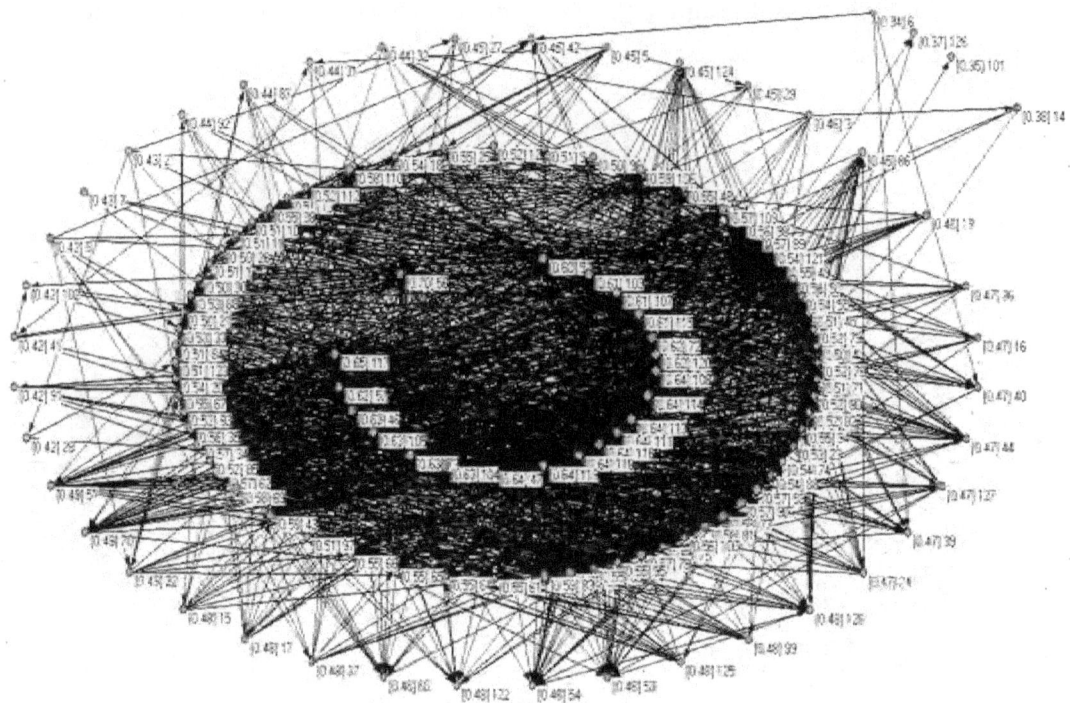

Fig. 6 Results of closeness centrality for the full CSM food web. The numbers in parentheses are closeness centralities and the numbers outside parentheses are species ID codes.

Table 3 Species with greater CC values in the full CSM food web and four sub-webs.

Predator-prey sub-web		Predator-parasite sub-web		Parasite-parasite sub-web		Parasite-host sub-web		Full CSM food web	
ID	Species	ID	Species	ID	Species	ID	Species	ID	Species
56	*Pachygrapsu scrassipes*	52	Dowitcher	116	Small cyathocotylid	119	*Pygidiopsoides spindalis*	56	*Pachygrapsus crassipes*
46	*Hemigrapsus oregonensis*	57	Willet	107	*Renicola buchanani*	116	Small cyathocotylid	117	*Stictodora hancocki*
47	*Funduluspar vipinnis*	58	Black-bellied Plover	109	*Catatropis johnstoni*	114	*Phocitremoides ovale*	116	Small cyathocotylid
73	*Gillycthys mirabilis*	59	California Gull	115	*Renicola cerithidicola*	117	*Stictodora hancocki*	119	*Pygidiopsoides spindalis*
18	*Anisogamma rusconfervicolus*	69	Clapper rail	120	Microphallid 1	118	*Mesostephanus appendiculatoides*	114	*Phocitremoides ovale*
38	Geonemertes	117	*Stictodora hancocki*	105	*Probolocorypheuca*	111	*Parorchis acanthus*	111	*Parorchis acanthus*
33	Macomanasuta	62	Marbled Godwit	118	*Mesostephanus appendiculatoides*	83	*Triakis semifasciata*	113	*Cloacitrema michiganensis*
72	*Leptocottusa rmatus*	63	Ring-billed gull	106	Himasthla species B	72	*Leptocottus armatus*	118	*Mesostephanus appendiculatoides*
1	Marine detritus	64	Western Gull	108	*Acanthoparyphium sp.*	57	Willet	108	*Acanthoparyphium sp.*

57	Willet	65	Bonaparte's Gull	117	*Stictodora hancocki*	113	*Cloacitremamic higanensis*	47	*Fundulus parvipinnis*
9	Oligochaete	111	*Parorchis acanthus*	113	*Cloacitrema michiganensis*	73	*Gillycthys mirabilis*		
11	Phoronid	114	*Phocitremoi des ovale*	114	*Phocitremoides ovale*				
		116	Small cyathocotylid	119	*Pygidiopsoides spindalis*				
		119	*Pygidiopsoi des spindalis*	111	*Parorchis acanthus*				
		111	*Parorchis acanthus*						

3.4 Clustering coefficient centrality

CU values of predator-parasite sub-web and parasite-host sub-web appear in two patterns: one for the degree values of some nodes are less than 2, and the CU values of these nodes are 999999998 in the Pajek; another for the neighboring nodes of one node are less than 2, and the CU values of these nodes are 0. From Table 4, we can find that the CU rankings of nodes in the full CSM food web and predator-prey sub-web are really different.

Table 4 Species with greater CU values in the full CSM food web and four sub-webs.

Predator-prey sub-web		Predator-parasite sub-web		Parasite-parasite sub-web		Parasite-host sub-web		Full CSM food web	
ID	Species	ID	Species	ID	Species	ID	Species	ID	Species
60	Whimbrel			104	*Himasthla rhigedana*			25	*Cerithidea californica*
81	Pied Billed Grebe			106	Himasthla species B			109	*Catatropis johnstoni*
38	Geonemertes			108	*Acanthoparyphium sp.*			70	Cooper's Hawk
78	Black-crowned Night heron			111	*Parorchis acanthus*			34	Protothaca
61	Mew Gull			113	*Cloacitrema michiganensis*			110	Large xiphideocercaria
63	Ring-billed gull			103	*Euhaplorchis californiensis*			35	*Tagelus spp.*
64	Western Gull			114	*Phocitremoides ovale*			106	Himasthla species B
65	Bonaparte's Gull			117	*Stictodora hancocki*			71	Northern Harrier
36	Cryptomya			119	*Pygidiopsoides spindalis*			115	*Renicola cerithidicola*
77	Snowy Egret			105	*Probolocoryphe uca*			103	*Euhaplorchis californiensis*
68	Bufflehead			107	*Renicola buchanani*			107	*Renicola buchanani*

3.5 Eigenvector centrality

Species with greater EC values in the full CSM food web are largely consistent with that in the predator-prey sub-web (Table 5; Fig. 7, 8). Species with greater EC values in the full CSM food web and predator-prey sub-web are free-living species, rather than parasites. Willet (species ID 57) has the largest EC value. Otherwise, species with larger EC values in predator-parasite sub-web and parasite-parasite sub-web are parasites.

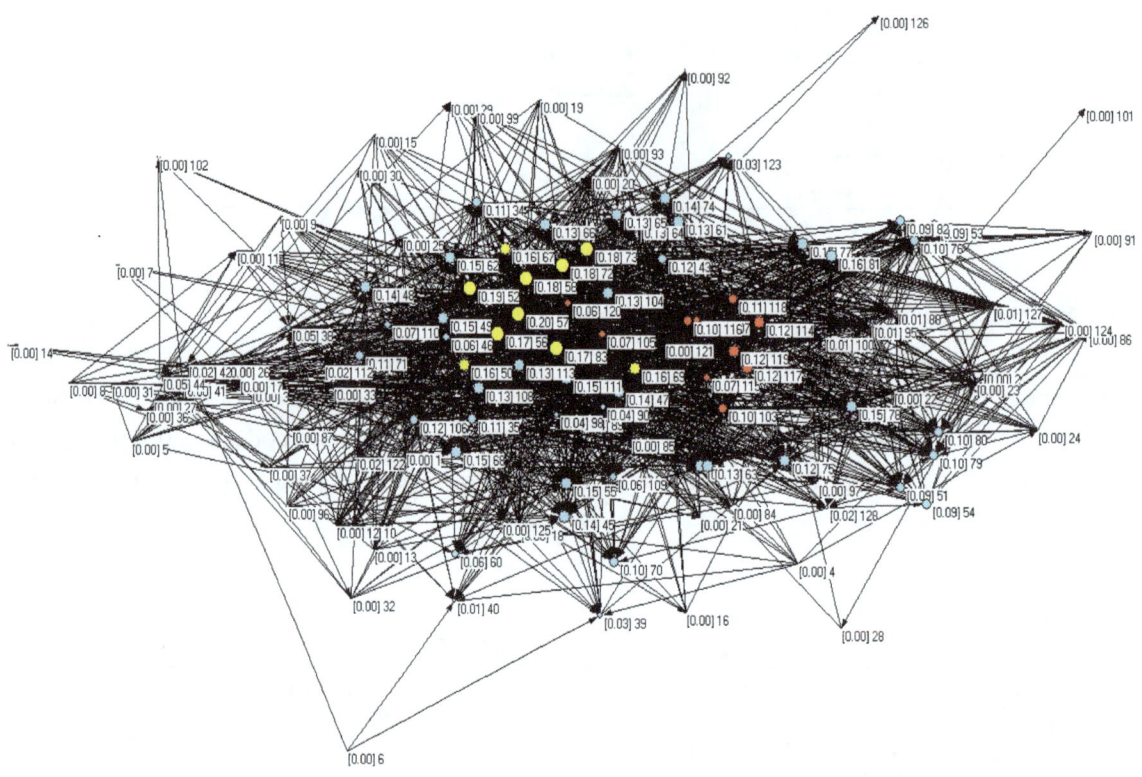

Fig.7 Results of eigenvector centrality for the full CSM food web. The numbers in parentheses are eigenvector centralities and the numbers outside parentheses are species ID codes.

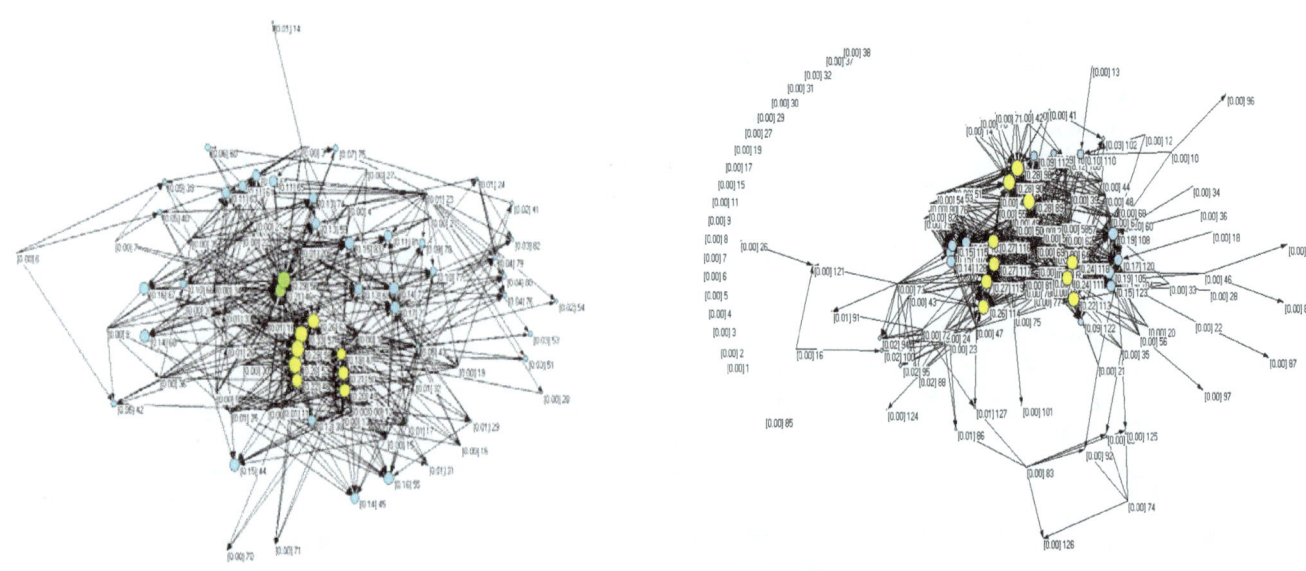

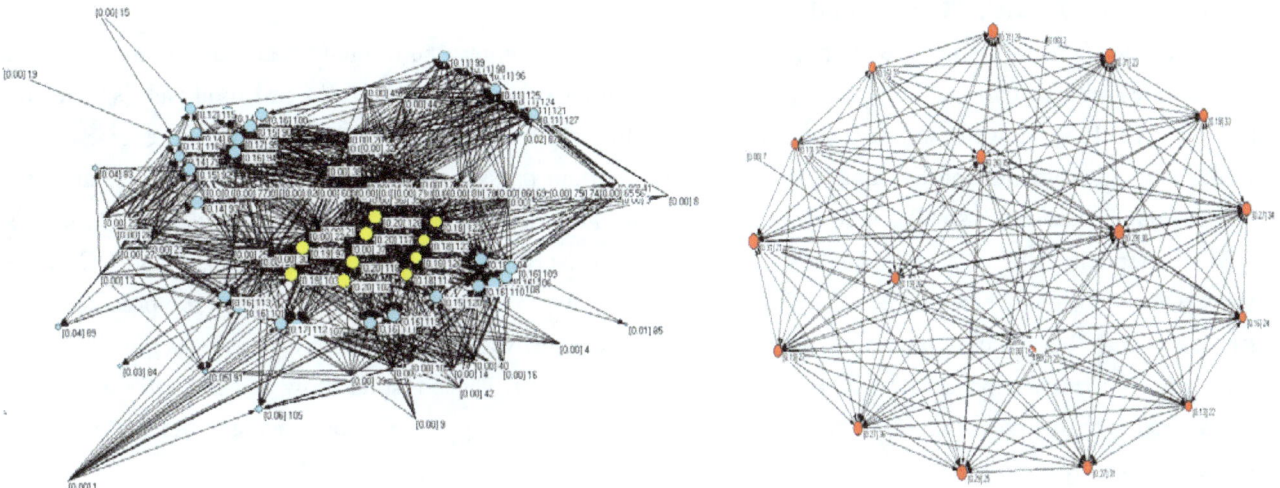

Fig. 8 Results of eigenvector centrality for the four sub-webs of CSM food web (upper left: predator-prey sub-web; upper right: predator-parasite sub-web; bottom right: parasite-parasite sub-web; bottom left: parasite-host sub-web). The numbers in parentheses are eigenvector centralities and the numbers outside parentheses are species ID codes. Species ID codes of different sub-webs are different from the original species.

Table 5 Species with greater eigenvector values in the full CSM food web and four sub-webs.

Predator-prey sub-web		Predator-parasite sub-web		Parasite-parasite sub-web		Parasite-host sub-web		Full CSM food web	
ID	Species	ID	Species	ID	Species	ID	Species	ID	Species
57	Willet	98	Plasmodium	111	*Parorchis acanthus*	83	*Triakis semifasciata*	57	Willet
58	Black-bellied Plover	90	*Culex tarsalis*	106	Himasthla species B	72	*Leptocottus armatus*	52	Dowitcher
56	*Pachygrapsus crassipes*	89	*Aedestaeniorhynchus*	104	*Himasthla Rhigedana*	73	*Gillycthys mirabilis*	58	Black-bellied Plover
52	Dowitcher	116	Small cyathocotylid	113	*Cloacitrema michiganensis*	57	Willet	72	*Leptocottus armatus*
62	Marbled Godwit	117	*Stictodora hancocki*	108	*Acanthoparyphium sp.*	52	Dowitcher	73	*Gillycthys mirabilis*
48	Western Sandpiper	119	*Pygidiopsoides spindalis*	119	*Pygidiopsoides Spindalis*	58	Black-bellied Plover	56	*Pachygrapsus crassipes*
46	*Hemigrapsus oregonensis*	114	*Phocitremoides ovale*	117	*Stictodora Hancocki*	77	Snowy Egret	83	*Triakis semifasciata*
50	Least Sandpiper	118	*Mesostephanus Appendiculatoides*	114	*Phocitremoides Ovale*	78	Black-crowned Night heron	67	Surf Scoter
59	California Gull	111	*Parorchis acanthus*	103	*Euhaplorchis californiensis*	81	Pied Billed Grebe	50	Least Sandpiper
47	*Fundulus parvipinnis*	113	*Cloacitrema michiganensis*	118	*Mesostephanus Appendiculatoides*	69	Clapper rail	69	Clapper rail

3.6 Analysis of DC, BC, CC, CU and EC

According to Table 6, the change of species ranking with CU is larger: the top ten species are totally different with species ranking by remaining four indices. The DC and CC analysis in the full CSM food web (species ID No. 1 to No. 128) showed that the parasites are more important than free-living species, while reverse results were obtained from BC and EC analysis. The more important parasites calculated from DC and CC analysis are *Stictodora hancocki,* small cyathocotylid, *Pygidiopsoides spindalis, Phocitremoides ovale* and *Parorchis acanthus* (species No. 117, 116, 119, 114, and 111, respectively). Species ranking by BC, DC and CC in the full CSM food web (species ID No. 1 to No. 83) are basically consistent with the species in the predator-prey sub-web, and the relative important species are *Pachygrapsus crassipes, Hemigrapsus oregonensis* and *Fundulus parvipinnis*(species ID No.56, 46, and 47, respectively). These results show that parasites in the full CSM food web do not change the relative importance of free-living species, but increase the DC value of free-living species.

Table 6 The top ten species (ID codes) ranking by DC, BC, CC, CU and EC in the full CSM food web and predator-prey sub-web, respectively.

	DC	BC	CC	CU	EC
Full CSM food web (Species ID No.1 to No. 128)	118	46	56	25	57
	117	56	**117**	109	52
	116	73	**116**	70	58
	119	83	**119**	34	72
	114	47	**114**	110	73
	111	72	**111**	35	56
	113	57	113	106	83
	105	108	118	71	67
	108	52	108	115	50
	56	113	47	103	69
Full CSM food web (Species ID No.1 to No. 83)	56	46	56	25	57
	57	56	47	70	52
	52	73	46	34	58
	47	83	57	35	72
	73	47	73	71	73
	58	72	72	43	56
	68	57	52	19	83
	50	52	58	16	67
	72	75	43	12	50
	46	74	83	23	69
Predator-prey sub-web (Species No.1 to No. 83)	**56**	56	**56**	60	57
	46	46	**46**	81	58
	47	47	**47**	38	56
	43	73	73	78	52
	57	72	18	61	62
	73	38	38	63	48
	33	43	33	64	46
	18	48	72	65	50
	1	50	1	36	59
	38	18	57	77	47

3.7 Pearson correlation of five topological indices

As can seen from Table 7, the Pearson's correlations of DC and CC are the largest in the full CSM food web and predator-prey sub-web (0.917 and 0.877, respectively), so DC and CC are strong correlated. DC mainly measures the importance of a node in the local scope, and thus denotes the self-correlation of the node. CC is a measure of the ability of one node for controlling the other nodes, and denotes the centralization extent of a node. Therefore, DC and CC analysis synthesizes the importance of a node locally and globally. Table 6 demonstrates that the keystone species in the CSM food web are *Stictodora hancocki*, small cyathocotylid, *Pygidiopsoides spindalis*, *Phocitremoides ovale* and *Parorchis acanthus* (species ID No. 117, 116, 119, 114, and 111, respectively).

Table 7 Pearson's correlation coefficients of five topological indices.

Pearson's correlation coefficient analysis		DC	BC	CC	CU	EC
DC	Full CSM food web	1.000	0.773	**0.917**	0.483	0.800
	predator-prey sub-web	1.000	0.789	**0.877**	0.053	0.498
BC	Full CSM food web	0.773	1.000	0.754	0.338	0.625
	predator-prey sub-web	0.789	1.000	0.595	-0.032	0.402
CC	Full CSM food web	**0.917**	0.754	1.000	0.525	0.695
	predator-prey sub-web	**0.877**	0.595	1.000	0.360	0.478
CU	Full CSM food web	0.483	0.338	0.525	1.000	0.307
	predator-prey sub-web	0.053	-0.032	0.360	1.000	0.205
EC	Full CSM food web	0.800	0.625	0.695	0.307	1.000
	predator-prey sub-web	0.498	0.402	0.478	0.205	1.000

3.8 Efficiency analysis of the full CSM food web

Table 8 indicates the changes of topological properties after removing different keystone species from the full CSM food web. The major topological changes before and after removing keystone species include

(1) Number of top species and basal species does not change. The top species are not necessarily the keystone species of the food web.

(2) Number of links and cycles reduces significantly. It means that the keystone species play an important role in the food web. There are less cycles between predators and preys due to the removal of parasites.

(3) Number of total links and maximum links, and link density and connectance decreases respectively.

(4) The maximum chain length did not change significantly.

Compared with the results of removing important species, the changes of the full food web are not significant in terms of all indices.

In conclusion, the topological structure of the full food web changed significantly after removing the keystone species, which further validates the results achieved previously.

Table 8 Comparison of topological properties of the full CSM food web with removed different keystone species.

	Removed species No.117	Removed species No.116	Removed species No.119	Removed species No.114	Removed species No.111	Removed species No.56	Full CSM food web
Number of species, S	127	127	127	127	127	127	128
Number of links, L	2197	2197	2198	2199	2205	2212	2290
Number of top species, T	3	3	3	3	3	3	3
Number of intermediate species, I	116	116	116	116	116	116	117
Number of basal species, B	8	8	8	8	8	8	8
Number of Chain cycles	71142	70472	71111	71526	74331	80450	85214
Link density, L/S	17.299	17.299	17.307	17.315	17.362	17.417	17.891
Connectance, L/S^2	0.13621	0.13621	0.13628	0.13634	0.13671	0.13714	0.13977
Mean connectance, D	34.598	34.598	34.614	34.630	34.724	34.835	35.781
Maximum chain length	No.1-5: 3 No.6: 5 No.7-8: 4	No.1-5, 7: 3 No.6: 5 No.8: 4	No.1-5, 7: 3 No.6: 5 No.8: 4	No.1-5, 7: 3 No.6: 5 No.8: 4	No.1-5, 7: 3 No.6: 5 No.8: 4	No.1,3-5,7: 3 No.2,6,8: 4	No.1-5,7: 3 No.6: 5 No.8: 4

4 Discussion

Since the concept of keystone species was first proposed by Paine (1969), the importance of them for conservation biology has been widely studied. However, due to the limitations of field experimental methods and the temporal and spatial variation (Menge et al., 1994; Paine, 1995; Estes et al., 1998), more and more researchers questioned the original concept of keystone species, and have developed various definitions of keystone species (Mills et al., 1993; Bond, 2001; Davic, 2003). So far, quantitative methods to identify keystone species remain to be little (Menge et al., 1994; Bond, 2001).

The traditional definitions of keystone species closely related to the richness and biomass of species, however, the definitions can be considered by combining the topological importance (Jordán et al., 1999, Jordán et al., 2003). Although the definitions of keystone species from network perspective and traditional definition are not fully consistent, they provide a quantitative and complementary view for the importance of species, and stress that the network theory and species conservation practices are highly correlated (Memmott, 1999; Dunne et al., 2002a). The identification of keystone species in food webs using network analysis depends on the topological characteristics of the network. In present study, we calculated the five centrality indices of nodes in the full CSM food web and its four sub-webs, and found that species rankings using different centrality indices were different. Species importance ranking by the degree centrality and betweenness centrality is based on their direct connection in the network. Degree considers the direct impact of a species with its neighboring species directly connected. Betweenness centrality represents the influence of a species in the "communication" process. On the other hand, closeness centrality, clustering coefficient centrality and eigenvector centrality take the influence of a species in the global network into consideration (Borgatti, 2005). In all of these indices, the importance of a species in the global or local network is equally important, so the different rankings using different centrality indices should be taken as the comprehensive measure of different topological properties, which are likely relevant to the direct target analysis of theoretical ecology and conservation ecology (Estrada, 2007).

Studies have indicated that there is a significant correlation between different topological parameters of a complex network (Wutchy and Stadler, 2003). Our results showed that DC and CC correlated significantly. Thus the combined use of DC and CC can better reflect the importance ranking of species in the global and local network.

Power et al. (1996) proposed that a quantitative and predictive generalization is a primary task for identifying keystone species. Research on complex networks will give us new thoughts and methods to further understand ecosystems (Abrams et al., 1996; Yodzis, 2001; Piraino et al., 2002). In this article, we identify keystone species by only using Pajek software, so the analytical method may be more unitary and lack of comparative study statistically. More methods, as Ecosim networks (Dunne et al., 2002b; Jordán et al., 2008), CosBiLaB Graph software (Jordán et al., 2008), etc., are suggested using in the future. In addition, we have used the conventional definition, i.e., taxonomical species, and simplify the life stages of species. In the further studies, we may distinguish species in different life stages and then integrate their relationship.

Acknowledgment

We thank Mr. WenJin Chen for his pre-treatment on part data in this article.

References

Abrams PA, Menge BA, Mittelbach GG, et al. 1996. The Role of Indirect Effects in Food Webs. In: Integration of Patterns and Dynamics. 371-395, Chapman and Hall, USA

Albert R, Barabási AL. 2002. Statistical mechanics of complex networks. Reviews of Modern Physics. 74: 47-97

Arii K, Derome R, Parrott L. 2007. Examining the potential effects of species aggregation on the network structure of food webs. Bulletin of Mathematical Biology, 69: 119-133

Bai KS, Gao RH, et al. 2011. Study on relationship between the keystone species of Larixgmelinii forest and rhododendron plant. Journal of Inner Mongolia Agricultural University, 32(2): 31-37

Beigrano A, Seharler UM, Dunne J, Ulanowicz RE. 2005. Aquatic Food Webs: An Ecosystem Approach. Oxford University Press, New York, USA

Berlow EL. 1999. Strong effects of weak interactions in ecological communities. Nature, 398: 330-334

Bonaeich P. 1987. Power and centrality: a family of measures. American Journal of Sociology, 92: 1170-1182

Bond WJ. 1989. The tortoise and the hare: ecology of angiosperm dominance and gymnosperm persistence. Biological Journal of the Linnean Society, 36(3): 227-249

Bond WJ. 2001. Keystone species - hunting the shark? Science, 292: 63-64

Borgatti SP. 2005. Centrality and network flow. Social Networks, 27: 55-71

Bustamante RH, Branch GM, Eekhout S. 1995. Maintenance of an exceptional intertidal grazer biomass in South Africa-subsidy by subtidal kelps. Ecology, 76: 2314-2329

Callaway DS, Newman ME, et al. 2000. Network robustness and fragility: Percolation on random graphs. Physical Review Letters, 85(25): 5468-5471

Davic RD. 2003.Linking keystone species and functional groups: a new operational definition of the keystone species concept. Conservation Ecology, 7(1)

Dunne JA, Williams RJ, Martinez ND. 2002a. Network structure and biodiversity loss in food webs: robustness increases with connectance. Ecology Letters, 5:558-567

Dunne JA, Williams RJ, Martinez ND, 2002b. Food-web structure and network theory: The role of connectance and size. Proceeding of the National Academy of Sciences of the United States of America, 99: 12917-12922

Eisenberg E, Levanon EY. 2003. Preferential attachment in the Protein network evolution. Physical Review Letters, 91: 1-4

Ernest SKM, Brown JH. 2001. Delayed compensation for missing keystone species by colonization. Science, 292: 101-104

Estes JA, Tinker MT, Williams TM, Doak DF. 1998. Killer whale predation on sea otters linking oceanic and nearshore ecosystems. Science, 282: 473-476

Estrada E. 2007. Characterization of topological keystone species Local, global and "meso-scale" centralities in food webs. Ecological Complexity, 4: 48-57

Go´mez D, González-Arangüena E, et al. 2003. Centrality and power in social networks: a game theoretic approach. Mathematical Social Sciences, 46: 27-54

Huang HB, Yang LM, et al. 2008. Identification technique of essential nodes in protein networks based on combined parameters. Acta Automatica Sinica, 34(11):1388-1395

Hurlbert SH. 1997. Functional importance vs keystoneness: reformulating some questions in theoretical biocenology. Australian Journal of Ecology, 22(4): 369-382

Ji LZ, Liu ZG, et al. 2002. Effect on cones picking in broad-leaved *pinuskoraiensis* forest in Changbai Mountain. Chinese Journal of Ecology, 21(3): 39-42

Jordán F. 2001. Trophic fields. Community Ecology, 2: 181-185

Jordán F, Liu WC, van Veen JF. 2003. Quantifying the importance of species and their interactions in a host-parasitoid community. Community Ecology, 4: 79-88

Jordán F, Okey TA, Bauer B, Libralato S. 2008. Identifying important species: a comparison of structural and functional indices. Ecological Modelling, 216: 75-80

Jordán F, Takács-Sánta A, Molnár I. 1999.A reliability for theoretical quest for keystones. Oikos, 86: 453-462

Khanina L. 1998. Determining keystone species. Conservation Ecology, 2(2): R2

Kuang WP, Zhang WJ. 2011. Some effects of parasitism on food web structure: a topological analysis. Network Biology, 1(3-4): 171-185

Lafferty KD, Allesina S, Arim M, et al. 2008. Parasites in food webs: the ultimate missing links. Ecology Letters, 11(6): 533-546

Lafferty KD, Dobson AP, Kuris AM. 2006a. Parasites dominate food web links. Proceedings of the National Academy of Sciences of the United States of America, 103(30): 11211-11216

Lafferty KD, Hechinger RF, Shaw JC, et al. 2006b. Food webs and parasites in a salt marsh ecosystem. In: Disease Ecology: Community Structure and Pathogen Dynamics (Collinge S, Ray C, eds). 119-134, Oxford University Press, UK

Libralato S, Christensen V, Pauly D. 2006. A method for identifying keystone species in food web models. Ecological Modelling, 1(195): 153-157

Memmott J. 1999. The structure of a plant-pollinator food web. Ecology Letters, 2: 276-280

Menge BA, Berlow EL, et al. 1994. The keystone species concept: variation in interaction strength in a rocky intertidal habitat. Ecological Monographs, 64: 249-286

Mills LS, Soule ME, Doak DF. 1993. The keystone-species concept in ecology and conservation. Bioscience, 43: 219-224

Mouquet N, Gravel D, Massol F, Calcagno V. 2013. Extending the concept of keystone species to communities and ecosystems. Ecology Letters, 16(1): 1-8

Naeem S, Li S. 1997. Biodiversity enhances ecosystem reliability. Nature, 390: 507-509

Navia AF, Cortes E, Mejia-Falla PA. 2010. Topological analysis of the ecological importance of elasmobranch fishes: A food web study on the Gulf of Tortugas, Colombia. Ecological Modelling, 221(24): 2918-2926

Newman ME. 2003. The structure and function of complex networks. Siam Review, 45: 167-256

Paine RT. 1966. Food web complexity and species diversity. American Naturalist, 100: 65-75

Paine RT. 1969. A note on trophic complexity and community stability. American Naturalist, 103: 91-93

Paine RT. 1992. Food-web analysis through field measurement of per capita interaction strength. Nature, 355: 73-75

Paine RT. 1995. A conversation on refining the concept of keystone species. Conservation Biology, 9(4): 962-964

Patten BC. 1991. Network ecology: indirect determination of the life-environment relationship in ecosystems. In: Theoretical Studies of Ecosystems: The Network Perspective. 288-351, Cambridge University, UK

Pimm SL. 1982. Food Webs. Chapman & Hall, London, UK

Piraino S, Fanelli G, Boero F. 2002. Variability of species' roles in marine communities: change of paradigms for conservation priorities. Marine Biology, 140: 1067-1074

Power ME, Tilman D, Estes JA, et al. 1996. Challenges in the quest for keystones. BioScience, 46: 609-620

Ravasz E, Somera AL, Mongru DA, et al. 2002. Hierarehieal organization of modularity in metabolic networks. Science, 297: 1551-1555

Simberloff D. 1998. Flagships, umbrellas, and keystones: is single-species management passé in the landscape area? Biological Conservation, 83: 247-257

Springer AM, Estes JA, van Vliet GB, et al. 2003. Sequential megafaunal collapse in the North Pacific Ocean: an ongoing legacy of industrial whaling? Proceeding of the National Academy of Sciences of the United States of America, 100: 12223–12228

Strogatz SH. 2001. Exploring complex networks. Nature, 410: 268-275

Soulé ME, Simberloff D. 1986. What do genetic sand ecology tell us about the design of nature reserve. Biological Conservation, 35(1): 19-40

Tanner JE, Hughes TP. 1994. Species coexistence, key stone species, and succession: a sensitivity analysis. Ecology, 75(8): 2204-2219

Tilman D. 2000.Causes, consequences and ethics of biodiversity. Nature, 405: 208-211

Ulanowicz RE, Puccia CJ. 1990. Mixed trophic impacts in ecosystems. Coenoses, 5: 7-16

Wassermann S, Faust K. 1994. Social Network Analysis. Cambridge University Press, USA

Watts D, Strogatz SH. 1998. Collective dynamics of "small-world" networks. Nature, 393: 440-442

West DB. 2001. Introduction to Graph Theory. Prentice Hall, USA

Williams RJ, Berlow EL, Dunne JA, et al. 2002. Two degrees of separation in complex food webs. Proceeding of the National Academy of Sciences of the United States of America, 99: 12913-12916

Wilson EO. 1987. The little things that run the world. Conservation Biology, 1: 344-346

Wootton JT. 1994. Predicting direct and indirect effects: an integrated approach using experiments and path analysis. Ecology, 75: 151-165

Wutchy S, Stadler PF. 2003. Centers of complex networks. Journal of Theoretical Biology, 223: 45-53

Yeaton RI. 1988. Porcupines, fires and the dynamics of the tree layer of the Burkeaafricana savanna. Journal of Ecology. 76(4): 1017-1029

Yodzis P. 2000. Diffuse effects in food webs. Ecology, 81: 261-266

Yodzis P. 2001. Must top predators be culled for the sake of fisheries? Trends in Ecology Evolution, 16: 78-84

Zhang WJ. 2011. Constructing ecological interaction networks by correlation analysis: hints from community sampling. Network Biology, 1(2): 81-98

Zhang WJ. 2012a. Computational Ecology: Graphs, Networks and Agent-based Modeling. World Scientific, Singapore

Zhang WJ. 2012b. How to construct the statistic network? An association network of herbaceous plants constructed from field sampling. Network Biology, 2(2): 57-68

Zhang WJ. 2012c. Modeling community succession and assembly: A novel method for network evolution. Network Biology, 2(2): 69-78

Zhang WJ. 2012d. Several mathematical methods for identifying crucial nodes in networks. Network Biology, 2(4): 121-126

6

Hierarchicalization of chaotic food webs using Interpretive Structural Modeling

Ping Liang[1], **WenJun Zhang**[1,2]
[1]School of Life Sciences, Sun Yat-sen University, Guangzhou 510275, China; [2]International Academy of Ecology and Environmental Sciences, Hong Kong
E-mail: zhwj@mail.sysu.edu.cn,wjzhang@iaees.org

Abstract

Ecologists always meet complex food webs without clear hierarchical structure. At certain degree it will retard further analysis of food webs. In present study we transferred chaotic food webs into hierarchicalized food webs using Interpretive Structural Modeling (ISM). As an example, the hierarchical structure of seven food webs was clearly identified and defined using ISM. ISM was thus proven to be effective.

Keywords Interpretive Structural Modeling; food web; hierarchicalization.

1 Introduction

Food webs are complex networks to describe the relationship between species. Food webs are mainly divided into two types, herbivorous food webs and saprophagous food webs. The earliest definition of the food web was made by Darwin in 1859, which was described as an entirety bounded together through a complex and well-connected network (Huang and Zhao, 1992). Food web studies had stagnated since MacArther published his famous study on the food web. Since mid-20th century, the studies on food webs have being quickly increased (Briand, 1983). Up till now, a large number of food web studies were conducted using various quantitative methods, such as network analysis (Cohen, 1978, 1988; Cohen et al., 1990; Matsuda and Namba, 1991; Martinez et al., 1992; Dunne et al., 2002; Montoya et al., 2006; Arii et al.,2007; Bascompte and Jordano, 2007; Vazquez et al., 2007; Hegland et al., 2009; Dormann, 2011; Zhang, 2011, 2012).

One of current focuses on food web research is trophic level-based food webs (Cohen, 1988). In a typical food web, various trophic levels form a pyramid structure (Lawton, 1989). Species, their abundance and stability in the same trophic level determine the function and structure of food webs (Paine, 1966; Polis and Winemiller, 1996). In such a food web, the loss of any species may reduce stability of the ecosystem. To study the hierarchical structure and trophic levels of the food web will facilitate the analysis of species diversity, ecosystem stability and functional characteristics.

After many years of observation studies, researchers have established various food web databases also. Among which Interaction Web Database is a detailed and open access database that includes seven types of food webs, Anemone-fish, Host-parasite, Plant-ant, Plant-herbivore, Plant-pollinator, Plant-seed disperser, and Predator-prey food webs.

Food web research starts from the analysis on hierarchical structure of food webs. Interpretive Structural Modeling (ISM) is an approach for analyzing complex socio-economic systems. It can be used to transfer a system with chaotic structure to a system with distinct hierarchical structure (Warfield et al., 1972). ISM is mostly used to analyze direct and indirect between-element relationships, and to find crucial factors, etc. Ecologists always meet food webs with a large number of species and complex interactions. Thus ISM will provide an effective tool for hierarchicalization of such food webs.

Based on the seven food webs in Interaction Web Database, the present study tried to transfer the chaotic food webs into the structural models with distinct hierarchical structure using ISM, in order to provide the basis for in-depth food web research.

2 Material and Methods

2.1 Data sources

Data were collated from Interaction Web Database (http://www.nceas.ucsb.edu/interactionweb/). Seven typical food webs, namely Anemone-fish, Host-parasite, Plant-ant, Plant-herbivore, Plant-pollinator, Plant-seed disperser, Predator-prey food webs in the database were used. Food webs selected are described as follows:

(1) Anemone-fish food web: recorded from the coral reef of Manado region, Sulawesi, Indonesia. It contains 8 species of anemones, 7 species of fish.

(2) Host-parasite food web: recorded from the freshwater reservoir, Ontario, Canada. It includes 6 species of hosts, 25 species of parasites.

(3) Plant-ant food web: recorded from the rainforest, Peru. It contains 8 species of plants, 18 species of ants.

(4) Plant-herbivore food web: recorded from Finland. It contains 5 species of plants, 64 species of herbivores.

(5) Plant-pollinator food web: recorded from rocks and open herbaceous communities, Azores Islands. It consists of 10 species of plants, 12 species of pollinators.

(6) Plant-seed disperser food web: recorded from the tropical rain forest, Central Panama. It contains 13 species of plants, 11 species of seed dispersers.

(7) Predator-prey food web: recorded from the pine forest, Otago, New Zealand.

Assign each species in seven food webs an ID number. Obtain between-species relationships from the database, in which 0 denotes no trophic connection, 1 denotes a trophic connection, and other values denote frequencies.

2.2 Interpretive Structural Modeling

Interpretive Structural Modeling (ISM) enables the structuring of 'elements' *vs.* any transitive relationship. A transitive relationship is one in which the following property holds (Wikipedia): If element 'A' → element 'B' AND 'B' → element 'C' THEN 'A' MUST → 'C' (where '→' stands for the transitive relationship under consideration).

ISM can be used to transfer a system with complex and ambiguous elements and relationships into several subsystems. A multi-level hierarchical and structural model is then constructed. Finally, the internal structure and hierarchy of the system can be determined according to the known relationships between the elements. It lays the basis for system optimization analysis.

ISM analysis of the food web follows these procedures:

(1) Identification of key species

Each species in seven food webs is recorded as a key species and assigned an ID number.

(2) Determine feeding relationship

In each of seven food webs, 0 denotes no trophic connection, 1 denotes a trophic connection, and other values denote frequencies. Feeding relationship is thus determined.

(3) Establish adjacency matrix

An element of adjacency matrix A shows feeding relationship between row species and column species (Zhang, 2012). Assume there are n species in the food web, a $n \times n$ adjacency matrix, $A=(a_{ij})_{n \times n}$, can be obtained. Species S_i is fed by species S_j, if $a_{ij}=1$; species S_i is not fed by species S_j, if $a_{ij}=0$.

(4) Establish reachability matrix

An non-zero element of reachability matrix M means that row species can reach column species. $M=(m_{ij})_{n \times n}$ $=(A+I)^n$. $m_{ij}=1$, means row species S_i can directly or indirectly affect column species S_j.

(5) Establish structural model (antecedent set and attainable set)

$P(S_i)$ is called the attainable set, i.e., the set of species that species S_i can reach. $P(S_i)$ can be obtained through examining all column species that the element are 1's in the i-th row of M. $Q(S_i)$ is called the antecedent set, i.e., the set of species that can reach species S_i. $Q(S_i)$ can be obtained through examining all row species that the element are 1's in the i-th column of M.

Calculate the species set at level 1, $L_1=\{S_i \mid P(S_i) \cap Q(S_i)=P(S_i)\}$, and then delete the rows and columns of matrix M corresponding to the species in L_1. The matrix M' can be thus obtained. Operate M' to find the species of L_2 at level 2. Repeat this procedure, $L_3, L_4, L_5...$, can thus be calculated. By doing so, all species are assigned to corresponding levels.

(6) Establish interpretive structural model

L_1 contains species at level 1; L_2 contains species at level 2, etc. Finally, rows and columns of matrix M are re-arranged according to this sequence. M is thus transferred to a partitioned and triangularized matrix. Connect species between adjacent levels and species at the same level with directed edges. The hierarchical structure of the food web can then be clearly determined.

3 Results and Discussion

3.1 Anemone-fish food web

Table 1 shows the species IDs in the Anemone-fish food web.

Table 1 Species IDs in the Anemone-fish food web.

Genus	Species	ID	Genus	Species	ID
Amphiprion	*Clarkia*	1	*Entacmaea*	*Quadricolor*	9
Amphiprion	*Melanopus*	2	*Heteractis*	*Magnifica*	10
Amphiprion	*Ocellaris*	3	*Stichodactyla*	*Mertensii*	11
Amphiprion	*Perideraion*	4	*Heteractis*	*Aurora*	12
Amphiprion	*Polymnus*	5	*Stichodactyla*	*Haddoni*	13
Amphiprion	*Sandaracinos*	6	*Macrodactyla*	*Doreensis*	14
Premnas	*Biaculeatus*	7	*Heteractis*	*Malu*	15
Heteractis	*Crispa*	8			

Construct the adjacency matrix, $A=(a_{ij})_{n \times n}$, based on Table 1, in which species S_i is fed by species S_j, if $a_{ij}=1$; species S_i is not fed by species S_j, if $a_{ij}=0$. The adjacency matrix A is

	1	2	3	4	5	6	7	8	9	10	11	12	13	14	15
1	0	0	0	0	0	0	0	0	0	0	0	0	0	0	0
2	0	0	0	0	0	0	0	0	0	0	0	0	0	0	0
3	0	0	0	0	0	0	0	0	0	0	0	0	0	0	0
4	0	0	0	0	0	0	0	0	0	0	0	0	0	0	0
5	0	0	0	0	0	0	0	0	0	0	0	0	0	0	0
6	0	0	0	0	0	0	0	0	0	0	0	0	0	0	0
7	0	0	0	0	0	0	0	0	0	0	0	0	0	0	0
8	1	0	0	1	0	0	0	0	0	0	0	0	0	0	0
9	1	1	0	0	0	0	1	0	0	0	0	0	0	0	0
10	0	0	1	1	0	0	0	0	0	0	0	0	0	0	0
11	1	0	0	0	0	1	0	0	0	0	0	0	0	0	0
12	1	0	0	0	0	0	0	0	0	0	0	0	0	0	0
13	0	0	0	0	1	0	0	0	0	0	0	0	0	0	0
14	1	0	0	0	1	0	0	0	0	0	0	0	0	0	0
15	1	0	0	0	0	0	0	0	0	0	0	0	0	0	0

Conducting ISM procedures above, the results are as follows:

Level 1   Species 8

Level 1   Species 9

Level 1   Species 10

Level 1   Species 11

Level 1   Species 12

Level 1   Species 13

Level 1   Species 14

Level 1   Species 15

Level 2   Species 1

Level 2   Species 2

Level 2   Species 3

Level 2   Species 4

Level 2   Species 5

Level 2   Species 6

Level 2   Species 7

Therefore, there are two functional levels (groups) in the Anemone-fish food web, which is coincident with the composition of the food web, i.e., anemones and fish. Level 1 includes the species No. 8-15, and Level 2 includes species No. 1-7. The interpretive structural model of the Anemone-fish food web is indicated in Fig. 1.

From Fig. 1, we conclude that the Anemone-fish food web is relatively stable. The loss of a species at a level can be replaced by other species at the same level. However, functional diversity of the food web is not ideal (only two functional groups).

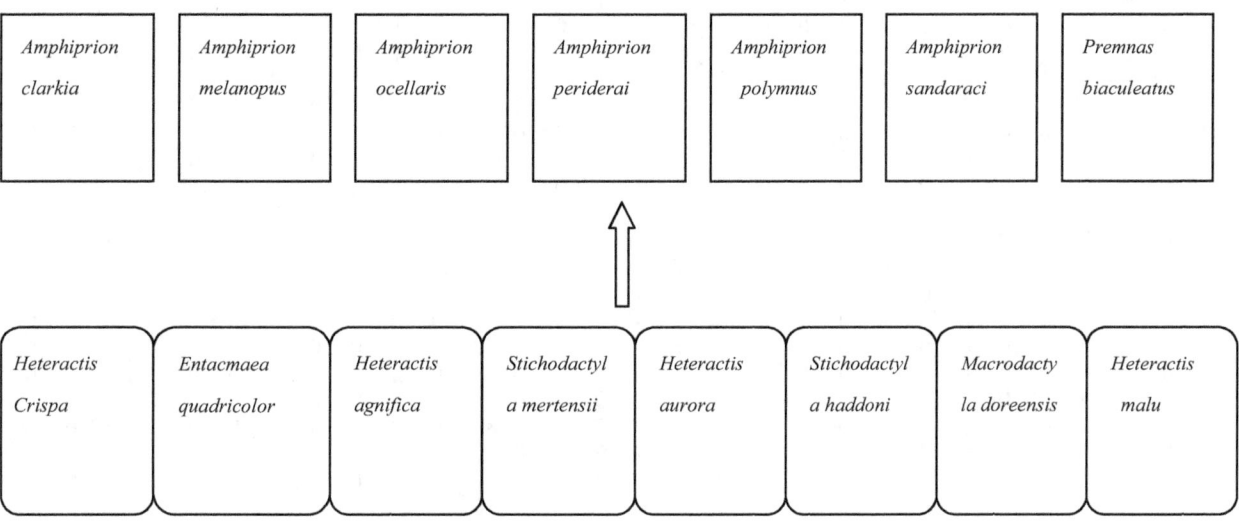

Fig. 1 Interpretive structural model of the Anemone-fish food web.

3.2 Host-parasite food web

Table 2 shows the species IDs in the Host-parasite food web.

Table 2 Species IDs in the Host-parasite food web.

Genus	Species	ID	Genus	Species	ID	Genus	Species	ID
Salmo	salar	1	Phyllodistomum	lachancei	12	Capillaria	salvelini	23
Salvelinus	fontinalis	2	Triaenophorus	Crassus	13	Metabronema	salvelini	24
Salvelinus	namaycush	3	Triaenophorus	crassus (L)	14	Metechinorhynchus	lateralis	25
Coregonus	clupeaformis	4	Eubothrium	salvelini	15	Neoechinorhynchus	crassus	26
Esox	lucius	5	Diphyllobothriu m	sp. (L)	16	Ergasilus	caeruleus	27
Catostomus	catostomus	6	Proteocephalus	pinguis	17	Salmincola	Salmincola	28
Tetraonchus	monenteron	7	Proteocephalus	tumidocollus	18	Salmincola	coregonorum	29
Discocotyle	sagittata	8	Proteocephalus	sp.1	19	Salmincola	edwardsii	30
Crepidostomum	farionis	9	Proteocephalus	sp.2	20	Salmincola	siscowet	31
Crepidostomum	cooperi	10	Raphidascaris	sp.	21			
Phyllodistomum	coregoni	11	Raphidascaris	canadensis	22			

Similar to the above procedure, we obtained the adjacency matrix *A* as the following:

```
0 0 0 0 0 0 0 1 0 1 0 1 0 0 1 1 0 0 0 1 0 0 0 0 0 0 0 1 0 0 0
0 0 0 0 0 0 0 1 1 0 0 1 0 0 1 1 0 1 1 1 1 0 1 1 1 0 0 0 0 1 0
0 0 0 0 0 0 1 0 1 1 0 1 0 1 1 1 0 1 1 1 1 0 1 0 0 0 0 0 0 0 1
0 0 0 0 0 0 0 1 1 1 1 0 0 1 0 1 0 1 1 1 1 0 1 0 0 0 0 1 1 0 0
0 0 0 0 0 0 1 0 0 0 0 0 1 0 0 0 1 0 0 0 0 1 0 0 1 0 0 0 0 0 0
0 0 0 0 0 0 0 0 0 0 0 0 0 0 0 0 0 0 0 0 0 0 0 0 0 1 1 0 0 0 0
0 0 0 0 0 0 0 0 0 0 0 0 0 0 0 0 0 0 0 0 0 0 0 0 0 0 0 0 0 0 0
0 0 0 0 0 0 0 0 0 0 0 0 0 0 0 0 0 0 0 0 0 0 0 0 0 0 0 0 0 0 0
0 0 0 0 0 0 0 0 0 0 0 0 0 0 0 0 0 0 0 0 0 0 0 0 0 0 0 0 0 0 0
0 0 0 0 0 0 0 0 0 0 0 0 0 0 0 0 0 0 0 0 0 0 0 0 0 0 0 0 0 0 0
0 0 0 0 0 0 0 0 0 0 0 0 0 0 0 0 0 0 0 0 0 0 0 0 0 0 0 0 0 0 0
0 0 0 0 0 0 0 0 0 0 0 0 0 0 0 0 0 0 0 0 0 0 0 0 0 0 0 0 0 0 0
0 0 0 0 0 0 0 0 0 0 0 0 0 0 0 0 0 0 0 0 0 0 0 0 0 0 0 0 0 0 0
0 0 0 0 0 0 0 0 0 0 0 0 0 0 0 0 0 0 0 0 0 0 0 0 0 0 0 0 0 0 0
0 0 0 0 0 0 0 0 0 0 0 0 0 0 0 0 0 0 0 0 0 0 0 0 0 0 0 0 0 0 0
0 0 0 0 0 0 0 0 0 0 0 0 0 0 0 0 0 0 0 0 0 0 0 0 0 0 0 0 0 0 0
0 0 0 0 0 0 0 0 0 0 0 0 0 0 0 0 0 0 0 0 0 0 0 0 0 0 0 0 0 0 0
0 0 0 0 0 0 0 0 0 0 0 0 0 0 0 0 0 0 0 0 0 0 0 0 0 0 0 0 0 0 0
0 0 0 0 0 0 0 0 0 0 0 0 0 0 0 0 0 0 0 0 0 0 0 0 0 0 0 0 0 0 0
0 0 0 0 0 0 0 0 0 0 0 0 0 0 0 0 0 0 0 0 0 0 0 0 0 0 0 0 0 0 0
0 0 0 0 0 0 0 0 0 0 0 0 0 0 0 0 0 0 0 0 0 0 0 0 0 0 0 0 0 0 0
0 0 0 0 0 0 0 0 0 0 0 0 0 0 0 0 0 0 0 0 0 0 0 0 0 0 0 0 0 0 0
0 0 0 0 0 0 0 0 0 0 0 0 0 0 0 0 0 0 0 0 0 0 0 0 0 0 0 0 0 0 0
0 0 0 0 0 0 0 0 0 0 0 0 0 0 0 0 0 0 0 0 0 0 0 0 0 0 0 0 0 0 0
0 0 0 0 0 0 0 0 0 0 0 0 0 0 0 0 0 0 0 0 0 0 0 0 0 0 0 0 0 0 0
0 0 0 0 0 0 0 0 0 0 0 0 0 0 0 0 0 0 0 0 0 0 0 0 0 0 0 0 0 0 0
0 0 0 0 0 0 0 0 0 0 0 0 0 0 0 0 0 0 0 0 0 0 0 0 0 0 0 0 0 0 0
0 0 0 0 0 0 0 0 0 0 0 0 0 0 0 0 0 0 0 0 0 0 0 0 0 0 0 0 0 0 0
0 0 0 0 0 0 0 0 0 0 0 0 0 0 0 0 0 0 0 0 0 0 0 0 0 0 0 0 0 0 0
0 0 0 0 0 0 0 0 0 0 0 0 0 0 0 0 0 0 0 0 0 0 0 0 0 0 0 0 0 0 0
0 0 0 0 0 0 0 0 0 0 0 0 0 0 0 0 0 0 0 0 0 0 0 0 0 0 0 0 0 0 0
```

The results indicate that Level 1 has 6 species, species No. 1-6, and Level 2 has 25 species, species No. 7-31. The interpretive structural model of the Host-parasite food web is indicated in Fig. 2. The host is largely responsible for the reproduction and growth of its parasites (Holling, 1973). In the Host-parasite food web, the species in Level 1 are less than the species in Level 2. Relatively, the matter and energy flow are easily blocked, and the food web is easily disturbed (Winter, 1990).

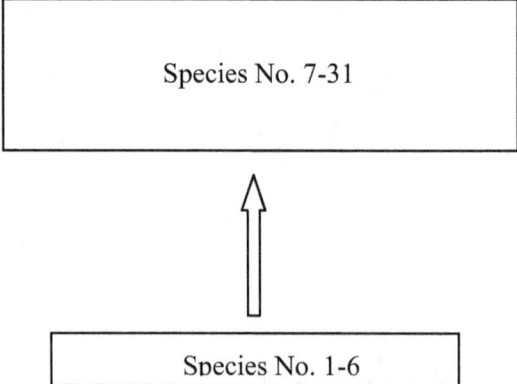

Fig. 2 Interpretive structural model of the Host-parasite food web.

3.3 Plant-ant, Plant-herbivore, Plant-pollinator, and Plant-seed disperser food webs

The ISM results for Plant-ant, Plant-herbivore, Plant-pollinator, and Plant-seed disperser food webs are concluded as follows.

Plant-ant food web: Level 1: 8 plant species; Level 2: 18 ant species.

Plant-herbivore food web: Level 1: 5 plant species; Level 2: 64 herbivore species.

Plant-pollinator food web: Level 1: 10 plant species; Level 2: 12 pollinator species.

Plant-seed disperser food web: Level 1: 13 plant species; Level 2: 11 seed disperser species.

For both Plant-ant and Plant-herbivore food webs, the number of species in Level 2 is much greater than that in Level 1. The feeding relationship between Level 1 and 2 is close. Level 2 is largely dependent upon Level 1 (Bai et al., 2004). In a sense, the two food webs are less stable. However in Plant-pollinator and Plant-seed disperser food webs, the number of species in Level 1 and 2 is similar to each other. In addition, the feeding relationship is not close. In a sense, the two food webs are relatively stable (Paine, 1992; McCann et al., 1998; Berlow, 1999).

3.4 Predator-prey food web

Table 3 shows the species IDs in the Predator-prey food web.

Table 3 Species IDs in the Predator-prey food web.

Species	ID	Species	ID	Species	ID
Unidentified detritus	1	*Austrosimulium*	18	*Paracalliope purple*	35
Plant material	2	*Cricotopus I*	19	*Paralimnophila*	36
Terrestrial invertebrates	3	*Cricotopus II*	20	*Podonomous*	37
Blue green algae	4	*Cristaperla*	21	*Polypedellum*	38
Auodinella	5	*Deleatidium*	22	*Psilachorema*	39
Rhoicosphenia curvata	6	*Eriopterini*	23	*Psychodid*	40
Navicula avenacea	7	*Eukiefidrella*	24	*Pycnocentria*	41
Unknown green algae	8	*Hydora nitida a*	25	*Scirtid brd*	42
Calothrix	9	*Hydora nitida l*	26	*Spaniocerca*	43
Cocconeis placentula	10	*Hydrobiosella stenocerca*	27	*Stenoperla prasinia*	44
Achnanthes lanceolata	11	*Hydrobiosis silvicola*	28	*Stictocladius*	45
Gomphonema angustatum	12	*Isopoda*	29	*Zelandobius*	46
Navicula pusio	13	*Lumbriculiid pink*	30	*Zelandoperla*	47
Synedra ulna	14	*Naonella*	31	*Crayfish*	48
Aotepsyche	15	*Nothodixa*	32	*Galaxias*	49
Aspectrotanypus	16	*Oligo I*	33		
Austroperla cyrene	17	*Paracalliope other*	34		

The results show that Level 1 has 15 species, Level 2 has 28 species, and Level 3 has 6 species. The interpretive structural model of the Predator-prey food web is indicated in Fig. 3.

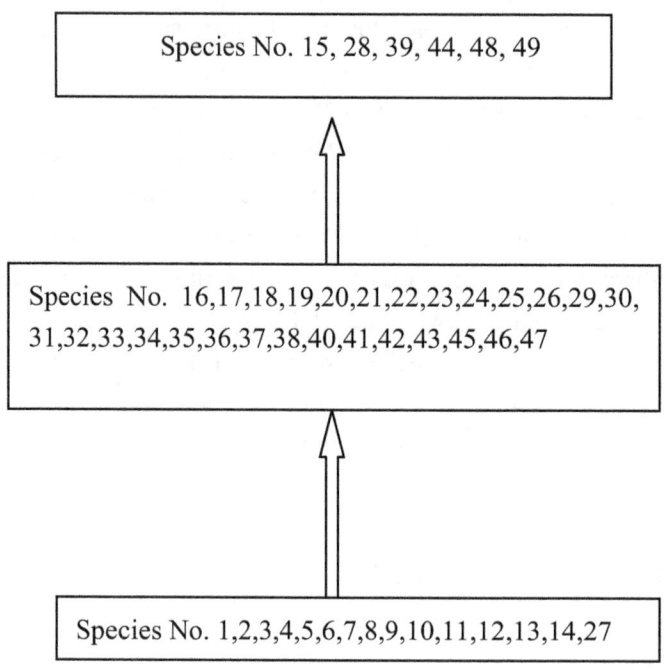

Species No. 15, 28, 39, 44, 48, 49

Species No. 16,17,18,19,20,21,22,23,24,25,26,29,30, 31,32,33,34,35,36,37,38,40,41,42,43,45,46,47

Species No. 1,2,3,4,5,6,7,8,9,10,11,12,13,14,27

Fig. 3 Interpretive structural model of the Predator-prey food web.

It is obvious that the functional diversity of the Predator-prey food web is larger than other food webs. More diverse functional groups make the food web more stable (Korner, 1993; Grime, 1997). However, species distribution among these groups is less ideal.

4 Remarks

In the Interaction Web Database, Anemone-fish, host-parasite, Plant-ant, Plant-herbivore, Plant-pollinator, and Plant-seed disperser food webs have two trophic levels respectively; Predator-prey food web is more complex. The ISM results of present study were coincident with the definition in Interaction Web Database. ISM was thus proven to be effective. However, it should be noted that seven food webs used in present study are parts of complete ecosystems. All species in a food web are only a few from full species list of the corresponding ecosystems. Therefore the resulted food webs did not show a pyramid structure. ISM is encouraged for use in more practical food webs.

References

Arii K, Derome R, Parrott L. 2007. Examining the potential effects of species aggregation on the network structure of food webs. Bulletin of Mathematical Biology, 69: 119-133

Bai YF. Han XG, Wu JG, et al.2004. Ecosystem stability and compensatory effects in the Inner Mongolia grassland. Nature, 431: 181-184

Bascompte J, Jordano P. 2007. The structure of plant-animal mutualistic networks: the architecture of biodiversity. Annual Review of Ecology, Evolution, and Systematics, 38: 567-593

Berlow EL. 1999. Strong effects of weak interactions in ecological communities. Nature, 398: 330-334

Briand F. 1983. Environmental control of food web structure. Ecology, 64: 253-263

Cohen JE. 1978. Food Webs and Niche Space. University Press, Princeton, New Jersey, USA

Cohen JE. 1988. Food webs and community structure. In: Perspectives in Theoretical Ecology (Roughgarden J, May RM, Levin SA, eds). Princeton University Press, Princeton, New Jersey, USA

Cohen JE, Briand F, Newman CM. 1990. Community Food Webs: Data and Theory. Biomathematics (Vol. 20), Springer, Berlin, Germany

Dormann CF. 2011. How to be a specialist? Quantifying specialisation in pollination networks. Network Biology, 1(1): 1-20

Dunne JA, Williams R, Martinez N. 2002. Conservation of species interaction networks. Ecology Letters, 5: 558

Grime TJ. 1997.Biodivebity and ecosystem function: the debate deepens. Science, 277: 1260-1261

Hegland SJ, Grytnes JA, Totland O. 2009. The relative importance of positive and negative interactions for pollinator attraction in a plant community. Ecological Research, 24: 929-936

Holling C S Resilience and stability of ecological system. 1973;

Huang DM, Zhao SL. 1992. Progress in food web researches. Journal of Gansu Agricultural University, 27(4): 277-288

Korner Ch. 1993. Scaling from species to vegetation: the useful of functional groups. Biodiversity and Ecosystem Function, 99: 117-140

Lawton JH. 1989. Food webs. In: Ecological Concepts (Cherett JM, ed). Blackwell Scientific, Oxford, UK

Martinez ND. 1992. Constant connectance in community food webs. American Naturalist, 139: 1208-1218

Matsuda H, Namba T. 1991. Food web graph of a coevolutionarily stable community. Ecology, 72: 267-276

McCann K, Hastings A, Huxel GR. 1998. Weak trophic interactions and the balance of nature. Nature, 395: 794-795

Montoya JM, Pimmental SL, Sole RV. 2006. Ecological networks and their fragility. Nature, 442: 259-264

Paine RT. 1966. Food web complexity and species diversity. American Naturalist, 100: 65-75

Paine RT. 1992. Food-web analysis through field measurement of per capita interaction strength. Nature, 355: 73-75

Polis GA, Winemiller KO. 1996. Food Webs: Integration of Patterns and Dynamics. Chapman & Hall, USA

Vazquez DP, Melian CJ, Williams NM, et al. 2007. Species abundance and asymmetric interaction strength in ecological networks. Oikos, 116: 1120-1127

Warfield JN, Hill JD, et al. 1972. A Unified Systems Engineering Concept. Monograph No. 1, Battelle Memorial Institute, Columbus, USA

Winter PA. 1990. A contemporary perspective on adapting modern planning and decision-making technologies to economic development. Economic Development Review, 8(2): 53-56

Zhang WJ. 2011. Constructing ecological interaction networks by correlation analysis: hints from community sampling. Network Biology, 1(2): 81-98

Zhang WJ. 2012. Computational Ecology: Graphs, Networks and Agent-based Modeling. World Scientific, Singapore

7

Improved methods for analyzing MRI brain images

Jyotsna Dogra, **Navdeep Prashar**, **Shruti Jain**, **Meenakshi Sood**

Department of Electronic and Communication Engineering, Jaypee Institute of Information Technology, Solan-173234, India

E-mail: jyotsnadogra1989@gmail.com, nav.prashar@gmail.com, shruti.jain@juit.ac.in, meenakshi.sood@juit.ac.in

Abstract

Image segmentation is a part of image processing for region or object extraction from the background area. Owing to the complex background, contrast of the infected portion, low intensity difference values, intricate inner body parts etc.; the problem of region extraction in segmentation is very challenging. Among various image segmentation techniques, thresholding is one of the simplest techniques, in which the region of interest is extracted from the background by comparing the pixel values with the threshold value. The threshold value is obtained from histogram of the image. The technique presented in the paper involves graph cut method in which the initial centroids are automatically selected by exploiting the symmetrical nature of the MRI images. The results obtained by the thresholding technique in this research work shows that any abnormality can be localized easily in horizontal divided MRI brain image rather than in vertical divided MRI image. Graph cut results show better segmentation than thresholding technique which is justified by PSNR and SSIM values.

Keywords segmentation; fuzzy c-mean clustering; k-mean clustering; split and merge; graph-cut.

1 Introduction

Since evolution, the most effective and convenient medium for the understanding of information is through images. With the advent of time the different techniques for different images got exploited for the image analysis as per the applications such as: medical application, satellite imaging for identification of various objects like land (Ballestores and Qiu, 2012; Hussain, 2014; Seljuq and Hussain, 2014), body, building and river, robot navigation, remote sensing, detecting tumors and malign tissue finger print recognition, face recognition, iris recognition etc (Pham et al., 2000). Image processing popularity resulted due to its versatile method and low cost (Sonka et al., 2014). In digital image processing the image undergoes with fast computers for signal analysis various phases: Data acquisition, Pre-processing, Image enhancement (Bhusri et al., 2016; Sharma et al., 2017; Wang et al., 1983), Information Extraction (Amandeep et al., 2017; Bhusri et al., 2016; Rana et al., 2016; Sharma et al., 2017), and Classification (Dhiman et al., 2016; Jain et al., 2011; Sood, 2017; Sood et al., 2015). Further, for information extraction images go through three levels: image analysis, image processing and image segmentation as shown in Fig 1. Image analysis involves the extraction of the meaningful information contained in the image. Image processing includes manipulation of images to enhance the quality of image: as removing noise, histogram equalization, processing RGB values etc (Sood et al., 2014). The process includes image as input in terms of pixels with intensity values and output may be an image or any characteristic or property of image (Soille, 2013). The various techniques that conclude for image segmentation are: detecting edges, counting objects, finding shapes and area of the objects in binary image (Pal et al., 1993).

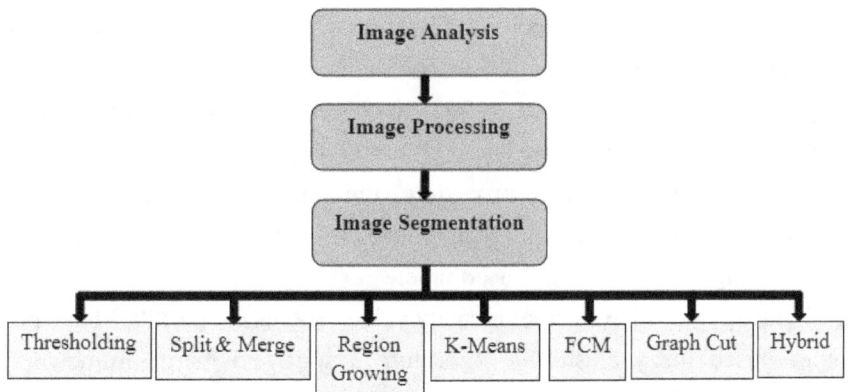

Fig. 1 Classification of image segmentation

In this research work different steps of image analysis, especially Medical Imaging is dealt with. Medical imaging comprises of techniques and process of creating visual representation of human body to reveal, diagnose and examine the disease for clinical application. It has been acknowledged to be most significant advancement over other medical technologies (Suetens, 2017). Some of the techniques encompassed by medical imaging are Positron emission tomography (PET), Medical resonance imaging (MRI), Computed tomography (CT) and more. Among all the MRI is one of the popular ways for non-invasive imaging of human body providing rich information about the human soft tissue anatomy. In comparison to X-ray, CT images, MRI shows more detailed image (Ferrari, 2005; Zhang, 2018).

Segmentation of brain MRI's is an important image processing procedure for both the physician and the brain researcher as it offers a valuable evaluation method for pre-and-post surgical after effects. A great deal of research is going on number of segmentation methods and lot of literature is available to segment images based on various criteria. Therefore, it is necessary to develop algorithms to obtain robust image segmentation. Thresholding is one of the oldest technique in image segmentation in which at an optimum threshold value the complete image is separated in object region and background image when the intensity of pixels is above or below the calculated threshold value. The limitation arises as it is not easy to find an appropriate threshold which can separate the image into two different groups.

Graph cut was first proposed byBoykov. It is an interactive segmentation. In this paper we have performed automatic centroid selection and returned the actual pixel value to the object region. The energy function responsible for the optimum cut in the graph comprises of the regional and the boundary term.

This research paper has been divided into following sections: Section II gives the description of Image Segmentation. Section III discusses the histogram thresholding and Section IV gives the results and discussions followed by conclusion.

2 Image Segmentation

Image segmentation plays a vital role in image processing by providing partitioning of image into region, measuring the area and shape, identification of region of interest in the image so that it becomes more meaningful and can be analysed (Comaniciu et al., 1997). Any image comprises background and the object region and image segmentation separates these two regions to provide the region of interest for its analysis. Main task of segmentation is the extraction of region of interest from the given image (Cheng et al., 2015). On similarity and discontinuity basis image segmentation is categorized in the categories as discussed in the following section.

a) *Thresholding*: One of the simplest and oldest techniques for segmentation of gray image is thresholding. An image described as $f(x, y)$ is composed of light object and a dark background, such that the pixels are distributed in two dominant modes. A threshold value is decided to group the pixels and extract the images for segmentation. Kallergi et al. (1992) have performed image segmentation in three levels: image pre-processing, local thresholding and region growing.

b) *Split and Merge*: In split and merge procedure, the image is divided into regions based on the homogeneity criterion. If the pixels in the region have different intensity value then the region is divided until no more splitting is required and a complete homogenous region is achieved. The regions are merged if the divided regions have same intensity value. Borges et al. (2000) have proposed a method in which prototype based fuzzy clustering algorithm in split and merge framework. Split and merge technique is also applied for bimodality detection and 3D image segmentation(Chaudhuri et al., 2010; Damiand et al., 2003).

c) *Region Growing*: In this technique the growth of region depends upon the homogeneous neighbouring pixels around the seed. The homogeneity criterion depends not only on the problem under consideration, but also on type of image data available. To increase the effectiveness of region growing, Pavlidis et al. (1990) and Gambotto (1993) have presented different approaches for integrating edge detection and region growing.

d) *Clustering*: Clustering is the organization of data with high intra cluster similarity and low inter cluster similarity. To find the similarity or dissimilarity between two data points i.e. the distance between image pixels is calculated. This distance can be the intensity difference between two pixels. It reflects the degree of separation or closeness among the data points. Coleman et al. (1979) presented the approach of clustering for image segmentation. The most commonly used clustering methods:

I. *K-mean Clustering*: K-mean Clustering also known as hard clustering is based on iterative process that divides the image into different clusters. The data points or the pixels are grouped in an exclusive way such that if a data point belongs to a certain cluster then it will not belong to any other cluster. K-mean assumes Euclidean distance on the basis of which the similarity and dissimilarity is measured and the clustering is performed. Chen et al. (1998) have proposed an adaptive K-mean algorithm for the segmentation of regions with the smooth varying intensity distribution which is applied on cardiac CT volumetric images.

II. *Fuzzy C mean Clustering*: In FCM clustering the clusters of similar pixel is formed by giving membership value to the assigned pixel, these membership values give the degree of similarity. But the algorithm lacked in the initialization as the final result is depended solely on the input or the initial values. FCM technique is also known as soft clustering because it does not put any hard constraints on the pixel in forming the clusters. Clark et al. (1994) proposed a hybrid method in which the fuzzy clustering has been used to detect the tumor and then label the clusters formed. Clark, et. al. provides an automatic segmentation of MR Images. Ahmed et al. (2002) proposed a method to modify the objective function of the conventional FCM so that the spatial information of the neighbourhood pixel could be used for the labelling of neighbourhood pixel (Chuang et al., 2006).

e) *Graph Cut method* Boykov et al. (2006) and Khanna et al. (2012) have given a technique in which the image is treated as binary and making a hard constraint for performing the image segmentation. Graph cut is an interactive technique which provides global optimal segmentation of N-dimensional images. The cost function defines the properties of boundary and region/object in the image. Felzenszwalb et al. (2004) have proposed an algorithm that measures the boundary between two regions in such a way that it satisfies two properties. Pauchard et al. (2016) have introduced graph cut for the interactive graph cut segmentation for fast creation of femur finite element models and observed that the proposed algorithm corresponded well to the manual segmentation.

f) *Hybrid Method*: Due to innovation in technology, the image segmentation techniques have become more application specific and the judgment of the most appropriate technique useful for a particular application is a difficult task. Therefore, hybrid or combined segmentation methods have been extensively used in different application specific biomedical image segmentation. The MRI segmentation method proposed by Ortiz et al. (2013, 2014) is based on Self-Organizing Maps (SOMs) and Genetic Algorithm (GAs). Sepas-Moghaddam et al. (2014) have presented a hybrid algorithm combining PSO, K-mean and learning automata based on multilevel thresholding are applied on standard test images.

3 Thresholding

Thresholding is the one of the simplest technique which has been based on the similarity index. Function f represents the complete image and $f(x, y)$ is the pixel intensity at the point (x, y). Single threshold value segments the image in two intensity values i.e. 0 (black) & 255 (white), resulting in extraction of region of interest inside the brain. Increase in the threshold values gives the increased number of intensity values in the segmented image. For application of thresholding based segmentation technique, it is required to apply the correct threshold values in order to achieve proper segmentation results.

Algorithm 1: Histogram Thresholding

Input: Image $f(x, y)$.
Output: Segmented image.

BEGIN

Step 1: Divide the MRI image in two halves (horizontal, vertical).

Step 2: Calculate the histogram for both images

Step 3: Difference between histograms (two halves) is calculated.

Step 4: Calculate the threshold value T from the difference value in step 3.

Step 5: Segment the image as per:

$$f(x, y) \geq T; \text{ Background}$$
$$f(x, y) \leq T; \text{ Object}$$

END

---

4 Graph Cut Segmentation

Graph cut is an interactive technique which provides the global optimal segmentation of N-Dimensional images. In this technique, image is expressed ina binary form where it is partitioned inobject and background regions, which restricts the segmentation to have a hard constraint. In Boykov et al. (2001), the authors have demonstrated the following hypothesis which are employed in this paper:

a. Combination of hard and soft constraint.

b. If the hard constraint are added or changed optimal segmentation can still be recalculated.

c. In case initial segmentation is not perfect the user can define additional seed points from the result and can be adjusted to the current segmentation without any re-computation. Hence it becomes time efficient.

As explained by Boykov et al. (2006), an image is represented in graphical form G, which comprises of set of nodes V and edges $\mathcal{E}$ which connect the nodes.

$$G = <V, \mathcal{E}>$$

Where nodes V consists of source s and the sink t; source s represents the set of nodes that belong to the segmented region of interest and sink t represents the set of nodes that belong to the segmented background region.

$$V = \{s, t\} \cup \mathcal{P}$$

An edge $\mathcal{E}$ is described as a $t-link$, if it connects the non terminal node with a terminal node; and as a $n-link$ if it connects two non-terminal nodes. Partition in an image is executed when edges which connect nodes are broken or cut; which require knowledge of boundary and region properties. The cost function defines these properties and provide optimal cut for the segmentation. Among the various strategies for optimizing energy function, in this paper authors have used the following energy function which was initially explored by Greig et al. (1989):

$$E(R) = \gamma U_p(R) + B_p(R)$$

$$U(R) = \sum_{p \epsilon P} U_p(r_p)$$

$$B(R) = B_{p,q} = e^{-\left(\frac{|I_p - I_q|}{\sigma}\right)}$$

where R gives the label to the image pixel, $U_p(R)$ is the regional term that gives the measure of assigning R to p (whether its object pixel or background pixel), $B_p(R)$ calculates the boundary term and γ is the relative importance factor.

Algorithm 2: Graph Cut Segmentation

Input: MRI Brain Image.

Output: Segmented image.

BEGIN

Step 1: Divide the MRI image in two halves.

Step 2: Calculate the pixel difference between the two halves to obtain maximum difference value.

Step 3: A range of is taken above and below the maximum value.

Step 4: From step 4. Obtain the centroids for the background and the object region

Step 5: Calculate object terminal and background terminal.

Step 6: Label the all pixel with the respect to the degree of belongingness.

Step7: Segment the region of interest by calculating the minimum cut.

Step 8: Segmented image is obtained.

END

---

5 Simulation Results and Discussions

In this section we have analyzed and compared the MRI image segmentation by two methods: a. Histogram Thresholding & b. Graph cut. Three MRI images are used among which thresholding is done on the first image and graph cut is implemented using other two MRI images. The two techniques proposed are explained in the following:

a. *Histogram Thresholding*:

MRI image used for experiment is infected with a tumor region in the temporal lobe. The MRI brain has been divided into two halves: vertical half and horizontal half. Brain is having the similar shape around the central axis, so any change in any half of the brain will result in high difference value in the pixel intensities. The histogram for the both upper and lower half portion of horizontally divided MRI image is plotted and compared. The optimum threshold value is obtained from this histogram. In our implementation we have used only one threshold value. The difference between both the histogram has been calculated and analyzed. Finally segmented image has been based on the threshold value, $T = 0.4980$ which has been calculated from the difference values of the histogram.

In this paper we have used only one threshold value so as to extract the infected portion from the brain. The final segmented image contains two intensity values and the infected portion has been clearly segmented.

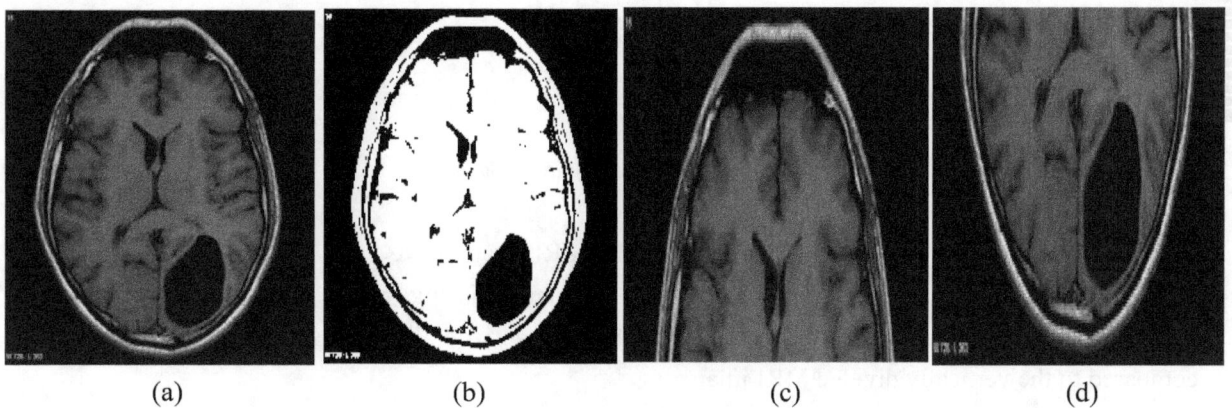

| (a) | (b) | (c) | (d) |

Fig. 2 (a) Original Image, (b) Output Image, (c) Upper half MRI, (d) Lower half MRI

Upper half and lower half of horizontally divided MRI image have been shown in Fig. 2(c) & 2(d) respectively, and histograms corresponding to Fig. 3 is shown in Fig. 3(a) & 3(b). In Fig. 4 it has been observed that number of pixel for the pixel values greater than 130 to 255 are zero. Higher difference has been seen for the pixel-value near to zero i.e. the tumor may contain these pixel intensity. Same procedure has been followed for the vertically divided MRI image.

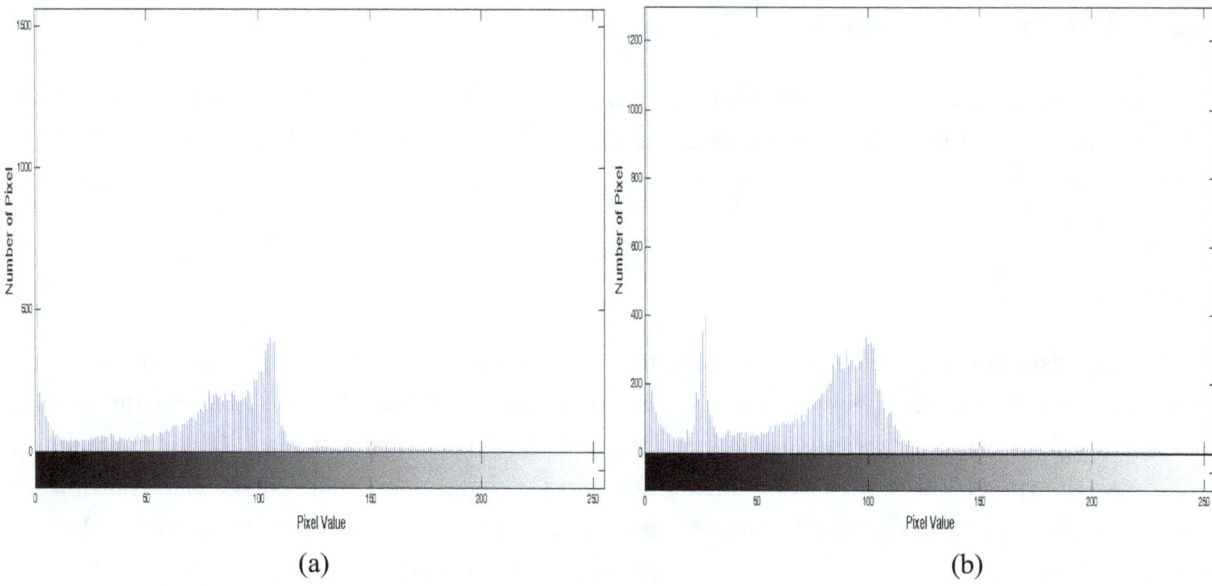

(a)   (b)

Fig. 3 (a) Histogram of Fig. 2 (c), Histogram of Fig. 2 (d).

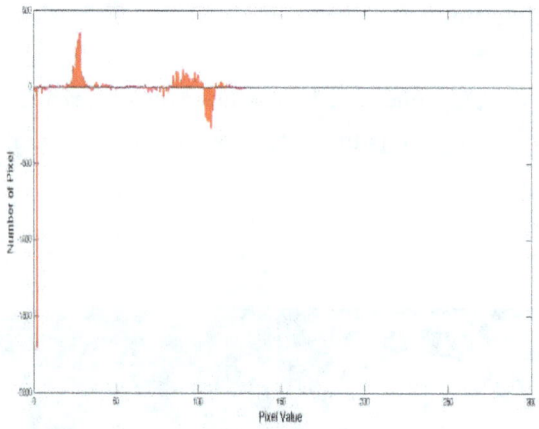

Fig. 4 Difference in histograms.

Right half and left half of the vertically divided MRI image have been shown in Fig. 5(a) & 5(b) respectively, and histograms corresponding to Fig. 6 is shown in Fig. 6(a) & 6(b). It has been observed from Fig. 4 to Fig. 7 that the difference values of the horizontally divided MRI image are low as compared to the vertically divided MRI image.

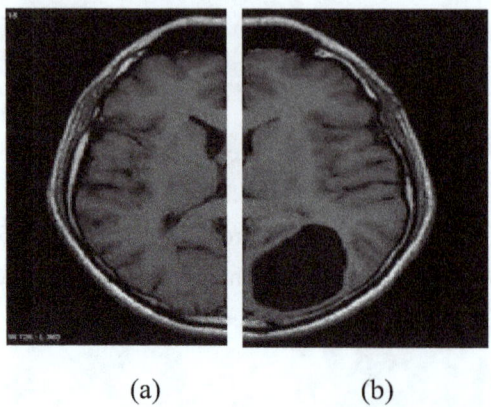

(a)   (b)

Fig. 5 (a) Right half MRI image, (b) Left half MRI image

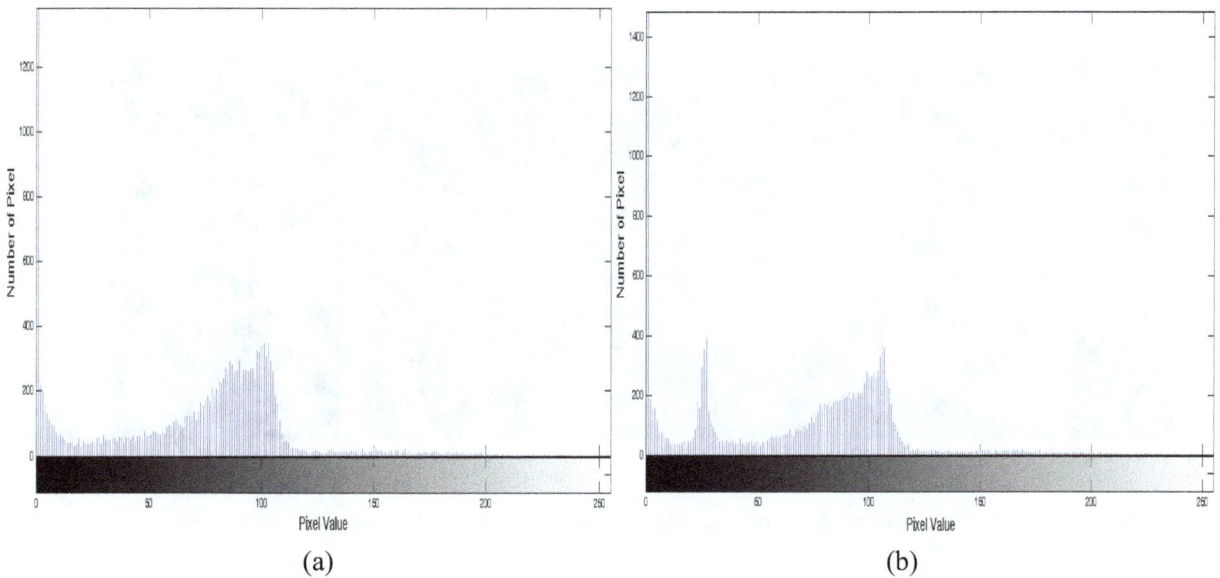

(a)   (b)

Fig. 6 Histogram of Fig. 6 (a), Histogram of Fig. 6 (b).

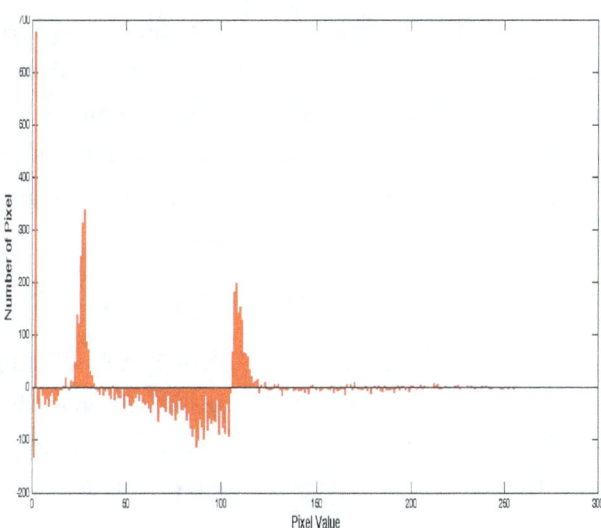

Fig. 7 Difference in histogram.

b. *Graph Cut segmentation:*

Initial centroids have been obtained by dividing the MRI Image in two halves (horizontal and vertical) and calculating the pixel difference. These difference values are responsible for evaluating the centroids from which the object and the background terminals are obtained. Hence the entire image is segmented in regions with homogeneous property. Image displayed in Fig. 8(a) is infected from tumor (URL: https://bigpictureeducation.com). Fig. 8(c) is astrocytoma infected brain and the image is taken from (URL: https://in.mathworks.com/matlabcentral.com). The corresponding segmented images are displayed in Fig. 8(b) and 8(d). The infected region is extracted from the background region and it is observed that they are not restricted to binary intensity values. The segmented region holds the original pixel values as in original image instead of intensity value 0. To measure the quality of the image parameters such as PSNR and SSIM are calculated for both the images. The values obtained for tumor affected segmented image is PSNR=22.28 dB, SSIM=0.75 and for astrocytoma affected segmented image is PSNR=18.80 dB, SSIM=0.90.

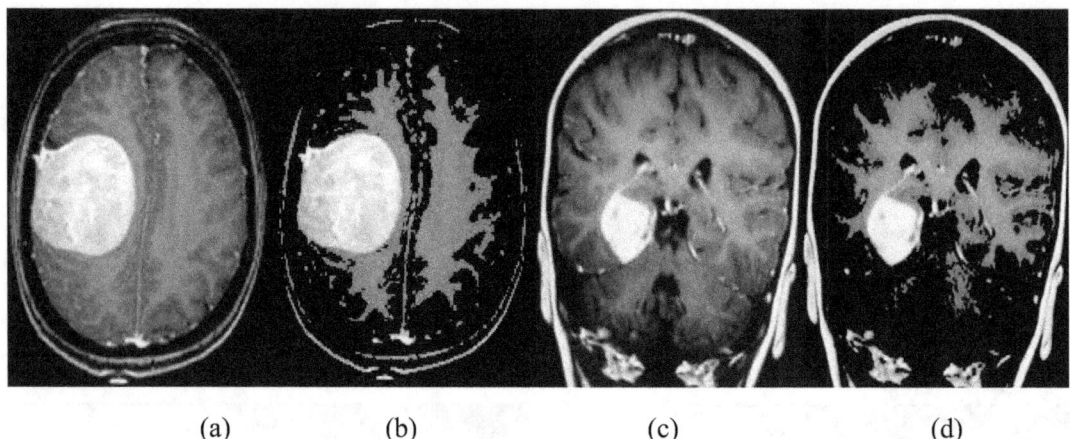

(a)   (b)   (c)   (d)

Fig. 8 Segmented output images: (a) and (c) original MRI; (b) and (d) segmented images using Graph cut.

6 Conclusion

In this paper two methods are presented and an improved segmentation is acquired by the graph cut method. The process of finding the difference in intensity values in thresholding method is blended in the graph cut method for evaluation of centroids in this research work. This leads to an automatic selection of centroids. The PSNR and SSIM parameter values also imply that presented graph cut segmentation method is an effective way to detect tumor of any irregular shape holding the properties of segmented region.

References

Ahmed MN, Yamany SM, Mohamed N, Farag AA, Moriarty T. 2002. A modified fuzzy c-means algorithm for bias field estimation and segmentation of mri data. IEEE transactions on medical imaging, 21(3): 193-199

Amandeep, Bhusri S, Jain S. 2017. CAD system for non small cell lung carcinoma using laws' mask analysis, Proceedings of the 11th INDIACom International Conference on Computing for Sustainable Global Development, BVICAM: 6285-6288

Ballestores Jr F, Qiu ZY. 2012. An integrated parcel-based land use change model using cellular automata and decision tree. Proceedings of the International Academy of Ecology and Environmental Sciences, 2(2): 53-69

Bhusri S, Jain S, Virmani J. 2016. Breast lesions classification using the amalagation of morphological and texture features. International Journal of Pharma and BioSciences, 7(2): 617-624

Bhusri S, Jain S, Virmani J. 2016. Classification of breast lesions using texture ratio vector technique, Proceedings of the 11th INDIACom International Conference on Computing for Sustainable Global Development: 6289-6293

Borges GA, Aldon MJ. 2000. A split-and-merge segmentation algorithm for line extraction in 2d range images, Pattern Recognition, 2000. Proceedings 15th International Conference, IEEE

Boykov Y, Funka-Lea G. 2006. Graph cuts and efficient nd image segmentation. International Journal of Computer Vision, 70(2): 109-131

Boykov YY, Jolly MP. 2001. Interactive graph cuts for optimal boundary & region segmentation of objects in nd images, Computer Vision, 2001. ICCV 2001. Proceedings Eighth IEEE International Conference, IEEE

Chaudhuri D, Agrawal A. 2010. Split-and-merge procedure for image segmentation using bimodality detection approach. Defence Science Journal, 60(3): 290

Chen CW, Luo J, Parker KJ. 1998. Image segmentation via adaptive k-mean clustering and knowledge-based morphological operations with biomedical applications. IEEE Transactions on Image Processing, 7(12): 1673-1683

Cheng M-M, Mitra NJ, Huang X, Torr PH, Hu SM. 2015. Global contrast based salient region detection. IEEE Transactions on Pattern Analysis and Machine Intelligence, 37(3): 569-582

Chuang K-S, Tzeng HL, Chen S, Wu J, Chen TJ. 2006. Fuzzy c-means clustering with spatial information for image segmentation. Computerized Medical Imaging and Graphics, 30(1): 9-15

Clark MC, Hall LO, Goldgof DB, Clarke LP, Velthuizen RP, Silbiger MS. 1994. MRI segmentation using fuzzy clustering techniques. IEEE Engineering in Medicine and Biology Magazine, 13(5): 730-742

Coleman GB, Andrews HC. 1979. Image segmentation by clustering. Proceedings of the IEEE, 67(5): 773-785

Comaniciu D, Meer P. 1997. Robust analysis of feature spaces: Color image segmentation, Computer Vision and Pattern Recognition. Proceedings IEEE Computer Society Conference, IEEE

Damiand G, Resch P. 2003. Split-and-merge algorithms defined on topological maps for 3d image segmentation. Graphical Models, 65(1): 149-167

Dhiman A, Singh A, Dubey S, Jain S. 2016. Design of lead ii ecg waveform and classification performance for morphological features using different classifiers on lead ii. Research Journal of Pharmaceutical, Biological and Chemical Sciences (RJPBCS), 7(4): 1226-1231

Felzenszwalb PF, Huttenlocher DP. 2004. Efficient graph-based image segmentation. International journal of computer vision, 59(2): 167-181

Ferrari M. 2005, Cancer nanotechnology: Opportunities and challenges. Nature Reviews Cancer, 5(3): 161-171

Gambotto JP. 1993. A new approach to combining region growing and edge detection. Pattern Recognition Letters, 14(11): 869-875

Greig DM, Porteous BT, Seheult AH. 1989. Exact maximum a posteriori estimation for binary images. Journal of the Royal Statistical Society, Series B (Methodological): 271-279

Hussain R. 2014. Multi Resolution Analysis (MRA) of satellite images of oil spill disasters. Computational Ecology and Software, 4(3): 193-204

Jain S, Naik PK, Bhooshan SV. 2011. Non linear modeling of cell survival/death using artificial neural network, Computational Intelligence and Communication Networks (CICN). 2011 International Conference, IEEE

Kallergi M, Woods K, Clarke LP, Qian W, Clark RA. 1992. Image segmentation in digital mammography: Comparison of local thresholding and region growing algorithms. Computerized Medical Imaging and Graphics, 16(5): 323-331

Khanna A, Sood M, Devi S. 2012. Us image segmentation based on expectation maximization and gabor filter. International Journal of Modeling and Optimization, 2(3): 230

Ortiz A, Gorriz J, Ramirez J, Salas-Gonzalez D. 2014. Improving mr brain image segmentation using self-organising maps and entropy-gradient clustering, Information Sciences, 262: 117-136

Ortiz A, Palacio AA, Górriz JM, Ramírez J, Salas-González D. 2013. Segmentation of brain MRI using som-fcm-based method and 3d statistical descriptors. Computational and Mathematical Methods in Medicine, 2013: 638563

Pal NR, Pal SK. 1993. A review on image segmentation techniques. Pattern Recognition, 26(9): 1277-1294

Pauchard Y, Fitze T, Browarnik D, Eskandari A, Pauchard I, Enns-Bray W, Pálsson H, Sigurdsson S,

Ferguson SJ, Harris TB. 2016. Interactive graph-cut segmentation for fast creation of finite element models from clinical CT data for hip fracture prediction. Computer Methods in Biomechanics and Biomedical Engineering, 19(16): 1693-1703

Pavlidis T, Liow YT. 1990. Integrating region growing and edge detection. IEEE Transactions on Pattern Analysis and Machine Intelligence, 12(3): 225-233

Pham DL, Xu C, Prince JL. 2000. Current methods in medical image segmentation. Annual Review of Biomedical Engineering, 2(1): 315-337

Rana S, Jain S, Virmani J. 2016. Svm-based characterization of focal kidney lesions from b-mode ultrasound images. Research Journal of Pharmaceutical, Biological and Chemical Sciences, 7(4): 837-846

Seljuq U, Hussain R. 2014. Synthetic Aperture Radar (SAR) image segmentation by fuzzy c-means clustering technique with thresholding for iceberg images. Computational Ecology and Software, 4(2): 129-134

Sepas-Moghaddam A, Yazdani D, Shahabi J. 2014. A novel hybrid image segmentation method. Progress in Artificial Intelligence, 3(1): 39-49

Sharma S, Jain S, Bhusri S. 2017. Classification of breast lesions using gabor wavelet filter for three classes. Proceedings of the 11th INDIA Com International Conference on Computing for Sustainable Global Development BVICAM: 6282-6284

Sharma S, Jain S, Virmani J. 2017. Two class classification of breast lesions using statistical and transform domain features. Journal of Global Pharma Technology, 9(7): 18-24

Soille P. 2013. Morphological Image Analysis: Principles and Applications. Springer Science & Business Media

Sonka M, Hlavac V, Boyle R. 2014. Image Processing, Analysis, and Machine Vision. Cengage Learning, USA

Sood M. 2017. Performance analysis of classifiers for seizure diagnosis for single channel EEG data. Biomedical Pharmacology, 10(2): 795-803

Sood M, Bhooshan SV. 2014. Design and development of prediction model to detect seizure activity utilizing higher order statistical features of EEG signals. Research Journal of Pharmaceutical, Biological and Chemical Sciences, 5(3): 1129-1145

Sood M, Bhooshan SV. 2015. Parameter-selective based CAD system for epileptic seizure classification, International Journal of Applied Engineering Research, 10(10): 25389-25408

Suetens P. 2017. Fundamentals of Medical Imaging. Cambridge University Press, UK

Wang DC, Vagnucci AH, Li C. 1983. Digital image enhancement: A survey. Computer Vision, Graphics, and Image Processing, 24(3): 363-381

Zhang WJ. 2018. Fundamentals of Network Biology. World Scientific, Singapore

8

Measurement and identification of positive plant interactions: Overview and new perspective

WenJun Zhang

School of Life Sciences, Sun Yat-sen University, Guangzhou 510275, China; International Academy of Ecology and Environmental Sciences, Hong Kong

E-mail: zhwj@mail.sysu.edu.cn, wjzhang@iaees.org

Abstract

Positive interactions play a key role in plant communities. The present study discusses measurement/identification methods of positive plant interactions. So far, some indices, e.g., RII, RCI, RNE and lnRR, and some models, e.g., site-based neighborhood models, individual-based models, etc., are usually used to measure and identify the type and strength of positive plant interactions. Most of these methods are based on interaction data of two species only. In a multi-species community or ecosystem, which occurs mostly in the nature, the interaction between two species is influenced by other species in the environment and may change as the time. Those indices and models may not exactly represent the true situations in the nature. Therefore, I argue that the inclusion of multi-species interactions in the network and utilization of theory and methods of network analysis and network evolution should be the focus in the future. The network evolution model, and correlation- and network-based methods in relation to species interactions were introduced and discussed. Finally, I think that network thinking and selforganizology are the basis for the future research of complex and dynamic species interactions.

Keywords positive plant interactions; measurement; identification; indices; models; sampling; network analysis; self-organization.

1 Introduction

Both positive and negative interactions between organisms are the most important factor for both construction of communities and maintenance of biodiversity (Grime, 1979; Bertness and Callaway, 1994; Zhang, 2007; Damgaard, 2011; Zhang, 2011, 2012a-b, 2014a-c, 2015a-b, 2016a; Maihait and Zhang, 2014; Zhang and Li, 2015a-b; Zhang and Liu, 2015). They determine population dynamics, community structure and ecosystem functions. Negative interactions refer to decreased growth, survival or reproduction of organisms due to limited resources and space. Negative interactions include competition (-/-) and amensalism (-/0) (Zhang,

2011). Positive interactions are interactions beneficial to an organism and at least harmless to another organism, i.e., the existence of an organism increases the growth, survival or reproduction of orther organism. Callaway (2007) and Zhang (2011) defined positive interactions are interactions beneficial to at least one of two interactive organisms, i.e., the existence of an organism increases the growth, survival or reproduction of another organism. Therefore, positive interactions include such types, (1) mutualism (+/+): both organisms co-exist and depend on each other for a long time, and the two sides profit from each other; (2) commensalism (+/0): symbiosis is beneficial to an organism and not effective to another organism; (3) protocooperation (+/+): it is similar to mutualism, however, their collaboration is loose and any one of the two organisms can live independently; (4) nurse effect, and (5) predation/parasitism (+/-), which mostly occurs across different trophic levels. Furthermore, Zhang (2011) revealed the quantitative mechanism of positive interactions based on foundamental models for predation/parasitism and competition. Nevertheless, in plant ecology, positive interaction can occur within the same species, and can take place between different species, but are limited at the same trophic level, i.e., the interaction types (1)-(4) above (i.e., facilitation).

Similar to the importance of positive interactions in animal communities (Zhang, 2011), positive interactions play a key role in plant communities. As early as in 1914, Pearson conducted the first experiment of positive interactions. He proved that Douglas fir *Pseudotsuga menziesii* showed greater surviveship under the canopy of aspen *Populus tremuloides*. Atsatt and O'Dowd (1976) found that the possibility of a plant feeding by animals was related to released chemicals, morphological composition, distribution and abundance of the surrounding plants. In 1986, DeAngelis et al. discussed the effects of positive interactions on ecosystems. Since then Hunter and Aarssen (1988) demonstrated that positive interactions are an important and common process in plant communities. In 2003, Bruno et al. maintained to include facilitation into ecological theory. In his comprehensive book, Callaway (2007) discussed all aspects of positive interactions, including the development history and research status, and the future direction. Further, Brooker et al. (2008) organized a symposium with the theme of Positive Interactions in Plant Communities, in University of Aberdeen, UK. The symposium proposed to include positive interactions into mainstream ecology based on competition.

2 Mechanisms and Hypothesis of Positive Plant Interactions

2.1 Mechanisms

Some mechanisms of positive plant interactions have been suggested as follows

2.1.1 Direct mechanisms

(1) Hydraulic hydraulic lift. Water is uplifted along root system from the deep layer to the surface dry soil through potential gradient of water, which makes a considerable amount of soil water reallocated inside the plant through its root system (Callaway and Walker, 1997).

(2) Canopy interception. Precipitation is intercepted by plant canopy, which makes soil humidity higher around the plant canopy and thus creates the more humid environment for plants (Brooker et al., 2008).

(3) Shading. Shading help to protect plant tissue, make it prevent from lethal temperature, cut down breathing consumption, reduce ultraviolet irradiation, reduce transpiration and increase soil water content (Brooker et al., 2008).

(4) Nutrients' feedback. Plants absorb nutrients from the soil and return nutrient to the soil by the fallen leaves, thus improving the physical and chemical properties of soil and benefitting the surrounding plants (Callaway and Walker, 1997).

2.1.2 Indirect mechanisms

Indirect mechanisms include using morphological structure (e.g., thorn) or chemical defense to resist herbivores or seed feeders, which prevents the surrounding plants from being eaten. Other examples include helping the surrenounding plants attract pollinators/distributors, or helping maintain soil microorganisms, etc (Brooker et al., 2008).

2.2 Hypothesis

Biomass, survivership, height and other plant traits usually represent the net effect of offset action by positive and negative interactions. In the superior environment, the net effect is competition, and in the stress environment, the net effect is positive interactions.

Stress gradient hypothesis argues that the strength of positive interactions enhances with the intensification of environment stress. Symmetric facilitation (mutualism) significantly increases the survival rate of plants in the extreme stressed environment. Asymmetric facilitation (commensalism) increases the survival rate of plants in the moderate stressed environment.

3 Indices, Models and Methods

3.1 Indices

Some indices are used to measure the strength of plant interactions, RII (Relative Interaction Index), RCI (Relative Competition Index), RNE (Relative Neighbor effect), and lnRR (Log Response Ratio) (Armas et al., 2004).

(1) RII

RII is represented by

$$RII=(\alpha B_0-B_0)/(\alpha B_0+B_0)=(\alpha-1)/(\alpha+1)$$

where $B_0=B_w/\alpha$ is the mass of plants with neighbors, B_w is the mass of isolated individuals. $\alpha>1$, facilitation prevails; $\alpha=1$, neutral interaction; $0\leq\alpha<1$, competition prevails.

(2) RNE

RNE is represented by

$$RNE=(B_0-B_w)/\max(B_0,\alpha B_0)=(B_0-\alpha B_0)/\max(B_0,\alpha B_0)$$
$$=1-\alpha, \text{ if } 0\leq\alpha<1$$
$$=(1-\alpha)/\alpha, \text{ if } \alpha>1,$$

(3) RCI

RCI is represented by

$$RCI=(B_0-B_w)/B_0=(B_0-\alpha B_0)/B_0=1-\alpha$$

(4) lnRR

lnRR is represented by

$$lnRR=\ln(B_0/B_w)=\ln(B_0/\alpha B_0)=-\ln\alpha$$

The criteria to determine the type of interactions using four indices are indicated in Table 1.

RII and lnRR are not so intuitionistic than RNE and RCI when competition prevails and than RNE when positive interactions prevail. RCI cannot be used to measure the strength of strong interaction similar to obligatory parasitism.

Table 1 Criteria to determine the type of interactions using four indices.

	RII	RNE	RCI	lnRR	Interactions
$B_w \rightarrow 0$	-1	1	1	∞	Competition
$B_w \rightarrow \infty$	1	-1	$-\infty$	$-\infty$	Facilitation
	0	0	0	0	Neutral

3.2 Models

Most models specifically containing positive interactions are derived from the competition models of plant individuals.

3.2.1 Site-based neighborhood models

The most popular site-based neighborhood model is the grid-based model, i.e., cellular automata (Chen et al., 2009; Xiao et al., 2009; Michalet et al., 2011; Ballestores Jr. and Qiu, 2012; Wang et al., 2012; Zhang et al., 2013; Zhang, 2012a, 2016a; Zhang et al., 2015).

3.2.2 Individual-based models

There are various individual-based models (Weiner et al., 2001; Xiao et al., 2009; Griebeler, 2011; Lin et al., 2012; Zhang et al., 2013), including tessellation model (McInnis et al., 2004) and distance model, and the latter futher includes fixed-radius-neighborhood model, zone-of-influence model (Chu et al., 2008; Weiner et al., 2001; Berger et al., 2008), and ecological field model (Wang, 1993).

Zone-of-influence model (ZOI) was proposed by Weiner et al. (2001). In ZOI, the plant obtains resources within the circular area around it (Weiner et al., 2001; Berger et al., 2008; Jia et al., 2011). The growth rate of the plant is

$$\mathrm{d}B/\mathrm{d}t = r(A - B^2/B_{max}^{4/3}) = r(CB^{2/3} - B^2/B_{max}^{4/3})$$

where B is the size of a plant, B_{max} is the maximum mass of the plant, r is the initial (maximum) growth rate, A is the area of the zone of influence of the plant, which represents the potential resources it can obtain, and C is a constant (e.g., $C=1$).

A plant competes for resources with its target plant within their overlapped zone of influence.

3.2.3 State space model

Damgaard (2011) proposed a novel method for measuring plant-plant interactions in undisturbed semi-natural and natural plant communities where it is difficult to distinguish individual plants. He assumed that the ecological success of the different plant species in the plant community may be adequately measured by plant cover and vertical density, the latter is correlated to the 3-dimensional space occupancy and biomass. In the outlined competition model the vertical density at the end of the growing season is assumed to be a function of the cover of all species at the start of the growing season, and the cover at the start of the growing season is assumed to be a function of the vertical density of all species at the end of the previous growing season. The method allows direct measurements of the competitive effects of neighbouring plants on plant performance and the estimation of parameters that describe the ecological processes of plant interactions during the growing season as well as the process of survival and recruitment between growing seasons. The presented method is suited for testing different ecological hypothesis on competitive interactions along environmental gradients, investigating the importance of competition, as well as predicting the likelihood of different ecological scenarios. In his state space model , the studied competitive processes are assumed to act on latent variables that model plant cover and vertical density by two process equations (or structural equations), and the latent variables are coupled to the observed plant cover and vertical density by two measuring equations.

3.3 New research area: network methods of multiple-species interactions

Michalet et al. (2006) argued that in a moderate or highly stressed environment, a species may expand the niche of its competitive species. According to stress gradient hypothesis, the strength of positive interactions increases with the intensification of environmental stress. In a multi-species community or ecosystem, which occurs mostly in the nature, the interaction between two species is influenced by other species in the environment (Atsatt and O'Dowd, 1976; Michalet et al., 2006), and may change as the time (Zhang, 2015a). Experimental and quantitative methods based on two species only, as mentioned mostly in the sections 3.1 and 3.2 (in exception of state space model) may not exactly represent the true situations in the nature. Therefore, inclusion of multi-species interactions in the network and utilization of theory and methods of network analysis and network evolution should be the focus in the future.

3.3.1 Network evolution model

In the nature, a community or ecosystem acts as a network. A variety of network evolution models have been proposed (Zhang, 2015a, 2016b). Zhang (2015a) developed a generalized network evolution model on community assembly. Different from differential equations with fixed state variables (nodes in the network), his model is the one with changed nodes (species). The model is based on difference equations with different number of species in different stages of evolution. It consists of pioneer rule, invasion and growth rule, extinction rule, connection (flow) rule and termination rule, etc. The invasion, establishment, growth, and extinction of species follow a series of rules. Coefficients of variables in the equations represent the type and strength of interactions. The model provides the basis to build self-organization models with various interactions.

3.3.2 Sampling based network analysis and measurement of multiple-species interactions

Jointly use of partial correlations of various correlation measures, e.g., Pearson correlation, Spearman correlation, Jaccard coefficient, Point correlation, etc., based on sampling data, are effective to identify positive/negative/neutral interactions in a network/system with multiple species (e.g., plant community). Details of these correlation- and network- based methods can be found in Zhang (2011, 2012a-b, 2014a, 2015b-c, 2016a), and Zhang and Li (2015a-b).

4 Discussion

As the increase of findings on positive plant interactions, more and more scientists strongly suggest to include positive interactions in mainstream ecology (Hunter and Aarssen, 1988; Callaway, 2007; Brooker et al., 2008). Nevertheless, in my view, we must focus on various interactions globally in the network/system with multiple species rather than interactions between two species only. We should treat biological systems as self-organing systems. As maintained by Zhang (2015a, 2016a), biological systems, including communities and ecosystems, are mostly self-organizing systems, and the community assembly and succession is a self-organization process. And according to the self-organization theory (Zhang, 2016a), positive interactions play an important role in self-organizing systems. Thus, it is not strange to find a large amount of positive plant interactions in the nature. Network thinking and selforganizology are the basis for the research of complex and dynamic species interactions.

Acknowledgment

We are thankful to the support of Discovery and Crucial Node Analysis of Important Biological and Social Networks (2015.6-2020.6), from Yangling Institute of Modern Agricultural Standardization, and High-Quality Textbook *Network Biology* Project for Engineering of Teaching Quality and Teaching Reform of Undergraduate Universities of Guangdong Province (2015.6-2018.6), from Department of Education of Guangdong Province.

References

Armas C, Ordiales R, Pugnaire FI. 2004. Measuring plant interactions: a new comparative index. Ecology, 85(10): 2682-2686

Atsatt PR, O' Dowd DJ. 1976. Plant defense guilds. Science, 193(4247): 24-29

Ballestores Jr. F, Qiu ZY. 2012. An integrated parcel-based land use change model using cellular automata and decision tree. Proceedings of the International Academy of Ecology and Environmental Sciences, 2(2): 53-69

Berger U, Piou C, Schiffers K, Grimm V. 2008. Competition among plants: concepts, individual-based modeling approaches, and a proposal for a future research strategy. Perspectives in Plant Ecology Evolution and Systematics, 9: 121-135

Bertness MD, Callaway R (1994). Positive interactions in communities. Trends in Ecology & Evolution, 9: 191-193

Brooker RW, Maestre FT, Callaway RM, et al. 2008. Facilitation in plant communities: the past, the present, and the future. Journal of Ecology, 96: 18-34

Bruno JF, et al. 2003. Inclusion of facilitation into ecological theory. Trends in Ecology and Evolution, 18: 119-125

Callaway RM. 2007. Positive Interactions and Interdependence in Plant Communities. Springer, Netherlands

Callaway RM, Walker LR. 1997. Competition and facilitation: a synthetic approach to interactions in plant communities. Ecology, 78: 1958-1965

Chen SY, Xu J, Maestre FT, et al. 2009. Beyond dual-lattice models: incorporating plant strategies when modeling the interplay between facilitation and competition along environmental severity gradients. Journal of Theoretical Biology, 258: 266-273

Chu CJ, Maestre FT, Xiao S, et al. 2008. Balance between facilitation and resource competition determines biomass-density relationships in plant populations. Ecology Letters, 11: 1189-1197

Damgaard C. 2011. Measuring competition in plant communities where it is difficult to distinguish individual plants. Computational Ecology and Software, 1(3): 125-137

DeAngelis DL, Post WM, Travis CC. 1986. Positive Feedback in Natural Systems. Springer-Verlag, NY, USA

Griebeler EM. 2011. Are individual based models a suitable approach to estimate population vulnerability? - a case study. Computational Ecology and Software, 1(1): 14-24

Grime JP. 1979. Plant Strategies and Vegetation Processes. Wiley, Chichester, UK

Hunter AF, Aarssen LW. 1988. Plants helping plants: New evidence indicates that beneficence is important in vegetation. BioScience, 38(1): 34-40

Jia X. 2011. Plant Competition and Facilitation and Their Role in Driving Population Dynamics Across Environmental Gradients: A Study Based on Zone-of-Influence (ZOI) Models. PhD dissertation. Zhejiang University, Hangzhou, China

Lin Y, Berger U, Grimm V, Ji QR. 2012. Differences between symmetric and asymmetric facilitation matter:

exploring the interplay between modes of positive and negative plant interactions. Journal of Ecology, 100: 1482-1491

Maihait M, Zhang WJ. 2014. A mini review on theories and measures of interspecific associations. Selforganizology, 1(3-4): 206-210

McInnis LM, Oswald BP, Williams HM, et al. 2004. Growth response of *Pinus taeda* L. to herbicide, prescribed fire, and fertilizer. Forest Ecology and Management, 199: 231-242

Michalet R, Xiao S, Touzard B, et al. 2011. Phenotypic variation in nurse traits and community feedbacks define an alpine community. Ecology Letters, 14: 433-443

Michalet R, Brooker RW, Cavieres LA, et al. 2006. Do biotic interactions shape both sides of the humped-back model of species richness in plant communities? Ecology Letters, 9: 767-773

Pearson GA. 1914. The role of aspen in the reforestation of mountain burns in Arizona and New Mexico. Plant World, 17: 249-260

Wang GX. 1993. On Ecological Field. Henan Science and Technology Press, Zhengzhou, China

Wang Y, Ellwood MDF, Maestre FT, et al. 2012. Positive interactions can produce species-rich communities and increase species turnover through time. Journal of Plant Ecology, 5: 417-421

Weiner J, Stoll P, Muller-Landau H, Jasentuliyana A. 2001. The effects of density, spatial pattern, and competitive symmetry on size variation in simulated plant populations. The American Naturalist, 158: 438-450

Xiao S, Michalet R, Wang G, Chen SY. 2009. The interplay between species' positive and negative interactions shapes the community biomass-species richness relationship. Oikos, 118: 1343-1348

Zhang WJ. 2011. Constructing ecological interaction networks by correlation analysis: hints from community sampling. Network Biology, 1(2):81-98

Zhang WJ. 2012a. Computational Ecology: Graphs, Networks and Agent-based Modeling. World Scientific, Singapore

Zhang WJ. 2012b. How to construct the statistic network? An association network of herbaceous plants constructed from field sampling. Network Biology, 2(2): 57-68

Zhang WJ. 2014a. Interspecific associations and community structure: A local survey and analysis in a grass community. Selforganizology, 1(2): 89-129

Zhang WJ, et al. 2014b. Interspecific associations of weed species around rice fields in Pearl River Delta, China: A regional survey. Selforganizology, 1(3-4): 143-205

Zhang WJ. 2014c. Research advances in theories and methods of community assembly and succession. Environmental Skeptics and Critics, 3(3): 52-60

Zhang WJ. 2015a. A generalized network evolution model and self-organization theory on community assembly. Selforganizology, 2(3): 55-64

Zhang WJ. 2015b. A hierarchical method for finding interactions: Jointly using linear correlation and rank correlation analysis. Network Biology, 5(4): 137-145

Zhang WJ. 2015c. Calculation and statistic test of partial correlation of general correlation measures. Selforganizology, 2(4): 65-77

Zhang WJ. 2016a. Selforganizology: The Science of Self-Organization. World Scientific, Singapore

Zhang WJ. 2016b. A random network based, node attraction facilitated network evolution method. Selforganizology, 3(1): 1-9

Zhang WJ, Li X. 2015a. General correlation and partial correlation analysis in finding interactions: with Spearman rank correlation and proportion correlation as correlation measures. Network Biology, 5(4): 163-168

Zhang WJ, Li X. 2015b. Linear correlation analysis in finding interactions: Half of predicted interactions are undeterministic and one-third of candidate direct interactions are missed. Selforganizology, 2(3): 39-45

Zhang WJ, Liu GH. 2015. Coevolution: A synergy in biology and ecology. Selforganizology, 2(2): 35-38

Zhang WJ, Qi YH, Zhang ZG. 2015. A cellular automaton for population diffusion in the homogeneous rectangular area. Selforganizology, 2(1): 13-17

Zhang WP, Pan S, Jia X, et al. 2013. Effects of positive plant interactions on population dynamics and community structures: a review based on individual-based simulation models. Chinese Journal of Plant Ecology, 37(6): 571-582

9

The modeling of predator-prey interactions

Muhammad Shakil[1], **H. A. Wahab**[1], **Muhammad Naeem**[2], **Saira Bhatti**[3], **Muhammad Shahzad**[1]

[1]Department of Mathematics, Hazara University, Manshera, Pakistan

[2]Department of Information Technology, Hazara University, Manshera, Pakistan

[3]Department of Mathematics, COMSATS Institute of Information Technology, Abbottabad, Pakistan

Email: wahabmaths@yahoo.com, wahab@hu.edu.pk

Abstract

In this paper, we aim to study the interactions between the territorial animals like foxes and the rabbits. The territories for the foxes are considered to be the simple cells. The interactions between predator and its prey are represented by the chemical reactions which obey the mass action law. In this sense, we apply the mass action law for predator prey models and the quasi chemical approach is applied for the interactions between the predator and its prey to develop the modeled equations for different possible mechanisms of the predator prey interactions.

Keywords predator-prey interactions; reaction diffusion systems; modeling; fox; rabbit.

1 Introduction

The Lotka Voltera equations are a pair of first order non linear differential equations, and these are also known as the predator prey equations .i.e. when growth rate of one population is decreased and the other increased then these populations are said to be in a predator-prey situation. The Lotka–Volterra equations are frequently used to describe the dynamics of biological system in which two species interact, one as a predator and the other as a prey (Murray, 2003).

The Lotka Volterra predator prey models were originally introduced by Alfred J. Lotka (Lotka, 1920) in the theory of autocatalytic chemical reactions. In 1920, the model to "organic systems", he made an extension while using plant species and an herbivorous animal species. In 1925, he utilized these equations for the possible analysis of predator-prey interactions and arrived at the equations which are well-known now a day.
In 1926, Vito Volterra (Volterra, 1926), made a statistical analysis of fish catches in the Adriatic independently investigated the equations. V. Voltera applied these equations to predator prey interactions; consist of a pair of first order autonomous ordinary differential equations.

Since that time the Lotka Voltera model has been applied to problems in chemical kinetics, population biology, epidemiology and neural networks. These equations also model the dynamic behavior of an arbitrary number of competitors. The Lotka–Volterra system of equations is an example of a Kolmogorov model, which is a more general framework that can model the dynamics of ecological systems with predator-prey interactions, competition, disease, and mutualism.

Lotka Volterra is the most famous predator prey model. According to Volterra if $x(t)$ is the prey population and $y(t)$ that of the predator at time t then Volterra's model is (Murray, 2003),

$$\frac{dx}{dt} = x(a - by); \tag{1}$$

$$\frac{dy}{dt} = y(cx - d). \tag{2}$$

where a, b, c and d are positive constants.

The reaction diffusion system is a mathematical model which clarifies how the concentrations of one or more substances which are distributed in the space, change under the influence of two processes. In the local chemical reactions, the substances are transformed into each other and the diffusion occur which causes the substances to spread out over a surface in space. Such systems are naturally applied in chemistry, however; the system can also explain dynamical processes of non chemical nature. Mathematically, the reaction diffusion system takes the form of the semi-linear parabolic partial differential equation.

2 Foxes and Rabbits Interaction: A Non-Linear System

In the existence of hundreds of different animals on the island, the modeling of the interaction among these animals is not an easy task. However, the interaction between two species is possible taking one as a predator and the other as a prey. The same single population growth model can be used for predators (foxes), if prey (rabbits) is considered as an unlimited food supply to foxes. But Foxes are territorial animals, and in general each fox claims its own territory by marking their territories with signals that other foxes will recognize e. g., by leaving their droppings in prominent positions but they also pair up in winter season. So it will be difficult for the foxes to cover the whole island without defining their own territories.

In the ecosystem there are various species and every one play a significant role in maintaining populations near the carrying capacity and in keeping the system in balance. In the ecosystems Predator and prey assist them through particular adaptations to compete for food resources. Predator and prey populations are directly related and they cannot survive without each other. Here this relationship is illustrated by using foxes and rabbits.

Between the foxes and rabbits populations here is a complicated and natural predator-prey relationship, since rabbits thrive in the absence of foxes and foxes thrive in the presence of rabbits. Foxes like to occupy a combination of forest and open fields. Foxes usually define their territory zones and use the transition zone or "edge" between these habitats as hunting areas. Foxes do not interfere in the defined areas of each other but they pair up in the winter season.

In case when more rabbits are there, the more foxes will prey on them, so that the number of foxes will increase, when more foxes will predate a lot such that the number of rabbits will start to decrease, swiftly it follows by falling fox numbers as there are a small amount of rabbits to sustain them, therefore it is observed that the populations of rabbits and foxes are locked together in an interactive population 'dance.' This system is non-linear in its behavior. In mathematics non-linear simultaneous equations are used to represent these interactions for which there exist an infinite number of solutions.

The cell-jump models may be considered as the proper diffusion models by themselves, the territorial animal like fox is given a simple cell as its territory. The sense in which the discrete equations for cells converge to the partial differential equations of diffusion is that the cell models give the semi-discrete approximation of the partial differential equations for diffusion which result in a system of ordinary differential equations in cells. There are jumps of concentrations on the boundary of cells. The system of semi discrete models for cells with "no-flux" boundary conditions has all the nice properties of the chemical kinetic equations for closed systems. Under the proper relations between coefficients, like complex balance or detailed balance, this system demonstrated globally stable dynamics. Here we specially refer this modeling by Gorban et al. (Gorban et al., 2011) who developed the idea and modeled diffusion equations for different mechanisms and proved the dissipation inequalities.

Let us suppose that our space is divided on two territories, a system represented as a chain of territories each with homogeneous composition and elementary acts of transfers on the boundary (for us there are only two territories). On each territory we have some concentrations (number of foxes and rabbits) for these processes. Let F^I (fox/ foxes in first territory) and R^I (rabbit/ rabbits in first territory) is the vector concentration in the first territory I and F^{II} (fox/ foxes in second territory) and R^{II} (rabbit/ rabbits in second territory) is the vector of concentration in the second territory II. Then in our study case the following mechanisms are possible between the predator (foxes) and its prey (rabbits).

3 Mechanism of Circulation

3.1 Foxes case

As the foxes are the territorial animals and they define their own territories to prey for, but they also pair up in winter seasons, so the following mechanism may or may not be possible in real, in study case of foxes, but theoretically this type of mutually inverted and mutually inverse processes can be written as,

$$F^I \rightarrow F^{II},$$
$$F^{II} \rightarrow F^I. \tag{3}$$

Let us suppose that c^I and c^{II} be the concentrations of foxes in the first and the second cell respectively, γ_F^I and γ_F^{II} be the stoichiometric vectors in the respective cells, w_r^I and w_r^{II} be the reactions rates in the respective cells and J_F be the total flux.

The mass action law is applicable here for the above mechanisms and we will calculate the stoichiometric vectors, reaction rates, vector of total flux and their diffusion equations as,

For mechanism, $F^I \rightarrow F^{II}$,

$$\gamma_{1F}^I = -1, \ w_r^I = kc^I.$$

For second mechanism, $F^{II} \rightarrow F^I$,

$$\gamma_{2F}^I = 1, \ w_r^{II} = kc^{II}.$$

Total flux will be calculated as, $J_F = w_r^{II} - w_r^I$,

$$J_F = k\left[c^{II} - c^I \right] = \frac{kl\left[c^{II} - c^I \right]}{l} = D\nabla c. \tag{4}$$

In this approximation, $kl = D$, where l is the cell size.

Then its modeled equation will be, $\dfrac{\partial c}{\partial t} = -div\left(J_F \right) = -D\Delta c. \tag{5}$

3.2 Rabbit's case

As the rabbits do not have moving restrictions like the foxes, so this mechanism for the case of rabbits will be written as,

$$R^I \rightarrow R^{II},$$
$$R^{II} \rightarrow R^I. \tag{6}$$

Now let d^I and d^{II} be the concentrations of rabbits in the first and the second cell respectively, γ_R^I and γ_R^{II} be the stoichiometric vectors in the respective cells, w_r^I and w_r^{II} are the reactions rates in the respective cells and J_R be the total flux. Then,

For mechanism: $R^I \rightarrow R^{II}$,

$$\gamma_{1R}^I = -1, \ w_r^I = kd^I.$$

For mechanism: $R^{II} \rightarrow R^I$,

$$\gamma_{2R}^I = 1, \ w_r^{II} = kd^{II}.$$

The flux in this case will be,

$$J_R = \frac{kl\left[d^{II} - d^I \right]}{l} = D\nabla d. \tag{7}$$

And modeled equation for this case will be,

$$\frac{\partial d}{\partial t} = -div\left(J_R\right) = -D\Delta d \tag{8}$$

4 Mechanism of Sharing Place
4.1 Foxes case

The mechanism of sharing place with interaction of n different foxes   i.e. when a fox F_i^I is in territory one and another fox F_j^{II} is in territory two and then fox F_i^I moves from territory one to territory two and vise versa. As the foxes are the territorial animals and they define their own territories to prey but they also pair up in winter seasons, so the above stated mechanism will have both the possibilities, that it may or may not be happens in real but in a multi component system it is given by stoichiometric equations of the form:

$$\begin{aligned}
F_i^I + F_j^{II} &\to F_j^I + F_i^{II}, \\
F_i^{II} + F_j^I &\to F_i^I + F_j^{II}.
\end{aligned} \tag{9}$$

For mechanism: $F_i^I + F_j^{II} \to F_j^I + F_i^{II}$,

$$\gamma_{F_i}^I = -1, \ \gamma_{F_j}^I = 1, \ w_r^I = kc_i^I c_j^{II}.$$

For mechanism: $F_i^{II} + F_j^I \to F_i^I + F_j^{II}$.

$$\gamma_{F_i}^I = 1, \ \gamma_{F_j}^I = -1, \ w_r^{II} = kc_i^{II} c_j^I.$$

Total flux will be calculated as,

$$J_F = k\left[c_i^{II}c_j^I - c_i^I c_j^{II}\right], \ J_F = k\left[\left(c_i^I + l\nabla c_i\right)c_j^I - c_i^I\left(c_j^I + l\nabla c_j\right)\right] = kl\left[c_j^I\nabla c_i - c_i^I\nabla c_j\right],$$

$$J_F = D\left[c_j^I\nabla c_i - c_i^I\nabla c_j\right]. \tag{10}$$

The modeled equation in this case will be,

$$\frac{\partial c}{\partial t} = -D\left[c_j^I\Delta c_i - c_i^I\Delta c_j\right]. \tag{11}$$

4.2 Rabbit's case
In the case study of rabbits the above stated mechanism may or may not be possible because rabbit's movement is naturally free without any restrictions. In this case the possible mechanisms will be as under,

$$R_i^I + R_j^{II} \rightarrow R_j^I + R_i^{II},$$
$$R_i^{II} + R_j^I \rightarrow R_i^I + R_j^{II}.$$

$$(12)$$

For the mechanism: $R_i^I + R_j^{II} \rightarrow R_j^I + R_i^{II}$,

$$\gamma_{R_i}^I = -1, \ \gamma_{R_j}^I = 1, \ w_r^I = kd_i^I d_j^{II}.$$

For the mechanism: $R_i^{II} + R_j^I \rightarrow R_i^I + R_j^{II}$,

$$\gamma_{R_i}^I = 1, \ \gamma_{R_j}^I = -1, \ w_r^{II} = kd_i^{II} d_j^I.$$

$$J_R = D\left[d_j^I \nabla d_i - d_i^I \nabla d_j \right].$$

$$(13)$$

Its modeled equation will be,

$$\frac{\partial d}{\partial t} = -D\left[d_j^I \Delta d_i - d_i^I \Delta d_j \right].$$

$$(14)$$

4.3 Mixed situation
Now in the case of mix situations i.e. by taking foxes and rabbits case together. When a fox is in territory one and a rabbit is in territory two, as foxes are territorial naturally but rabbits movement is free, so if a rabbit that moves from territory two to territory one (or from territory one to territory two), the fox in that area will prey that rabbit (territory change is possible only between (among) the foxes but with respect to any other animal (here rabbits) change of territory for foxes will not be possible). In this case the possible mechanisms will be,

$$F_i^I + R_j^{II} \rightarrow F_i^I + R_j^I \rightarrow F_i^I,$$
$$F_i^{II} + R_j^I \rightarrow F_i^{II} + R_j^{II} \rightarrow F_i^{II}.$$

$$(15)$$

In mechanism: $F_i^I + R_j^{II} \rightarrow F_i^I + R_j^I \rightarrow F_i^I$,

If we take the first possibility. i.e. $F_i^I + R_j^{II} \to F_i^I + R_j^I$,

Then $\gamma_F^I = 0$, $w_r^I = kc^I d^{II}$, $\gamma_R^I = 1$, $w_r^{II} = kc^{II} d^I$..

Equation of flux will be,

$$J = k\left[c^{II} d^I - c^I d^{II}\right] = k\left[\left(c^I + l\nabla c\right)d^I - c^I\left(d^I + l\nabla d\right)\right], \tag{16}$$

Modeled equation for foxes in this case will be, $\dfrac{\partial c}{\partial t} = -D\left[d^I \Delta c - c^I \Delta d\right].$ (17)

Now if we take the second possibility i.e. $F_i^I + R_j^I \to F_i^I$,

Then, $\gamma_F^I = 0$, $w_r^I = kc^I d^I$, $\gamma_R^I = -1$, $w_r^{II} = kc^{II} d^{II}$.

Equation of flux will be, $J = k\left[c^{II} d^{II} - c^I d^I\right] = k\left[\left(c^I + l\nabla c\right)\left(d^I + l\nabla d\right) - c^I d^I\right],$

$$J = kl\left[c^I \nabla d + d^I \nabla c + l\nabla c\nabla d\right] = D\left[c^I \nabla d + d^I \nabla c + l\nabla c\nabla d\right]. \tag{18}$$

Modeled equation for foxes in this case will be,

$$\frac{\partial c}{\partial t} = -D\left[c^I \nabla d + d^I \nabla c + l\nabla c\nabla d\right]. \tag{19}$$

5 Mechanism of Attraction

5.1 Foxes case

Foxes are solitary and necessitate quite huge hunting areas. A fox constantly patrols its territory looking for food, using its urine to mark places it has completed searching. Foxes are territorial and fight other foxes that they find on their territory but they also pair up   in winter seasons so the mechanism of attraction in the case study of foxes will take place only in winter seasons, so in a multi component system it is given by stoichiometric equations of the form:

Attraction:

$2F^I + F^{II} \to 3F^I,$

$F^I + 2F^{II} \to 3F^{II}.$

(20)

For mechanism: $2F^I + F^{II} \rightarrow 3F^I$,

$$\gamma_F^I = 1, \ w_r^I = k\left(c^I\right)^2 c^{II}.$$

For mechanism: $F^I + 2F^{II} \rightarrow 3F^{II}$,

$$\gamma_F^I = -1, \ w_r^{II} = kc^I \left(c^{II}\right)^2.$$

Flux from cell one to cell two will be,

$$J_F = kc^I c^{II}\left(c^{II} - c^I\right) = klc^I c^{II}\frac{\left(c^{II} - c^I\right)}{l} = klc^I\left(c^I + l\nabla c\right)\nabla c,$$

To the first order we have that,

$$J_F = klc^2\nabla c = \frac{1}{3}kl\left(3c^2\nabla c\right) = \frac{1}{3}kl\nabla c^3. \tag{21}$$

And its modeled equation will be,

$$\frac{\partial c}{\partial t} = -kldiv\left(c^2\nabla c\right) = \frac{\partial c}{\partial t} = -kl\frac{1}{3}\Delta c^3. \tag{22}$$

5.2 Rabbit's case

In the study case of rabbits the mechanisms of attraction may or may not be possible in real because rabbits move freely in nature without any restrictions. If two rabbits are there in territory one and one rabbit is in territory two and vise versa, if they all move in the same territory then these are mechanisms of attraction, so in a multi component system it is given by stoichiometric equations of the form:

Attraction:

$$2R^I + R^{II} \rightarrow 3R^I,$$
$$R^I + 2R^{II} \rightarrow 3R^{II}.$$

$$\tag{23}$$

For mechanism: $2R^I + R^{II} \rightarrow 3R^I$,

$$\gamma_R^I = 1, \ w_r^I = k\left(d^I\right)^2 d^{II}.$$

For mechanism: $R^I + 2R^{II} \rightarrow 3R^{II}$,

$$\gamma_R^I = -1, \ w_r^{II} = kd^I\left(d^{II}\right)^2.$$

Equation of flux to the first order will be,

$$J_R = kld^2\nabla d = \frac{1}{3}kl\left(3d^2\nabla d\right) = \frac{1}{3}kl\nabla d^3. \tag{24}$$

And modeled equation in this case will be,

$$\frac{\partial d}{\partial t} = -kl\frac{1}{3}\Delta d^3.$$

(25)

6 Mechanism of Repulsion

6.1 Foxes case

As it is stated in the early lines that foxes are solitary and necessitate quite huge hunting areas. Foxes are territorial and fight other foxes that they find on their territory but they also pair up in the winter seasons, so mechanism of attraction in the study case of foxes will take place only in winter seasons while in other months there will be only repulsion so in a multi component system it is given by stoichiometric equations of the form: Repulsion:

$$3F^I \rightarrow 2F^I + F^{II},$$
$$3F^{II} \rightarrow F^I + 2F^{II}.$$

(26)

For mechanism: $3F^I \rightarrow 2F^I + F^{II}$,

$$\gamma_F^I = -1, \ w_r^I = k\left(c^I\right)^3.$$

For mechanism: $3F^{II} \rightarrow F^I + 2F^{II}$,

$$\gamma_F^I = 1, \ w_r^{II} = k\left(c^{II}\right)^3.$$

Equation of flux for this mechanism will be calculated as,

$$J_F = k\left[\left(c^{II}\right)^3 - \left(c^I\right)^3\right],$$

$$J_F = k\left[\left(c^{II} - c^I\right)\left(c^I + l\nabla c\right)^2 + \left(c^I\right)^2\left(c^I + l\nabla c\right)c^I\right],$$

$$J_F = k\left[\left(c^{II} - c^I\right)\left(3\left(c^I\right)^2 + l^2\left(\nabla c\right)^2 + 2l\nabla c + lc^I\nabla c\right)\right],$$

$$J_F = k\left[\left(c^{II} - c^I\right)\left(3\left(c^I\right)^2 + l\nabla c\left(l\nabla c + c^I + 2\right)\right)\right],$$

$$J_F = kl\left[\left(\frac{c^{II} - c^I}{l}\right)\left(3\left(c^I\right)^2 + l\nabla c\left(l\nabla c + c^I + 2\right)\right)\right],$$

$$J_F = kl\left[\left(\nabla c\right)\left(3\left(c^I\right)^2 + l\nabla c\left(l\nabla c + c^I + 2\right)\right)\right],$$

So to the first order we will have,

$$J_F = kl\nabla c.3\left(c^I\right)^2 = kl.c^2\nabla c = kl\nabla c^3.$$

(27)

And its modeled equation will be,

$$\frac{\partial c}{\partial t} = -kl\,div\left(c^2 \nabla c\right) = -kl\Delta c^3.$$

(28)

6.2 Rabbit's case

In this study case of rabbits the above stated mechanism of repulsion may or may not be possible because rabbits move freely in nature without any restrictions. If three rabbits are there in the same territory and one of them changes his territory then this will be a case of repulsion so in a multi component system it is given by stoichiometric equations of the form:

Repulsion:

$$3R^I \rightarrow 2R^I + R^{II},$$
$$3R^{II} \rightarrow R^I + 2R^{II}.$$

(29)

For mechanism: $3R^I \rightarrow 2R^I + R^{II}$,

$$\gamma_R^I = -1, \ w_r^I = k\left(d^I\right)^3.$$

For mechanism: $3R^{II} \rightarrow R^I + 2R^{II}$,

$$\gamma_R^I = 1, \ w_r^{II} = k\left(d^{II}\right)^3.$$

Vector of total flux will be,

$$J_R = kl\nabla d.3\left(d^I\right)^2 = kl.d^2\nabla d = kl\nabla d^3.$$

(30)

And its modeled equation will be,

$$\frac{\partial d}{\partial t} = -kl\,div\left(d^2 \nabla d\right) = -kl\Delta d^3.$$

(31)

7 The Observations

The sense in which the discrete equations for cells converge to the partial differential equations is that the cell models give the semi-discrete approximation of the partial differential equations. They result in a system of ordinary differential equations in cells. Such approximation appears often in finite element methods and cells are discontinuous finite elements. There are jumps of concentrations on the boundary of cells. The Taylor expansion of the right hand sides of the discrete system of ordinary differential equations for cells produces the second order in the cell size approximation to the continuous diffusion equation (the standard result for the central differences).

   Significant difference from the classical finite elements is in construction of right hand sides of the ordinary differential equations for concentrations in cells: there are flows in both directions: from cell 1 to cell II and from cell II to cell I. These flows have a simple mass action law construction and the resulting diffusion flow is the difference between them. The kinetic constants should be scaled with the cell size to keep $kl = D$ (k is the kinetic constant, l is the cell size and D is the diffusion coefficient for the particular mechanism).

This is the approximation of the right hand sides. The approximation of solutions is a more difficult problem and depends upon the properties of solutions of PDEs. It seems a good hypothesis that for obtained diffusion systems with convex Lyabunov functional and "no-flux" boundary conditions in bounded areas with smooth boundaries the cell model gives the uniform approximation to the solution of the correspondent PDE.

The system of semi-discrete models for cells with "no-flux" boundary conditions has all the nice properties of the chemical kinetic equations for closed systems. Under the proper relations between coefficients, like complex balance or detailed balance, this system demonstrated globally stable dynamics. This global stability property can help with the study of the related PDE.

Finally, the cell-jump models may be considered as the proper diffusion models by themselves, for the finite physically reasonable cell size, without limit. This size may be quite large for the coarse-grained models (it depends on the medium microstructure and on the smoothness of the concentration fields).

8 Summary

Our goals for the present study are:

- ❖ To build up a brief study of complex biological systems by taking a study case of foxes and rabbits.
- ❖ To tackle key research questions about our study case by proposing new techniques and algorithms that are inspired by those complex biological systems.
- ❖ Further in our study case we want to study and aim to extend these ideas to all other possible mechanisms complete in all aspects between foxes and rabbits. We also aim to check the stability of these interactions along with finding the solutions of the above determined modeled equations.

References

Alabdullatif M, Abdusalam HA, Fahmy ES. 2007. Adomian decomposition method for nonlinear reaction diffusion system of Lotka-Voltera type. International Mathematical Forum, 2: 87-96

Berezovskaya F, Karev G, Arditi R. 2001. Parametric analysis of the ratio-dependent predator-prey model. Journal of Mathematical Biology, 43(3): 221-246

Gorban AN, Sargsyan HP, Wahab HA. 2011. Quasichemical models of multicomponent nonlinear diffusion. Mathematical Modelling of Natural Phenomena, 6(5): 184-262

Lotka AJ. 1920. Undamped oscillations derived from the law of mass action. Journal of American Chemical Society, 42: 1595

Murray JD. 2003. Mathematical Biology I: An Introduction. Springer-Verlag, Germany

Voltera V. 1926. Variations and fluctuations of the number of the individuals in animal species living together. In: Animal Ecology (Chapman RN, ed). McGraw-Hill, New York, USA

Shakil M, Wahab HA, Naeem M, Bhatti S, 2014. A quasi chemical approach for the modeling predator-prey interactions. Network Biology 4(3): 130-150

Wahab HA, Shakil M, Khan T, Bhatti S, 2013.A comparative study of a system of Lotka-Voltera type of PDEs through perturbation methods. Computational Ecology and Software, 3(4): 110-125

10

Network matrix based methods for between-network comparison

WenJun Zhang

School of Life Sciences, Sun Yat-sen University, Guangzhou 510275, China; International Academy of Ecology and Environmental Sciences, Hong Kong

E-mail: zhwj@mail.sysu.edu.cn, wjzhang@iaees.org

Abstract

In present article, I introduced some network matrix based methods for comparing and testing between-network difference/similarity, including the methods for interval weights based network matrix, including between-network similarity, randomization test of between-network difference, and statistic test of between-network difference, and the method for Boolean weights based network matrix. In addition, degree change index, weight change index, and eigenvector matrix change index were presented also. Matlab codes of the methods were provided.

Keywords network matrix; network comparison; difference; similarity; algorithms; Matlab.

1 Introduction

The structure of network covers node degree, network connectance, aggregation strength, etc (Dormann,2011; Zhang and Zhan, 2011). Various statistics may be used in the difference comparison of network structure (Solow, 1993; Manly, 1997; Zhang, 2012a, 2018). For example, a Java algorithm has been presented to statistically compare between-network structure difference (Zhang, 2011a). In addition, the Java algorithm was also developed to statistically compare between-community structure difference (Zhang, 2011b). In present article, I introduce some network matrix based methods to compare and test between-network difference/similarity.

2 Methods

Suppose we compare the two networks A and B, based on their network matrices $A=(a_{ij})$ and $B=(b_{ij})$, $i, j=1, 2, ..., m$, where m is the number of nodes, a_{ij} and b_{ij} are the weights between two nodes i and j in two networks respectively.

2.1 Interval weights based network matrix

Assume the weights a_{ij} and b_{ij} are all interval values. Several methods bellow can be used to compare two networks. In these methods, network matrices A and B are firstly transformed to vectors A_1 and B_1 respectively, i.e., $(A_1, B_1) = \{(a_{ij}, b_{ij}) \mid i, j = 1, 2, ..., m\}$, and the corresponding methods are thus used.

2.1.1 Between-network similarity

By using Pearson correlation, etc (Zhang, 2012a, 2012c, 2015, 2018; Zhang and Li, 2015a, 2015b), we can achieve the test results for network similarity/correlation. Generalized Matlab codes for Pearson correlation between two network matrices, A and B, are as follows (Zhang, 2012a, 2018)

```
AA=input('Input the excel file name of network matrix A: ','s');
A=xlsread(AA);
BB=input('Input the excel file name of network matrix B: ','s');
B=xlsread(BB);
m=size(A,1); mm=size(A,2); n=size(B,1); nn=size(B,2);
if ~((m==mm) & (n==nn) & (m==n))
error('Network matrices A and B should be the square matrices of the same size.');
end
% Matrices A and B are the square matrices of the same size.
sig=input('Input significance level(e.g., 0.01): ');
A1=reshape(A, m*m, 1);    %Transform matrix A to vector A1
B1=reshape(B, m*m, 1);    %Transform matrix B to vector B1
r=corr(A1,B1);
fprintf(['Pearson correlation r=' num2str(r) '\n']);
tvalue=abs(r)/sqrt((1-r^2)/(m*m-2));
p=(1-tcdf(tvalue,m*m-2))*2;
sigma=p<sig;
if (sigma==1) fprintf(['Pearson correlation is statistically significant (p=' num2str(p) ')\n']); end
if (sigma==0) fprintf(['Pearson correlation is not statistically significant (p=' num2str(p) ')\n']); end
```

2.1.2 Randomization test of between-network difference

By using randomization test for between-network difference based on Euclidean distance, Manhattan distance, Chebyshov distance, and Pearson correlation-based distance (Zhang, 2012a, 2015, 2018; Manly, 1997; Schoenly and Zhang, 1999; Solow, 1993; Zhang and Schoenly, 1999), we can achieve the test results for network difference/similarity. Generalized Matlab codes for randomization test between two network matrices, A and B, are as follows (Zhang, 2012a, 2018)

```
AA=input('Input the excel file name of network matrix A: ','s');
A=xlsread(AA);
BB=input('Input the excel file name of network matrix B: ','s');
B=xlsread(BB);
m=size(A,1); mm=size(A,2); n=size(B,1); nn=size(B,2);
if ~((m==mm) & (n==nn) & (m==n))
error('Network matrices A and B should be the square matrices of the same size.');
end
% Matrices A and B are the square matrices of the same size.
sim=input('Input the maximum number of simulations (e.g., 100): ');
sel=input('Choose distance measure (1: Euclidean distance; 2: Manhattan distance; 3: Chebyshov distance; 4: Pearson
correlation): ');
```

```
sig=input('Input the significance level (e.g., 0.05): ');
A1=reshape(A, m*m, 1);    %Transform matrix A to vector A1
B1=reshape(B, m*m, 1);    %Transform matrix B to vector B1
pvalue=randTest(A1,B1,sim,sel);
if (pvalue<sig) fprintf(['Difference is statistically significant (p=' num2str(pvalue) ')\n']); end
if (pvalue>=sig) fprintf(['Difference is not statistically significant (p=' num2str(pvalue) ')\n']); end
```

The functions, randTest.m, euclideandis.m, manhattandis.m, chebyshovdis.m, and correcoeffdiff.m, used in the algorithm above are as follows (Zhang, 2018)

```
functionpvalue=randTest(x,y,sim,sel)              %pvalue: calculated p value.
if ((min(size(x))~=1) | (min(size(y))~=1))        %sim: times of randomizations.
error('Both x and y are vectors');
end                                               %x and y: two vectors to be tested. x and y are row vectors.
m=max(size(x));
if (max(size(y))~=m)
error('Vector sizes do not match.');
end
if (sim<=1)
error('No. randomizations are too less.');
end
dum=min(min(x),min(y));
if (dum<0)
x=x-dum;
y=y-dum;
end
ma=-1e10;
for j=1:2
for i=1:m
in=1;
if (j==1) dum=x(i);
else dum=y(i);
end
while (m~=0)
if ((abs(dum-floor(dum))<1) & (~(abs(dum-floor(dum))<=1e-10)))
in=in*10;
dum=dum*10;
if ((floor(dum+1e-10))~=(floor(dum))) break; end
else break; end
end
if (in>ma) ma=in; end
end; end
x=x.*ma.*1.0;
y=y.*ma.*1.0;
switch sel
```

```
case 1
dxy=euclideandis(x,y);
case 2
dxy=manhattandis(x,y);
case 3
dxy=chebyshovdis(x,y);
case 4
dxy=correcoeffdiff(x,y);
end
nrx=sum(x);
nrxy=sum(sum(x+y));
fr=0;
for sm=1:sim
ar=floor(x+y);
col=sum(ar);
br(1)=ar(1);
for i=2:m
br(i)=br(i-1)+ar(i);
end
cols=randperm(nrxy);
p1(1:m)=0;
for j=1:m
if (ar(j)==0) continue; end
if (j==1) temp=0;
else temp=br(j-1);
end
for i=1:nrx
if ((cols(i)>temp) & (cols(i)<=br(j))) p1(j)=p1(j)+1; end
end; end
p2=ar'-p1;
switch sel
case 1
dum=euclideandis(p1,p2);
case 2
dum=manhattandis(p1,p2);
case 3
dum=chebyshovdis(p1,p2);
case 4
dum=correcoeffdiff(p1,p2);
end
if (abs(dum)>=abs(dxy))
fr=fr+1;
end
end
pvalue=fr/sim;
```

```matlab
function distance = euclideandis(x,y)        %x and y: two vectors to be tested.
if (max(size(x))~=max(size(y)))
error('Array sizes do not match.');
end
if ((min(size(x))~=1) | (min(size(y))~=1))
error('Both x and y are vectors');
end
distance =sqrt(sum((x-y).^2))/max(size(x));


function distance = manhattandis(x,y)        %x and y: two vectors to be tested.
if (max(size(x))~=max(size(y)))
error('Array sizes do not match.');
end
if ((min(size(x))~=1) | (min(size(y))~=1))
error('Both x and y are vectors');
end
distance =sum(abs(x-y))/max(size(x));


function distance = chebyshovdis(x,y)        %x and y: two vectors to be tested.
if (max(size(x))~=max(size(y)))
error('Array sizes do not match.');
end
if ((min(size(x))~=1) | (min(size(y))~=1))
error('Both x and y are vectors');
end
distance =max(abs(x-y));


function diff = correcoeffdiff(x,y)        %x and y: two vectors to be tested.
m=max(size(x));
if (m~=max(size(y)))
error('Array sizes do not match.');
end
if ((min(size(x))~=1) | (min(size(y))~=1))
error('Both x and y are vectors');
end
xbar=mean(x);
ybar=mean(y);
aa=sum(x.*y)-ybar*sum(x)-xbar*sum(y)+m*xbar*ybar;
bb=sum(x.^2)-2*xbar*sum(x)+m*xbar^2;
cc=sum(y.^2)-2*ybar*sum(y)+m*ybar^2;
diff =1-aa/sqrt(bb*cc);
```

2.1.3 Statistic test of between-network difference

Known the weight pairs (a_{ij}, b_{ij}), $i, j=1, 2, ...,m$, i.e., (A_1, B_1). Assume the difference A_1-B_1 between two vectors A_1 and B_1 follows the normal distribution. If there is not statistic difference between network matrices A and B, the mean of A_1-B_1 should be zero. The t-test can be used to test the statistic difference (Zhang, 2018). The Matlab codes are as follows

```
AA=input('Input the excel file name of network matrix A: ','s');
A=xlsread(AA);
BB=input('Input the excel file name of network matrix B: ','s');
B=xlsread(BB);
m=size(A,1); mm=size(A,2); n=size(B,1); nn=size(B,2);
if ~((m==mm) & (n==nn) & (m==n))
error('Network matrices A and B should be the square matrices of the same size.');
end
% Matrices A and B are the square matrices of the same size.
alpha=input('Input significance level(e.g., 0.01)');
A1=reshape(A, m*m, 1);    %Transform matrix A to vector A1
B1=reshape(B, m*m, 1);    %Transform matrix B to vector B1
[h,p,ci,stats]=ttest(A1-B1,0,alpha,0);
if (h==1) fprintf(['Difference is statistically significant (p=' num2str(p) ')\n']);
else fprintf(['Difference is not statistically significant (p=' num2str(p) ')\n']); end
```

2.2 Boolean weights based network matrix

If network matrices A and B are Boolean type, i.e., the weights a_{ij} and b_{ij} take 0 or 1, point correlation can be used to calculate between-network similarity/correlation (Zhang, 2015, 2017, 2018). In this method, network matrices A and B are firstly transformed to vectors A_1 and B_1 respectively, i.e., $(A_1, B_1)=\{(a_{ij}, b_{ij}) \mid i, j=1, 2, ..., m\}$, and point correlation is thus calculated and tested.

Generalized Matlab codes for Point correlation between two network matrices, A and B, are as follows (Zhang, 2015, 2017, 2018)

```
AA=input('Input the excel file name of network matrix A: ','s');
A=xlsread(AA);
BB=input('Input the excel file name of network matrix B: ','s');
B=xlsread(BB);
m=size(A,1); mm=size(A,2); n=size(B,1); nn=size(B,2);
if ~((m==mm) & (n==nn) & (m==n))
error('Network matrices A and B should be the square matrices of the same size.');
end
% Matrices A and B are the square matrices of the same size.
sig=input('Input the significance level (e.g., 0.05): ');
x=reshape(A, m*m, 1);    %Transform matrix A to vector x
y=reshape(B, m*m, 1);    %Transform matrix B to vector y
n=size(x,1);
aa=sum((x==0) & (y==0));
bb=sum((x==0) & (y~=0));
```

cc=sum((x~=0) & (y==0));

dd=sum((x~=0) & (y~=0));

pointcorr=(aa*dd-bb*cc)/sqrt((aa+bb)*(cc+dd)*(aa+cc)*(bb+dd));

chi2=n*(aa*dd-bb*cc)^2/((aa+bb)*(cc+dd)*(aa+cc)*(bb+dd));

chi2test=chi2>chi2inv(1-sig,1);

%chi2=10.8 for sig=0.001; chi2=6.635 for sig=0.01; chi2=7.87 forsig=0.005

fprintf(['Point correlation=' num2str(pointcorr) '\n']);

if (chi2test==1) fprintf(['Point correlation is statistically significant (p=' num2str(sig) ')\n']); end

if (chi2test==0) fprintf(['Point correlation is not statistically significant (p=' num2str(sig) ')\n']); end

2.3 Other generalized methods

2.3.1 Degree change index

Degree change index, proposed by Zhang (2012b), can be popularized to compare two networks

$$D=\sum_{j=1}^{m}(|O_{Aj}-O_{Bj}|+|I_{Aj}-I_{Bj}|)$$

where D: value of degree change index; m: total number of nodes in the network; O_{Aj}, O_{Bj}: out-degree of node for networks A and B respectively; I_{Aj}, I_{Bj}: in-degree of node j for networks A and B respectively.

2.3.2 Weight change index

Weight change index (S), popularized from parameter robustness (Zhang, 2016), can be used to compare two networks

$$S=\sum_{i}\sum_{j}|a_{ij}-b_{ij}|$$

2.3.3 Eigenvector matrix change index

Suppose the eigenvector matrices of the network matrices A and B are V_a and V_b respectively. The eigenvector matrix change index is defined as

$$V=\sum_{i}\sum_{j}|V_{aij}-V_{bij}|$$

where V_{aij} and V_{bij} are elements of the eigenvector matrices V_a and V_b respectively, $i,j=1, 2, ..., m$.

AA=input('Input the excel file name of network matrix A: ','s');

A=xlsread(AA);

BB=input('Input the excel file name of network matrix B: ','s');

B=xlsread(BB);

m=size(A,1); mm=size(A,2); n=size(B,1); nn=size(B,2);

if ~((m==mm) & (n==nn) & (m==n))

error('Network matrices A and B should be the square matrices of the same size.');

end

% Matrices A and B are the square matrices of the same size.

S=sum(sum(abs(A-B)));

[VA,D] = eig(A,'nobalance');

[VB,D] = eig(B,'nobalance');

```
V=sum(sum(abs(VA-VB)));
fprintf([' Weight change index S=' num2str(S) '\n']);
fprintf(['Eigenvector matrix change index V=' num2str(V) '\n']);
```

3 Discussion

In practical uses, several methods above can be jointly used to analyze various differences or similarities between two networks. It should be noted that all methods above can be improved, in particular the degree change index, weight change index, and eigenvector matrix change index. For instance, these absolute indices can be standardized as some forms of proportions.

Finally, I assume that the two network matrices are known, which should be defined and specified by researchers.

Acknowledgment

We are thankful to the support of The National Key Research and Development Program of China (2017YFD0201204), and Discovery and Crucial Node Analysis of Important Biological and Social Networks (2015.6-2020.6), from Yangling Institute of Modern Agricultural Standardization.

References

Dormann CF. 2011. How to be a specialist? Quantifying specialisation in pollination networks. Network Biology, 1(1): 1-20

Manly BFJ. 1997. Randomization, Bootstrap and Monte Carlo Methods in Biology (2nd Edition). Chapman & Hall, London, UK

Schoenly KG, Zhang WJ. 1999. IRRI Biodiversity Software Series. V. RARE, SPPDISS, and SPPANK: programs for detecting between-sample difference in community structure. IRRI Technical Bulletin No. 5. International Rice Research Institute, Manila, Philippines

Solow AR. 1993. A simple test for change in community structure. Journal of Animal Ecology, 62: 191-193

Zhang WJ. 2011a. A Java algorithm for non-parametric statistic comparison of network structure. Network Biology, 1(2): 130-133

Zhang WJ. 2011b. A Java program for non-parametric statistic comparison of community structure. Computational Ecology and Software, 1(3): 183-185

Zhang WJ. 2012a. Computational Ecology: Graphs, Networks and Agent-based Modeling. World Scientific, Singapore

Zhang WJ. 2012b. Several mathematical methods for identifying crucial nodes in networks. Network Biology, 2(4): 121-126

Zhang WJ. 2012c. How to construct the statistic network? An association network of herbaceous plants constructed from field sampling. Network Biology, 2(2): 57-68

Zhang WJ. 2015. Calculation and statistic test of partial correlation of general correlation measures. Selforganizology, 2(4): 65-77

Zhang WJ. 2016. Network robustness: Implication, formulization and exploitation. Network Biology, 6(4): 75-85

Zhang WJ. 2017. Network pharmacology of medicinal attributes and functions of Chinese herbal medicines: (II) Relational networks and pharmacological mechanisms of medicinal attributes and functions of Chinese herbal medicines. Network Pharmacology, 2(2): 38-66

Zhang WJ. 2018. Fundamentals of Network Biology. World Scientific Europe, London, UK

Zhang WJ, Li X. 2015a. General correlation and partial correlation analysis in finding interactions: with Spearman rank correlation and proportion correlation as correlation measures. Network Biology, 5(4): 163-168

Zhang WJ, Li X. 2015b. Linear correlation analysis in finding interactions: Half of predicted interactions are undeterministic and one-third of candidate direct interactions are missed. Selforganizology, 2(3): 39-45

Zhang WJ, Schoenly KG. 1999. IRRI Biodiversity Software Series III. BOUNDARY: A program for detecting boundaries in ecological landscapes. IRRI Technical Bulletin No. 3.International Rice Research Institute, Manila, Philippines

Zhang WJ, Zhan CY. 2011. An algorithm for calculation of degree distribution and detection of network type: with application in food webs. Network Biology, 1(3-4): 159-170

11

Network robustness: Implication, formulization and exploitation

WenJun Zhang

School of Life Sciences, Sun Yat-sen University, Guangzhou 510275, China; International Academy of Ecology and Environmental Sciences, Hong Kong

E-mail: zhwj@mail.sysu.edu.cn,wjzhang@iaees.org

Abstract

Robustness refers to a system's capacity for maintaining some performance when the system's internal structure is perturbed. In previous study, network robustness, i.e., network structure robustness, includes both resistance capacity (connection robustness) of network structure to perturbation and restoration capacity (restoration robustness) of network structure if it is perturbed. Besides network structure robustness, in present paper I defined two more categories of robustness, network parameter robustness, and comprehensive robustness. Network parameter robustness refers to a network's capacity, without any structural changes, for maintaining between-node flows (fluxes) / link weights if it is perturbed. Comprehensive robustness refers to the network's capacity that not only the topological structure of the network, e.g., nodes and links, are not or less changed, but also between-node flows (fluxes), link weights, nodes' state values are maintained also if the network is perturbed. Comprehensive robustness considers both structure and parameter changes of a network. Furthermore, some new indices for network parameter robustness, and comprehensive robustness were proposed. In addition to specialized indices for network robustness, the inverse of various indices of global sensitivity analysis were suggested as indices for network robustness. Differences between robustness and stability were discussed. Misuse or inaccurate use of robustness / stability in ecology was clarified. In addition, I proposed methods to facilitate network robustness. Parameters / properties of some robust bio-networks were analyzed and summarized.

Keywords network; structure robustness; parameter robustness; comprehensive robustness; stability; sensitivity analysis; biological networks

1 Introduction

1.1 Definition and implication

Perturbation of a system's structure / parameters occurs occasionally. System perturbation occurs due to (1) external disturbances, and (2) internal factors such as unknown variables or mechanisms, structural catastrophe, or slow drift of parameters / properties, etc. Robustness refers to a system's capacity for maintaining some performance (stability, topological structure, functionality, or flux, etc.) if the system's internal structure (or parameters) is perturbed. It denotes the insensitivity of system's performance to perturbation of its structure / parameters (Zhang, 2016d). According to the specific types of system's performances, robustness is divided into stability robustness and performance robustness.

Robustness is the key for the survival of the system in the case of abnormal and dangerous situations. Robust control refers to control theory and methodology that maintains satisfied performance when system model encounters perturbation or other uncertain disturbances (Ferrarini, 2011, 2013, 2014, 2015).

In terms of robustness of biological systems, Alon (2006) defined robustness as that a biological network can almost make its basic functions irrelevant to original biochemical parameters, while non-robustness was defined as fine-turned, i.e., system properties greatly change when biochemical parameters are perturbed (Alon et al., 1999). Robustness is related to the survival of organisms, which is a response to internal perturbation and which reflects the capacity of an organism's internal organization (Zhang and Zhang, 2009).

Exploiting the mechanism of robustness is very important for understanding biological networks, through which we can understand how biological networks maintain its own features under various disturbances, such as changes in the environment (lack of nutrition level, chemical induction, temperature), internal faults (DNA damage, genetic failure of metabolic pathways), etc (Zhang and Wang, 2009; Gao and Guo, 2011).

1.2 Robustness and stability

Stability is divided into state stability and structure stability. State stability is a system's capacity to maintain its operation state after it is disturbed by external factors, while the system's structure / parameters is / are maintained. State stability includes uniform stability, and asymptotic stability, etc. Structure stability denotes a system's capacity to maintain its structure / parameters after it / they is / are disturbed by external factors.

Robustness is a mapping from system's structure / parameters to system's performance, driven by the system's perturbation, while stability is driven by external disturbance.

Both robustness and stability are determined by a system's structure / parameters.

2 Network Robustness

Complex networks are characterized by some or all of the properties of self-organization, self-similarity, attractors, small-world, and scale-free. Some of these properties, however, are occasionally disturbed or even destroyed.

Network structure robustness is the capacity of the network for maintaining its functionality when network nodes or links are damaged (by random failures, malicious attacks, etc.) (Newman et al., 2006; Barabasi and Albert, 2000; Zhu and Liu, 2012). It includes both resistance capacity (connection robustness) of network structure to external damage and restoration capacity (restoration robustness) of network structure if it is damaged (Du et al., 2010). Research on network robustness have been widely reported (Kwon and Cho, 2007; Ash and Newt, 2007; Gao et al., 2006; Wang et al., 2006a, 2006b). Kwon and Cho (2007) studied the relationship between feedback structure and network structure robustness, and found that scale-free network model may evolve more feedback structures than random graph model and its network structure robustness enhanced considerably. Ash and Newt (2007) optimized network structure using evolutionary algorithm and found that clustering, modularity and length of long paths all have important effects on the network structure robustness. Gao et al. (2006) demonstrated that for most food webs the attacks based on betweenness centrality are more effective than that based on node degree. Wang et al. (2006a, b) argued that network structure robustness to random failures can be improved by optimizing the efficiency of the network (average inverse length of paths). So far, network structure robustness is usually represented by network connectivity (connection robustness). Restoration robustness is seldom studied.

Besides network structure robustness, here I define two more categories of robustness, network parameter robustness, and comprehensive robustness. Network parameter robustness refers to a network's capacity, without any structural changes, for maintaining between-node flows (fluxes) / link weights if it is perturbed. Comprehensive robustness refers to a network's capacity that not only topological structure of the network, e.g., nodes and links, are not or less changed, but also between-node flows (fluxes), link weights, nodes' state values are maintained also if the network is perturbed. Comprehensive robustness considers both structure and parameter changes of a network.

In addition to conventional indices for network robustness, I propose to use the inverse of some indices of global sensitivity analysis such as Sobol index (Sobol, 1993), and Extended Fourier Amplitude Sensitivity Test, etc (Tarantola et al., 2002; Xu et al., 2004; Zhang, 2012c, 2016d) as indices of network robustness.

As a basic formula, for instance, we may define network robustness as

$$R=1/(dN/dp/N)$$

or

$$R=1/(dN/dp)$$

where N: network measure, dN: final variation of network measure due to perturbation; p: strength of perturbation, which is usually expressed by variation of network measure itself, e.g., number of removed nodes / links, or amount of reduced flux, etc. The greater R value means the stronger robustness.

There are many network measures for uses, i.e., total number of nodes or links, network fluxes, etc.

2.1 Structure robustness

2.1.1 Connection robustness

Connection robustness refers to the capacity of a network to maintain connectivity among remaining nodes if some nodes are attacked and destroyed. Suppose N_r nodes with maximal degree, along with their links are simultaneously removed from a network X. Connection robustness is thus (Dodds et al., 2003)

$$R=c/(n-n_r)$$

where n: total number of nodes in the original network, n_r: number of nodes removed, and c: number of nodes in the maximal connected subgraph (i.e., components; Zhang, 2012a) after n_r nodes have been removed.

2.1.2 Restoration robustness

The full information of a node with unknown or incomplete information can be achieved by consulting the information of its adjacent nodes, as used in detection of key terrorist in the terrorist organization (Bohannon, 2009). Restoration robustness refers to the capacity of a network to restore missed nodes / links if they are destroyed.

Restoration robustness in terms of nodes (D) and links (L) are

$$D=1-(n_r-n_s)/n$$
$$L=1-(l_r-l_s)/l$$

respectively, where n: total number of nodes in the original network, l: total number of links in the original network, n_s: number of nodes restored, n_r: number of nodes removed, l_s: number of links restored, and l_r: number of links removed.

2.2 Parameter robustness

Suppose the topological structure of a network, e.g., nodes and links, are not changed if the network is perturbed. Parameter robustness refers to a network's capacity, without any structural changes, for maintaining between-node flows (fluxes) / link weights if it is perturbed. Here I propose to use the following index, revised from adjacency matrix index (Zhang, 2012c), to represent parameter robustness. Following the definition of Zhang (2012c), suppose flows (fluxes) / link weights matrix of a network with n nodes is $w=(w_{ij})_{n \times n}$, where w_{ij} is the flow (flux) or the weight of the link between nodes v_i and v_j; $w_{ij}=0$ if there is not a link between nodes v_i and v_j; $i, j=1,2,\ldots, n$. The index is

$$S=\sum_i \sum_j |w_{ij0}-w_{ijt}|$$

where w_{ijt}, w_{ij0}: flow (flux) or link weight between nodes v_i and v_j after and before a network is perturbed. The less S value means the stronger robustness.

2.3 Comprehensive robustness

Following the definition of Zhang (2012c), suppose adjacency matrix of a network with n nodes is $d=(d_{ij})_{n \times n}$, If $d_{ij}=d_{ji}=0$, then there is not link between nodes v_i and v_j; if $d_{ij}=d_{ji}=1$, then there is a link between nodes v_i and v_j. Suppose w_{ij} is the flow (flux) or the weight of the link between nodes v_i and v_j, $i, j=1,2,\ldots, n$. The index for comprehensive robustness is

$$S=\sum_i \sum_j |w_{ij0}d_{ij0}-w_{ijt}d_{ijt}|$$

In addition to the measures on network robustness, other robustness measures can be defined according to various performance indices of a network, e.g., network type, variation coefficient, network entropy, etc (Zhang and Zhan, 2011). For instance, robustness of network entropy is defined as a network's capacity to maintain network entropy if the network's structure (or parameters) is perturbed.

For specific networks, e.g., ecosystems, network performance indices may be additionally defined as total productivity, number of functional groups, etc.

3 Theoretical Analysis of Structure Robustness of Typical Networks

In the random network of n nodes, two nodes are connected at a certain probability p (Zhang, 2012a). In the regular network, nodes are connected following certain rules. For instance, the nearest neighbors coupling network (Wang et al., 2006a, 2006b), namely for a given k (k is an even number), n nodes in the network are linked to generate a ring, in which each node is only connected with its $k/2$ neighborhood nodes. Scale-free network is the most popular network (Barabasi and Albert, 1999; Zhang, 2012a). Small-world network was proposed by Watts and Strogatz (1998). Methods for network generation / evolution are from Barabasi and Albert (1999) (scale-free network), Watts and Strogatz (1998) (small-world network), Wang et al. (2006a, 2006b) (regular network), and Zhang (2012a) (random network).

3.1 Connection robustness

For the random network, increasing connection probability p enhances connection robustness. Network connectivity will not be destroyed if $p \geq 0.3$ (the threshold density for connection robustness).

As increase of the nodes removed from scale-free network, the decrease of network connection capacity produces "emergence" phenomena. Connection robustness enhances as the increase of network density.

In general, the connection robustness of regular network and random network is stronger than scale-free network. In terms of connection robustness, small-world network is the interim between regular network and random network. Connection robustness of all networks produces "emergence" as change of number of nodes removed and network density.

3.2 Restoration robustness

3.2.1 Node restoration

Suppose that node i is adjacent to node j. If node i is removed from node j, we may try to restore node i and the link to node j according to the information of node j.

For the random network, if network density is less than 0.3, the restored nodes decline as increase of the nodes removed. Removed nodes can be restored if network density is not less than 0.3 (threshold density for node restoration robustness). Similar to threshold density for connection robustness, the threshold density for node restoration robustness declines as the increase of network size.

The scale-free network can be thoroughly restored if only a few of nodes are removed. After the removed nodes have reached a threshold, the number of restored nodes will decrease quickly. However, node restoration robustness increases as network density.

In the small-world network, for a fixed re-connection probability p, the decline of node restoration rate shows "emergence" as increase of nodes removed. The increase of node degree or re-connection probability p will enhance node restoration robustness.

Overall, node restoration robustness of all networks produces "emergence" as change of number of nodes removed and network density.

3.2.2 Link restoration

For the random network, link restoration robustness increases as network density. However, link restoration robustness will be almost irrelevant to network density after network density has reached a threshold.

Link restoration robustness of scale-free network decreases linearly as the number of nodes removed, and has not significant relationship with network density.

For the small-world network, with a fixed re-connection probability p, the decline of link restoration rate shows "emergence" as increase of nodes removed. The increase of node degree or re-connection probability p will enhance link restoration robustness.

In summary, both link and node restoration robustness of the random network are the best among four types of networks, and the regular network is the worst. Small-world network is the interim between the random network and the regular network. Node restoration robustness of scale-free network is better but its link restoration robustness is as worse as the regular network.

In addition to network evolution methods and resultant networks above, more complex network evolution model has demonstrated that network density (connectance; Zhang, 2011, 2012a, 2012b) increases as time (Zhang, 2016a), which means that the network evolves to the stronger robustness according to the conclusions above.

As theoretical models, all of network evolution methods and resultant networks above do not limit the number of links connected to a node, i.e., node degree is limitless in these methods and networks.

4 Facilitation of Network Structure Robustness

Different from theoretical networks above, practical networks sometimes demonstrate different mechanisms (e.g., food webs with random degree distributions were highly fragile to removals of species, see Montoya and

Sole, 2003) for robustness-maintaining due to the limitation of node degree (e.g., many species have only one or two links in food webs), etc. Based on both theoretical analysis and practical observations, here I summarize the following methods to enhance structure robustness of a given network

(1) Maintaining a certain network density / connectance / connectivity (Dunne et al., 2002; Allesina et al., 2005; Zhang, 2011, 2012a). It has been reported that a food web with the higher connectance has more numerous reassembly pathways and can thus restore faster from perturbation (MacArthur, 1955; Law and Blackford, 1992; Zhang, 2012a). However, some models suggested that food webs with the lower connectance restored faster after a disturbance (May, 1973; Pimm, 1991; Chen and Cohen, 2001; Cohen et al., 1990; Zhang, 2012a). Therefore, for a network with fixed number of nodes, maintaining (increasing or reducing) a certain number of links may improve network structure robustness.

(2) Deploying network circuits (Alon, 2006; Zhang, 2012a, 2016f). Deploying some circuits in network means the existence of some feedback controls, which help to improve network robustness. Feedback control is the basis of robustness of bio-systems and bio-processes, e.g., the chemotaxis and heat shock response of *Escherichia coli*, biological rhythms, and cell cycles, etc (Oleksiuk et al., 2011).

(3) Constructing hierarchical sub-networks / modules / connected components (Zhang, 2012a, 2016e). A large network is always organized from various small mosaics (modules). Organizing mosaics to a large network will probably influence the robustness of entire network (May, 1973). Pinnegar et al. (2005) used a detailed Ecopath with Ecosim model to examine the impacts of food web aggregation and the removal of weak linkages. They found that aggregation of a 41-compartment food web to 27 and 16 compartment systems greatly affected system properties (e.g. connectance, system omnivory and ascendancy) and influenced dynamic stability. Highly aggregated webs restored more quickly following disturbances compared to the original disaggregated model.

Existence of hierarchical sub-networks / modules / connected components may prevent failures diffuse across over a network, simplify the evolution and update of nodes and links, and thus help to enhance network structure robustness. In addition, utilization of several hierarchical sub-networks / modules / connected components may avoid malfunction of the major components in a network. A biological cell is a typical example. In a cell, mitochondria, ribosomes, chloroplasts, etc., are sub-networks / modules / connected components.

(4) Incorporating redundancy. Besides useful nodes and links, adding redundant (i.e., temporarily not useful, or candidate) nodes, links and circuits in a network will help to enhance structure robustness. Redundant nodes / links / circuits are expected to play key roles if some original nodes, links and circuits are destroyed. In biological networks, repeated genes, homeotic genes (McAdams and Arkin, 1999), redundant metabolites, multi-pathway signaling, and similar metabolic circuits (Edwards et al., 2001), etc., are all examples of network redundancy.

Some methods above can be used to facilitate parameter robustness and comprehensive robustness. Actually comprehensive robustness is more reasonable than structure robustness, because network structure is usually determined by the strengths of between-node interactions (i.e., link weights, between-node fluxes, etc.). As an example, past studies have demonstrated that under the condition of constant structure robustness, to increase nodes and links in a network must be at the cost of weakening the added links (weak interactions, i.e., weak weights, weak fluxes, etc.; McCann, 1988; Paine, 1992; Zhang, 2011, 2012a).

Computational simulation, e.g., the network evolution model (Zhang, 2015) can be used to exploit the relationship between network structure and network robustness.

5 Biological Networks and Robustness

5.1 Misuse or inaccurate use of stability in biological studies

So far, most research used stability, etc., to describe the robustness of various biological systems. This includes such topics as "relationship between biodiversity and stability", "stability of ecosystems", "stability of metabolic networks", etc. It is unsurprising because the terminology "robustness" was defined as late as about 30 years ago, while "stability" had been used by ecologists for more than 40 years (May, 2973). Obviously, "stability" in these topics was substantially robustness in most situations. Definition and implication of robustness and stability, as discussed above, are soundly distinctive. We concern both external disturbance and internal perturbation in biodiversity and ecosystems. According to the substantial implication of so-called stability in these topics, I argue to use robustness to replace misused or inaccurately used stability in these situations in exception of the fewer cases for stability-specific topics. In most of the situations, the exact and correct designate should be "relationship between biodiversity and robustness", "robustness of ecosystems", "robustness of metabolic networks", etc.

5.2 Robust structure and parameters of some biological networks (systems)

I assume that naturally existing and sustainable networks (food webs, biochemical networks, etc.) are robust. I summarize some network structures and parameters from references in Table 1, in which S is the number of species, L is the number of actual links, and C is connectance. λ is the parameter of power-law distribution of node degrees, $p(x)=x^{-\lambda}$. It can be concluded that overall power-law λ is around 1.5 (1.5±0.4), the mean node degree is around 2~3 (Zhang and Li, 2016).

Table 1 Network structure and parameters of food webs, metabolic pathways, etc.

Structures	Parameters	Values			Sources
$L=CS^2$	C	0.14			
		food webs			Martinez, 1991, 1992
$S=C^a$	a	-0.5			
		food webs			Montoya and Sole, 2003
$L=aS^b$	a, b	a=1.3, b=1.1			
		40 food webs			Cohen and Briand, 1984
$L=aS^b$	b	1.5			
		food webs			Sugihara et al., 1989; Schoenly et al., 1991; Havens, 1992; Martinez, 1994
$L=aS^b$	a, b	a=2, b=1			
		food webs			Cohen et al., 1990; Martinez, 1992
$C=L/S^2$	C	0.1~0.15			
		food web			Martinez, 1992; Warren, 1994
		1.63	1.71	1.58	

Formula	Param				Reference
$p(x)=x^{-\lambda}$	λ	transcription network	signaling network	metabolic network	Goemann et al., 2011
$p(x)=x^{-\lambda}$	λ	1.64 arthropod family networks	1.43 arthropod family	1.11 arthropod family	Zhang, 2011
$m= L/S$	m	2.16	4.07	3.14	
$m= L/S$	m	2.84 arthropod species networks	2.19		Zhang, 2011
$m= L/S$	m	2.1 tumor pathway	2.9 tumor pathway		Huang and Zhang, 2012
$m= L/S$	m	2.35 transcription network	2.18 signaling network	3.09 metabolic network	Goemann et al., 2011
$m= L/S$	m	~2 food webs			Cohen et al., 1990; Martinez, 1992
$m= L/S$	m	4~15 immunization network			Shams and Khansari, 2014
$m= L/S$	m	4.68 normal pathway	10.58 cancer pathway		Rahman et al., 2013

Note: S is the number of species, L is the number of actual links, and C is connectance. λ is the parameter of power-law distribution of node degrees, $p(x)=x^{-\lambda}$.

Acknowledgment

I am thankful to the support of High-Quality Textbook *Network Biology* Project for Engineering of Teaching Quality and Teaching Reform of Undergraduate Universities of Guangdong Province (2015.6-2018.6), from Department of Education of Guangdong Province, Discovery and Crucial Node Analysis of Important Biological and Social Networks (2015.6-2020.6), from Yangling Institute of Modern Agricultural Standardization, and Project on Undergraduate Teaching Reform (2015.7-2017.7), from Sun Yat-sen University, China.

References

Allesina S, Bodini A, Pascual M. 2005. Functional links and robustness in food webs. Philosophical Transactions of the Royal Society B, 364(1524): 1701-1709

Alon U. 2006. An Introduction to Systems Biology: Design Principles of Biological Circuits. CRC Press, USA

Alon U, Surette MG, Barkai N, et al. 1999. Robustness in bacterial chemotaxis. Nature, 397(6715): 168-171

Ash J, Newt HD. 2007. Optimizing complex networks for resilience against cascading failure. Physical Statistical Mechanics and its Applications, 380(7): 673-683

Barabasi AL, Albert R. 1999. Emergence of scaling in random networks. Science, 286(5439): 509-512

Bohannon J. 2009. Counter terrorism's new tool: "metanetwork" analysis. Science, 325(5939): 409-411

Chen X, Cohen JE. 2001. Global stability, local stability and permanence in model food webs. Journal of Theoretical Biology, 212: 223-235

Cohen JE, Briand F. 1984. Trophic links of community food webs. Proceedings of the National Academy of Sciences of USA, 81: 4105-4109

Cohen JE, Briand F, Newman CM. 1990. Community Food Webs: Data and Theory. Springer, Berlin, Germany

Dodds SP, Watts DJ, Sabel FC. 2003. Information exchange and robustness of organizational networks. Proceedings of the National Academy of Sciences of USA, 100(21): 12516-12521

Du W, Cai M, Du HF. 2010. Study on indices of network structure robustness and their application. Journal of Xi'an Jiao Tong University, 44(4): 93-97

Dunne JA, Williams RJ, Martinez ND. 2002. Food-web structure and network theory: the role of connectance and size. Ecology, 99(20): 12917-12922

Edwards JS, Ibarra RU, Palsson BO. 2001. In *silico* predictions of *Escherichia coli* metabolic capabilities are consistent with experimental data. Nature Biotechnology, 19(2): 125-130

Ferrarini A. 2011. Some thoughts on the controllability of network systems. Network Biology, 1(3-4): 186-188

Ferrarini A. 2013. Exogenous control of biological and ecological systems through evolutionary modelling. Proceedings of the International Academy of Ecology and Environmental Sciences, 3(3): 257-265

Ferrarini A. 2014. Local and global control of ecological and biological networks. Network Biology, 4(1): 21-30

Ferrarini A. 2015. Evolutionary network control also holds for nonlinear networks: Ruling the Lotka-Volterra model. Network Biology, 5(1): 34-42

Gao L, Guo JL. 2011. Review in research on biological networks. China Journal of Bioinformation, 9(2): 113-119

Gao L, Li MH, Wu JS. 2006. Betweenness - based attacks on nodes and edges of food weds dynamics of continuous Discrete and Impulsive Systems Series B, 13(3): 421-428

Goemann B, Wingender E, Potapov AP. 2011. Topological peculiarities of mammalian networks with different functionalities: transcription, signal transduction and metabolic networks. Network Biology, 1(3-4): 134-148

Havens K. 1992. Scale and structure in natural food webs. Science, 257: 1107-1109

Huang JQ, Zhang WJ. 2012. Analysis on degree distribution of tumor signaling networks. Network Biology, 2(3): 95-109

Kwon YK, Cho KH. 2007. Analysis of feedback loops and robustness in network evolution based on Boolean models. BMC Bioinformatics, 8(9): 430-438

Law R, Blackford JC. 1992. Self-assembling food webs. A global view-point of coexistence of species in Lotka-Volterra communities. Ecology, 73: 567-578

MacArthur R. 1955. Fluctuation of animal populations and a measure of community stability. Ecology, 36(3): 533-536

Martinez ND. 1991. Artifacts or attributes? Effects of resolution on the Little Rock Lake food web. Ecological

Monographs, 61: 367-392

Martinez ND. 1992. Constant connectance in community food webs. American Naturalist, 139: 1208-1218

Martinez ND. 1994. Scale-dependent constraints on food web structure. American Naturalist, 144: 935-953

May RM. 1973. Stability and Complexity in Model Systems. Princeton University Press, USA

McAdams HH, Arkin A. 1999. It's a noisy business! Genetic regulation at the nanomolar scale. Trends in Genetics, 15(2): 65-69

McCann K, et al. 1998. Weak trophic interactions and the balance of nature. Nature, 395: 794-798

Montoya JM, Sole RV. 2003. Topological properties of food webs: from real data to community assembly models. Oikos, 102: 614-622

Newman EJ, Barabasi AL, Watts DJ. 2006. The Structure and Dynamic of Networks. Princeton University Press, Princeton, NJ, USA

Oleksiuk O, Jakovljevic V, Vladimirov N, et al. 2011. Thermal robustness of signaling in bacterial chemotaxis. Cell, 145(2): 312-321

Paine RT. 1992. Food-web analysis through field measurement of per capita interaction strength. Nature, 355: 73-75

Pimm SL. 1991. The balance of nature? Ecological issues in the conservation of species and communities. University of Chicago Press, USA

Pinnegar JK, Blanchard JL, Mackinson S, et al. 2005. Aggregation and removal of weak-links in food-web models: system stability and recoveryfrom disturbance. Ecological Modelling, 184: 229-248

Rahman KMT, Md. Islam F, Banik RS, et al. 2013. Changes in protein interaction networks between normal and cancer conditions: Total chaos or ordered disorder? Network Biology, 3(1): 15-28

Schoenly K, Beaver RA, Heumier TA. 1991. On the trophic relations of insects: a food web approach. American Naturalist, 137: 597-638

Shams B, Khansari M. 2014. Using network properties to evaluate targeted immunization algorithms. Network Biology, 4(3): 74-94

Sobol IM. 1993. Sensitivity Estimates for nonlinear mathematical models. Mathematical Model and Computational Experiment, 1: 407-414

Sugihara G, Schoenly K, Trombla A. 1989. Scale invariance in food web properties. Science, 245: 48-52

Tarantola S, Giglioli N, Jesinghaus J, et al. 2002 .Can global sensitivity analysis steer the implementation of models for environmental assessments and decision-making? Stochastic Environmental Research and Risk Assessment, 16: 63-76

Wang B, Tang HW, Guo CH. 2006a. Optimization of network structure to random failures. Physical Statistical Mechanics and its Applications, 368(2): 607-614

Wang XF, Li X, Chen GR. 2006b. The Theory of Complex Network and Its Application. Tsinghua University Press, Beijing, China

Warren PH. 1994. Making connections in food webs. Trends in Ecology and Evolution, 4: 136-140

Watts DJ, Strogatz SH. 1998. Collective dynamics of small-world' networks. Nature, 393: 440-442

Xu CG, Hu YM, Chang Y, et al. 2004. Sensitivity analysis in ecological modeling. Chinese Journal of Applied Ecology, 15(6): 1056-1062

Zhang WJ. 2011. Constructing ecological interaction networks by correlation analysis: hints from community sampling. Network Biology, 1(2): 81-98

Zhang WJ. 2012a. Computational Ecology: Graphs, Networks and Agent-based Modeling. World Scientific, Singapore

Zhang WJ. 2012b. How to construct the statistic network? An association network of herbaceous plants constructed from field sampling. Network Biology, 2(2): 57-68

Zhang WJ. 2012c. Several mathematical methods for identifying crucial nodes in networks. Network Biology, 2(4): 121-126

Zhang WJ. 2015. A generalized network evolution model and self-organization theory on community assembly. Selforganizology, 2(3): 55-64

Zhang WJ. 2016a. A random network based, node attraction facilitated network evolution method. Selforganizology, 3(1): 1-9

Zhang WJ. 2016b. Screening node attributes that significantly influence node centrality in the network. Selforganizology, 3(3): 75-86

Zhang WJ. 2016c. Selforganizology: The Science of Self-Organization. World Scientific, Singapore

Zhang WJ. 2016d. Some methods for sensitivity analysis of systems / networks. Network Pharmacology, 1(3): 74-81

Zhang WJ. 2016e. A method for identifying hierarchical sub-networks / modules and weighting network links based on their similarity in sub-network / module affiliation. Network Pharmacology, 1(2): 54-65

Zhang WJ. 2016f. A node-similarity based algorithm for tree generation and evolution. Network Biology, 6(3): 55-64

Zhang WJ, Li X. 2016. Generate networks with power-law and exponential-law distributed degrees: with applications in link prediction of tumor pathways. Network Pharmacology, 1(1): 15-35

Zhang WJ, Zhan CY. 2011. An algorithm for calculation of degree distribution and detection of network type: with application in food webs. Network Biology, 1(3-4): 159-170

Zhang SY, Zhang X. 2009. Biological networks and some of their developments. Journal of System Simulation, 17: 5300-5305

Zhang ZL, Wang ZY. 2009. Sytems Biology. Science Press, Beijing, China

Zhu YH, Liu J. 2012. A research on the robustness of complex networks. Science and Technology Information, 32: 6

12

Networks control: Introducing the degree of success and feasibility

Alessandro Ferrarini

Department of Evolutionary and Functional Biology, University of Parma, Via G. Saragat 4, I-43100 Parma, Italy

E-mail: sgtpm@libero.it, alessandro.ferrarini@unipr.it

Abstract

Taming ecological and biological networks is a key-issue. It could be used to: a) neutralize damages to ecological and biological networks, b) safeguard rare and endangered species, c) manage ecological systems at the least possible cost, and d) counteract the impacts of climate change. While I recently showed that ecological and biological networks can be efficaciously controlled both from inside (inside-control model) and outside (outside-control model), here I propose a solution to the choice of the most feasible solution to network control. To do this, I introduce the concepts of control success and feasibility.

Keywords edges control feasibility; control success; control uncertainty; genetic algorithms; network control; stochastic simulations.

1 Introduction

Recently, I proposed that ecological and biological networks can be controlled by coupling network dynamics and evolutionary modelling (Ferrarini, 2011). They can be efficaciously tamed from outside (Ferrarini, 2013a), but also through the use of endogenous controllers (Ferrarini, 2013b). These two approaches are different from both a theoretical and methodological viewpoint. The endogenous control requires that the network is optimized at the beginning of its dynamics (by acting upon nodes, edges or both) so that it will then go inertially to the desired state. Instead, the exogenous control requires that exogenous controllers act upon the network at each cycle. *A priori*, it's hard to say which of the two approaches is more effective, it mainly depends on the kind of ecological or biological network one is dealing with.

In another paper (Ferrarini, 2013c), I have faced a further important question: how reliable is the achieved solution? In other words, which is the degree of uncertainty about getting the desired result if values of edges and nodes were a bit different from optimized ones? This is a pivotal question, because it's not assured that while managing a certain system we are able to impose to nodes and edges exactly the optimized values we would need in order to achieve the desired results. In order to face this topic, I have coined a 3-parts framework (network dynamics - genetic optimization - stochastic simulations).

Here I propose a solution to the choice of the most feasible solution to network control. To do this, I introduce the concepts of control success and feasibility.

2 Mathematical Formulation

Most real systems' dynamics can be modelled and simulated as follows (Liu et al., 2011; Slotine and Li, 1991):

$$
\begin{cases}
\dfrac{dS_1}{dt} = a_{11}S_1 + \ldots + a_{1n}S_n + I_1 + O_1 \\
\ldots \\
\dfrac{dS_n}{dt} = a_{n1}S_1 + \ldots + a_{nn}S_n + I_n + O_n
\end{cases}
\tag{1}
$$

where S_i is the number of individuals (or the total biomass, or the covered surface in case of plant species) of the generic i-th species, while I and O represent inputs and outputs from/to outside.

Biological and ecological systems can be tamed from outside using the following 1-external-controller model (Ferrarini 2013a):

$$
\begin{cases}
\left(\dfrac{dS_1}{dt}\right)_{OPT} = a_{11}S_1 + \ldots + a_{1n}S_n + I_1 + O_1 + c_{11*}C_{1*} \\
\ldots \\
\left(\dfrac{dS_n}{dt}\right)_{OPT} = a_{n1}S_1 + \ldots + a_{nn}S_n + I_n + O_n + c_{n1*}C_{1*} \\
\dfrac{dC_1}{dt} = f_1S_1 + \ldots + f_nS_n
\end{cases}
\tag{2}
$$

where asterisks stand for the genetic optimization (Holland 1975) of exogenous node's edges (i.e., coefficients of interaction with the inner system) and exogenous node's stock, i.e. the modification of such values at the beginning of network dynamics in order to get a certain goal (e.g., maximization of the final value of a certain variable). There's 1 controller C_1 that, in some cases, can also receive feedbacks from the network, It's clear that also the feedback dC_1/dt to the controller could be subject to genetic control by taming $<f_1...f_n>$.

In case 1 controller is not enough, the model in (2) must be expanded to the following k-external-controllers model (Ferrarini 2013a):

$$\begin{cases} \dfrac{dS_1}{dt} = a_{11}S_1 + \ldots + a_{1n}S_n + I_1 + O_1 + c_{11*}C_{1*} + \ldots + c_{1k*}C_{k*} \\[2ex] \ldots \\[1ex] \dfrac{dS_n}{dt} = a_{n1}S_1 + \ldots + a_{nn}S_n + I_n + O_n + c_{n1*}C_{1*} + \ldots + c_{nk*}C_{k*} \\[2ex] \dfrac{dC_1}{dt} = f_{11}S_1 + \ldots + f_{1n}S_n \\[2ex] \ldots \\[1ex] \dfrac{dC_k}{dt} = f_{k1}S_1 + \ldots + f_{kn}S_n \end{cases} \qquad (3)$$

Alternatively, an ecological or biological network can be controlled from inside using the following control model (Ferrarini, 2013b):

$$\begin{cases} \left(\dfrac{dS_1}{dt}\right)_{OPT} = a_{11*}S_1^{*} + \ldots + a_{1n*}S_n^{*} + I_{1*} + O_{1*} \\[2ex] \ldots \\[1ex] \left(\dfrac{dS_n}{dt}\right)_{OPT} = a_{n1*}S_1^{*} + \ldots + a_{nn*}S_n^{*} + I_{n*} + O_{n*} \end{cases} \qquad (4)$$

where asterisks stand for the optimization of edges (i.e., coefficients of interaction among variables) or nodes (i.e., initial stocks), that is the modification of their values at the beginning of the network dynamics in order to get a certain goal.

After optimization is reached, the degree of uncertainty of (2), (3) and (4) about getting the desired result can be computed as (Ferrarini, 2013c):

$$\begin{cases} \left(\dfrac{dS_1}{dt}\right)_{OPT} = \underline{a_{11*}S_1^{*}} + \ldots + \underline{a_{1n*}S_n^{*}} + \underline{I_{1*}} + \underline{O_{1*}} \\[2ex] \ldots \\[1ex] \left(\dfrac{dS_n}{dt}\right)_{OPT} = \underline{a_{n1*}S_1^{*}} + \ldots + \underline{a_{nn*}S_n^{*}} + \underline{I_{n*}} + \underline{O_{n*}} \end{cases} \qquad (5)$$

where:

$$\begin{cases} 0.95 * a_{ij*} \le \underline{a_{ij}} \le 1.05 * a_{ij*} \\[1ex] 0.95 * S_j^{*} \le \underline{S_j} \le 1.05 * S_j^{*} \end{cases} \qquad (6)$$

or alternatively:

$$\begin{cases} 0.9 * a_{ij*} \leq \underline{a_{ij}} \leq 1.1 * a_{ij*} \\ 0.9 * S_j^* \leq \underline{S_j} \leq 1.1 * S_j^* \end{cases} \tag{7}$$

Hence, $\underline{a_{ij}}$ represents a 5% (or 10%) uncertainty about a_{ij*}, while $\underline{S_j}$ represents a 5% (or 10%) uncertainty about S_j^*. If, after genetic optimization, we stochastically vary n times (e.g. 10,000 times) a_{ij*} and S_j^*, we are able to compute how many times such uncertainty makes the optimization procedure useless. Hence, uncertainty about network control can be computed as (Ferrarini, 2013c):

$$U_\% = 100 * \frac{k}{n} \tag{8}$$

where k is the number of stochastic simulations acting upon optimized parameters that make the optimization procedure useless (i.e. the goal of optimization is not reached).

Now, let's assign to each species a weight of importance σ_i:

$$\sigma_i : \begin{cases} > 0 & \text{for benefit species (or network actors)} \\ 0 & \text{for species (or network actors) of no interest} \\ < 0 & \text{for cost species (or network actors)} \end{cases} \tag{9}$$

I suggest here that the degree of success DS_i of network control for each i-th species can be computed as the weighted difference between the optimized dynamic of the species (S_i^{opt}: how it goes, at equilibrium, after optimization) and the inertial one (S_i^{in}: how it would go, at equilibrium, without optimization):

$$DS_i = \sigma_i * \Delta_i = \sigma_i * (S_i^{opt} - S_i^{in}) \tag{10}$$

DS_i is positive if a benefit species has increased thanks to the network control or a cost species has decreased. Instead it's negative in case a benefit species has decreased due to the network control or a cost species has increased. The overall degree of success of network control for n species (or network actors) can be hence calculated as:

$$DS = \sum_{i=1}^{n} \sigma_i * \Delta_i \tag{11}$$

Now I define the degree of feasibility F of network control as:

$$F = \frac{DS}{1 + U_\%} \tag{12}$$

where the constant 1 has been added to avoid that the denominator goes to 0. As a first approximation, I suggest that the weight of importance σ_i should go from -1 to +1. But, in order to give DS and $1 + U_\%$ the same order of magnitude, I suggest that σ_i should be set so that:

$$\begin{cases} 1 \le \dfrac{DS}{1+U_{\%}} \le 10 \\ or \\ 1 \le \dfrac{1+U_{\%}}{DS} \le 10 \end{cases} \quad (13)$$

It's clear that F is a 3D surface equation in the form:

$$Z = \frac{x}{1+y} \quad (14)$$

Hence the feasibility surface is like in Fig. 1.

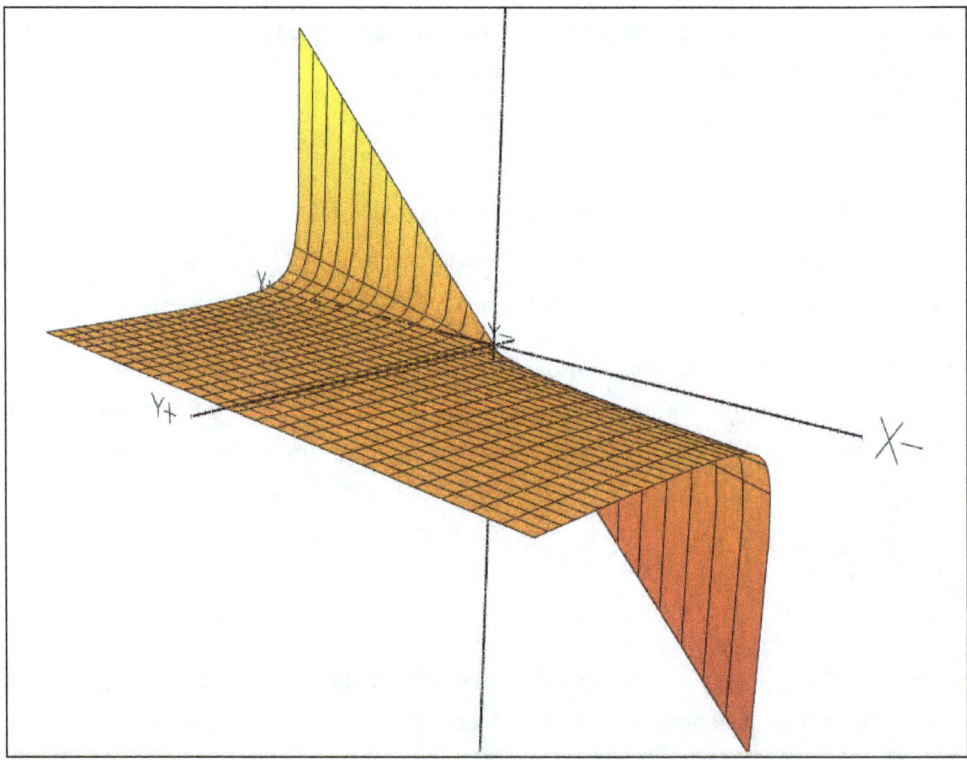

Fig. 1 Feasibility surface as a function of control success (X-axis) and control uncertainty (Y-axis).

Since many solutions to network control can be found using the previous control models (2), (3) and (4), each *j-th* solution will receive its degree of feasibility

$$F_j = \frac{DS_j}{1+U_{j\%}} \quad (15)$$

and the best solution to network control will be the one with

$$max(F_j) \quad (16)$$

3 Conclusions

Taming ecological and biological networks is a key-issue. It could be used to: a) neutralize damages to ecological and biological networks, b) safeguard rare and endangered species, c) manage ecological systems at the least possible cost, and d) counteract the impacts of climate change.

While I recently showed that ecological and biological networks can be efficaciously controlled both from inside (inside-control model) and outside (outside-control model), here I have proposed a solution to the choice of the most feasible solution to network control. To do this, I have introduced the concepts of control success and feasibility.

References

Ferrarini A. 2011. Some thoughts on the controllability of network systems. Network Biology, 1(3-4): 186-188

Ferrarini A. 2013a. Exogenous control of biological and ecological systems through evolutionary modelling. Proceedings of the International Academy of Ecology and Environmental Sciences, 3(3): 257-265

Ferrarini A. 2013b. Controlling ecological and biological networks *via* evolutionary modelling. Network Biology, 3(3): 97-105

Ferrarini A. 2013c. Computing the uncertainty associated with the control of ecological and biological systems. Computational Ecology and Software, 3(3): 74-80

Holland JH. 1975. Adaptation in Natural and Artificial Systems: An Introductory Analysis with Applications to Biology, Control and Artificial Intelligence. University of Michigan Press, Ann Arbor, USA

Liu YY, Slotine JJ, Barabasi AL. 2011. Controllability of complex networks. Nature, 473:167-173

Slotine JJ, Li W. 1991. Applied Nonlinear Control. Prentice-Hall, USA

13

A quasi chemical approach for the modeling of predator-prey interactions

Muhammad Shakil[1], **H. A. Wahab**[1], **Muhammad Naeem**[2], **Saira Bhatti**[3]

[1]Department of Mathematics, Hazara University, Manshera, Pakistan

[2]Department of Information Technology, Hazara University, Manshera, Pakistan

[3]Department of Mathematics, COMSATS Institute of Information Technology, Abbottabad, Pakistan

E-mail: wahabmaths@yahoo.com,wahab@hu.edu.pk

Abstract

We aim to develop the reaction diffusion equation for different types of mechanism of the predator-prey interactions with quasi chemical approach. The chemical reactions representing the interactions obey the mass action law. Since the cell-jump models may be considered as the proper diffusion models by themselves, the territorial animal like fox is given a simple cell as its territory. Under the proper relations between coefficients, like complex balance or detailed balance, this system demonstrated globally stable dynamics.

Keywords quasi-chemical approach; predator-prey interactions; modeling fox and rabbit interactions.

1 Introduction

The Lotka Voltera equations are a pair of first order non linear differential equations these are also known as the predator prey equations .i.e. when the growth rate of one population is decreased and the other increased then the populations are in a predator-prey situation. The Lotka–Volterra equations are frequently used to describe the dynamics of biological in which two species interact, one as a predator and the other as prey.

In this review paper, we aim to study the interactions between the territorial animals like foxes and the rabbits. The territories for the foxes are considered to be the simple cells. The interactions between predator and its prey are represented by the chemical reactions which obey the mass action law. In this sence, we aim to study the mass action law for Lotka-Volterra predator prey models and consider the quasi chemical approach for the interactions between the predator and its prey. We also aim to develop the reaction diffusion equation for different types of mechanism of the predator prey interactions. The principle of detailed balance, however can also be introduced in some cases. Since the cell-jump models may be considered as the proper diffusion models by themselves, the territorial animal like fox is given a simple cell as its territory. The sense in which the discrete equations for cells converge to the partial differential equations of diffusion is that the cell models give the semi-discrete approximation of the partial differential equations for diffusion which result in a system of ordinary differential equations in cells. There are jumps of concentrations on the boundary of cells. The system of semi discrete models for cells with "no-flux" boundary conditions has all the nice properties of the chemical kinetic equations for closed systems. Under the proper relations between coefficients, like complex balance or detailed balance, this system demonstrated globally stable dynamics.

Here in this review paper, we will give background of the development of the different models of Lotka-Volterra types of reaction-diffusion equation along with different approaches that were acquired by the different scientist in the ecology and mathematical biology. Then we come to our goals, where we borrowed the properties of the chemical reactions for our approach to develop the reaction-diffusion equation of Lotka Volterra type.

The Lotka Volterra predator prey models were originally introduced by Alfred J. Lotka (1920) in the theory of autocatalytic chemical reactions in 1910 this was effectively the logistic equation which was originally derived by Pierre Francois Verhulst. In 1920 Lotka extended the model to "organic systems" using a plant species and an herbivorous animal species as an example and in 1925 he utilized the equations to analyze predator-prey interactions in his book on biomathematics arriving at the equations that we know today.

In 1926, Vito Volterra (Voltera, 1926), made a statistical analysis of fish catches in the Adriatic independently investigated the equations. V. Voltera applied these equations to predator prey interactions; consist of a pair of first order autonomous ordinary differential equations.

Since that time the Lotka–Voltera model has been applied to problems in chemical kinetics, population biology, epidemiology and neural networks. These equations also model the dynamic behavior of an arbitrary number of competitors. The Lotka–Volterra system of equations is an example of a Kolmogorov model, which is a more general framework that can model the dynamics of ecological systems with predator-prey interactions, competition, disease, and mutualism.

Lotka Volterra is the most famous predator prey model. According to Volterra if $x(t)$ is the prey population and $y(t)$ that of the predator at time t then Volterra's model is (Murray, 2003),

$$\frac{dx}{dt} = x(a - by) \tag{1}$$

$$\frac{dy}{dt} = y(cx - d) \tag{2}$$

where a, b, c and d are positive constants. Often some necessary modifications are possible here e.g. for an absent predator a limited growth of prey can be introduced. The assumptions in the model are:

•   In the absence of any predation the prey grows unboundedly in a Malthusian way; this is the ax term in equation (1).

•   The effect of the predation is to reduce the prey's per capita growth rate by a term proportional to the prey and predator populations; this is the $-bxy$ term.

•   In the absence of any prey for sustenance the predator's death rate results in exponential decay, that is, the $-dy$ term in equation (2).

•   The prey's contribution to the predators' growth rate is cxy; i.e. it is proportional to the available prey as well as to the size of the predator population.

•   The xy terms are representing the conversion of energy from one source to another.

• bxy is taken from the prey and cxy accrues to the predators.

The model (1) and (2) are known as the *Lotka–Volterra model* but this model has serious drawbacks. Nevertheless it has been of considerable value in posing highly relevant questions and is a jumping-off place for more realistic models; this is the main motivation of study here (Murray, 2003).

2 Interacting Population Models

The population dynamics of each species is affected whenever two or more species interact. Generally there consist whole webs of interacting species; those webs which consist structurally for complex communities are called a trophic web. The dynamics outcomes of the interactions are very sensitive to initial data and parameter values. The interactions lead to the following possible outcomes,

- Competitive exclusion;
- Total extinction, i.e., collapse of the whole system;
- Coexistence in the form of positive steady state;
- Coexistence in the form of oscillatory solutions;
- A better and friendly competitor can save a otherwise doomed prey species.

Two or more species models consist in case of concentrating particularly on two-species systems. There are three main types of interactions.

- If the growth rate of one population is decreased and the other increased the populations are in a predator–prey situation.
- If the growth rate of each population is decreased then it is competition.
- If each population's growth rate is enhanced then it is called mutualism or symbiosis (Murray, 2003).

3 The Effect of Complexity on Stability

To explain the relationship between structural complexity and stability of ecosystems an extensive literature is available. Theoretical study reviews shows that the term stability is mostly used to mean the condition that whereby species densities, when perturbed from equilibrium, again return to that equilibrium. Analytically this condition can be made tractable by assuming that the populations are perturbed slightly so that linear approximations can be made to the nonlinear equations.

To know the effect of complexity on stability briefly let us take the generalized Lotka–Volterra predator–prey system where there are n prey species and n predators, which prey on all the prey species but with different severity. Then in place of (1) and (2) it can be written as,

$$\frac{dx_i}{dt} = x_i \left[a_i - \sum_{j=1}^{n} b_{ij} y_j \right]$$
$$\frac{dy_i}{dt} = y_i \left[\sum_{j=1}^{n} c_{ij} x_j - d_i \right] \qquad i = 1, 2, 3, ..., n \qquad (3)$$

and all of the a_i, b_{ij}, c_{ij} *and* d_i are positive constants (Murray, 2003).

4 Realistic Predator–Prey Models

Though the Lotka–Volterra model is unrealistic because one of the unrealistic assumption is that growth of the prey population is unbounded in the absence of predation and the other is that there is no limit to the prey consumption but it suggests that simple predator–prey interactions can result in periodic behaviour of the populations. Since it is not unexpected that if a prey population increases then it encourages growth of its predator and more predators consume more prey due to which its population starts to decline. With the less food around the predator population declines and when it is low enough, this allows the prey population to increase and the whole cycle starts over again. Depending on the detailed system such oscillations can grow or decay or go into a stable *limit cycle* oscillation or even exhibit chaotic behaviour, although in the latter case there must be at least three interacting species, or the model has to have some delay terms. A limit cycle solution is a closed trajectory in the predator–prey space which is not a member of a continuous family of closed trajectories. A stable limit cycle trajectory is such that any small perturbation from the trajectory decays to zero.

One of the unrealistic assumptions in the Lotka–Volterra models, (1) and (2), and generally (3), is that the prey growth is unbounded in the absence of predation.

In the form we have written the model (1) and (2) the bracketed terms on the right are the density-dependent per capita growth rates. To be more realistic these growth rates should depend on both the prey and predator densities as,

$$\frac{dx}{dt} = x\,F(x,y), \quad \frac{dy}{dt} = y\,G(x,y). \tag{4}$$

where the forms of F and G depend on the interaction, the species and so on. A more realistic prey population equation might take the form,

$$\frac{dx}{dt} = x\,F(x,y), \quad F(x,y) = r\left(1 - \frac{x}{k}\right) - y\,R(x). \tag{5}$$

where $R(x)$ is one of the predation term and K is the constant carrying capacity for the prey when $P=0$.

The predator population equation, the second of (5), should also be made more realistic than simply having $G = -d + cx$ as in the Lotka–Volterra model (2). Possible forms are,

$$G(x,y) = k\left(1 - \frac{hy}{x}\right), \quad G(x,y) = -d + eR(x) \tag{6}$$

where k, h, d and e are positive constants. The first of (6) says that the carrying capacity for the predator is directly proportional to the prey density (Murray, 2003).

5 Competitioning Models

Here, for a common food source two species compete to each other. For example, competition may be for territory which is directly related to food resources. Mathematically, carrying capacity of one species is reduced by the other species and vice versa. When two or more species compete for the same limited food source or in some way inhibit each other's growth, a very simple competition model which demonstrates a fairly general principle which is observed to hold in nature is that in the competition for the same limited resources one of the species usually becomes extinct. The basic two species Lotka–Volterra competition model with each species n_1 and n_2 have logistic growth in the absence of the other. Inclusion of logistic growth in the Lotka–Volterra systems makes them much more realistic, but to highlight the principle the simpler model which nevertheless reflects many of the properties of more complicated models, particularly as regards stability.

$$\frac{dx_1}{dt} = r_1 x_1 \left[1 - \frac{x_1}{k_1} - b_{12} \frac{x_2}{k_1} \right] \tag{7}$$

$$\frac{dx_2}{dt} = r_2 x_2 \left[1 - \frac{x_2}{k_2} - b_{21} \frac{x_1}{k_2} \right] \tag{8}$$

where $r_1, k_1, r_2, k_2, b_{12}$ and b_{21} are positive constants and r's are the linear birth rates and the k's are the carrying capacities. The b_{12} and b_{21} measure the competitive effect of n_2 on n_1 and n_1 on n_2 respectively and they are not equal generally (Murray, 2003).

Note that the competition models (7) and (8) are not a conservative system like its Lotka–Volterra predator–prey counterpart and if we non dimensionalise this model by writing

$$u_1 = \frac{x_1}{k_1}, \ u_2 = \frac{x_2}{k_2}, \ \tau = r_{1t}, \ \rho = \frac{r_1}{r_1},$$

$$a_{12} = b_{12} \frac{k_2}{k_1}, \ a_{21} = b_{21} \frac{k_1}{k_2} \tag{9}$$

(6) and (8) become

$$\frac{du_1}{d\tau} = u_1 (1 - u_1 - a_{12} u_2) = f_1(u_1, u_2),$$

$$\frac{du_2}{d\tau} = p u_2 (1 - u_2 - a_{21} u_1) = f_2(u_1, u_2), \tag{10}$$

The steady states, and phase plane singularities are the solutions of $f_1(u_1, u_2) = f_2(u_1, u_2) = 0$ (Murray, 2003).

6 Mutualism or Symbiosis

In these cases, the two species benefit from each other. To some extent it is the opposite of the competition model here carrying capacity of each species is increased by the other species.

There are many cases where the interaction of two or more species is to the advantage of all. In promoting and even in maintaining such species mutualism or symbiosis often plays a very crucial role. Plant and seed dispersal is an example. If the survival is not at stake even in those situations the advantage of mutualism or symbiosis can have its own importance. As a topic of theoretical ecology, even for two species, this area has not been as widely studied as the others even though its importance is comparable to that of predator–prey and competition interactions (Lotka, 1920). This is in part due to the fact that simple models in the Lotka–Volterra vein give silly results. The simplest mutualism model equivalent to the classical Lotka–Volterra predator–prey one is,

$$\frac{dx_1}{dt} = r_1 x_1 + a_1 x_1 x_2. \ \frac{dn_2}{dt} = r_2 n_2 + a_2 n_2 n_1,$$

where r_1, r_2, a_1 and a_2 are positive constants.

Since $\frac{dx_1}{dt} > 0$ and $\frac{dx_2}{dt} > 0$, n_1 and n_2 simply grow unboundedly (Murray, 2003).

7 Threshold Phenomena

With the exception of the Lotka–Volterra predator–prey model, the two species models, which have either a stable steady states where small perturbations die out, or unstable steady states where perturbations from them grow unboundedly or result is produced in limit cycle periodic solutions. There is an interesting group of models which have a nonzero stable state such that if the perturbation is sufficiently large or it is of the right kind, then before returning to the steady state population densities undergo large variations. Such models are said to exhibit a threshold effect (Murray, 2003). One such group of models is studied here. The model predator–prey system is

$$\frac{dx}{dt} = x\left[F(x) - y\right] = f(x, y),$$
$$\frac{dy}{dt} = y\left[x - G(y)\right] = g(x, y) \tag{11}$$

where for convenience all the parameters have been incorporated in the F and G by a suitable rescaling.

8 Discrete Growth Models for Interacting Populations

Now taking the two interacting species, each have the non overlapping generations and each species affect the other's population dynamics. In the continuous growth models, there are some main types of interaction, such as, predator–prey, mutualism and competition. Nowhere near to the same extent as for continuous models for which, in the case of two species, there is a complete mathematical treatment of the equations. In view of the complexity of solution behaviour with single-species discrete models it is not surprising that even more complex behaviour is possible with coupled discrete systems (Murray, 2003).

The interaction between the prey (x) and the predator (y) to be governed by the discrete time (t) system of coupled equations as,

$$x_{t+1} = rx_t f(x_t, y_t), \tag{12}$$

$$y_{t+1} = x_t g(x_t, y_t), \tag{13}$$

where $r > 0$ is the net linear rate of increase of the prey and f and g are functions which relate the predator-influenced reproductive efficiency of the prey and the searching efficiency of the predator respectively.

9 Detailed Analysis to Predator Prey Models

Predator-dependent predator-prey model is of the form

$$\frac{dx}{dt} = xg\left(\frac{x}{k}\right) - yP(x, y), \quad x(0) > 0,$$
$$\frac{dy}{dt} = cyP(x, y) - dy, \quad y(0) > 0, \tag{14}$$

when $P(x, y) = \left(\frac{x}{y}\right)$ then model (14) is strictly ratio dependent. The traditional or prey-dependent model takes the form

$$\frac{dx}{dt} = xg\left(\frac{x}{k}\right) - yP(x,y), \quad x(0) > 0,$$

$$\frac{dy}{dt} = cyP(x) - dy, \quad y(0) > 0,$$

(15)

Mathematically both the traditional prey-dependent and ratio-dependent models as a limiting cases i.e. for the former $c = 0$ and for the later $a = 0$ of the general predator-dependent functional response

$$P(x,y) = \frac{\theta x}{a + bx + cy}$$

when $P(x) = \dfrac{\theta x}{x+m}$ and $g\left(\dfrac{x}{k}\right) = r\left(1 - \dfrac{x}{k}\right)$ becomes

$$\frac{dx}{dt} = rx\left(1 - \frac{x}{k}\right) - \frac{\theta xy}{m+x}, \quad x(0) > 0$$

where r, k, θ, m, f, d are positive constants and the population density of prey and predator at time t is

represented by $x(t), y(t)$ respectively.

In the absence of predation the prey carrying capacity K *and* grows with intrinsic growth rate r. The

predator consumes the prey with functional response of type $\dfrac{cxy}{(m+x)}$ and contributes to its growth with rate

$\dfrac{fxy}{(m+x)}$. The constant d is the death rate of predator. According to model (15) in a predator-prey system

enrichment will cause an increase in the equilibrium density of the predator but not in that of the prey, and will destabilize the positive equilibrium, that is according to (15) a low and stable prey equilibrium density is not possible. Another prediction that can be made from model (15) is that the mutual extinction between the prey and predator cannot extinct simultaneously.

Recently by both mathematicians and ecologists the rich dynamics provided at the boundary and close to the origin by the strict ratio-dependent models is ignored. If there is a positive steady state in ratio dependent models both prey and predator will go extinct i.e., the collapse of the system. The extinction may occur in two different ways. One of the way is that regardless of the initial densities both species become extinct and the other way is that both species will die out if the initial prey/predator ratio is too low. In the first case, extinction often occurs as a result of high predator efficiency in catching the prey and in the second way there are interesting implications. For example, it indicates that altering the ratio of prey to predators through over-harvesting of prey species may lead to the collapse of the whole system and the extinction of both species may occur. In many aspects the richest dynamics is provided by the ratio-dependent models while in the prey-dependent models the least in dynamical behavior is provided.

In ratio dependent models there are still some controversial aspects. A specific main controversial aspect is that in the ratio dependent models the high population densities of both prey and predator species are required while the most interesting dynamics of ratio-dependent models occurs near the origin. If the area of the population interaction is large certainly it is a valid concern because in such cases rather than interfering each other most efforts are spent by the predators in searching the prey hence the functional response is likely to be much more sensible to prey density than predator density. However, if the area of the population is small for the prey and even the densities are low since predators can remain interfering effectively each other then certainly in such cases ratio dependence formulation may remain valid for very small field or patch, even when the numbers of individuals of prey and predators are low while their densities may remain high. In such cases, ratio dependence models are a valid mechanism which suggests that there is a high possibility of mutual extinction.

Some interesting and new dynamics are revealed by the analysis of the ratio dependent models. For competing predators competitive exclusion principle still holds for most parameter values, it is very often that through the result of the parameter values both can go extinct. In fact, for certain choices of initial values and parameters even the prey species can go extinct it in turn cause the extinction of both predators. Coexistence is possible for some parametric values in both the forms of positive steady state and oscillatory solutions. Most surprisingly when a predator is in a position of driving the prey and itself to extinction, the introduction of a predator which is more friendly to prey and is a stronger as compared to the existing one competitor the prey species may be saved (Berezovskaya et al., 2001).

10 The Law of Mass Action. Basic Concepts (Gorban et al., 2011)

Heat energy flows from a higher temperature region to a lower temperature region. There is no net heat energy flow in the case when both the regions have a same temperature. For example a covered cup of tea will not be colder or warmer then the room temperature after it has been there for a few hours. This phenomenon is known as equilibrium. Equilibrium happens in phase transition. The chemical equilibrium and the law of mass action are the two fundamental concepts of classical chemical kinetics. In a chemical process chemical equilibrium is the state in which concentrations of the reactants and the products have no net change over time. The law of mass action is the mathematical model that explains and predicts the behaviours of the solutions in dynamic equilibrium. This law provides an expression for a constant for all reversible reactions and concerns with the composition of reactions at equilibrium and the rate equitions for elementery reactions. Gulberg and waage (1864-1879) has proved that chemical equilibrium is a dynamic process in which rate of reactions for the forward and backward reactions must be equal. For a chemical reaction the law of mass action was first stated as follows:

" when two reactants A and B react together at a given temperature in a sub situatin reaction the affinity or chemical force between them is proportional to the active masses $[A]$ and $[B]$ each raise to a particular power".

$$\text{Affinity} = \alpha[A]^a[B]^b \tag{16}$$

Here α, a, b were regarded as emperical constants to be determined. In 1867, the rate expression were simplified as the chemical force was assumed to be directly proportional to the product of the active masseses of the reactants.

$$\text{Affinity} = \alpha[A][B] \tag{17}$$

In 1879 this assumption was explained in trems of collision theory so that the general condition for the equilibrium could be applied to any orbitrary chemical equilibrium.

Affinity $= k[A]^{\alpha}[B]^{\beta}$ $\qquad\qquad\qquad\qquad\qquad\qquad\qquad\qquad$ (18)

The exponents α, β, σ and τ are explicitly defined as the "stoichiometric coefficients" for the reactions so that for a general reaction of the type,

$$\alpha A + \beta B + \ldots \rightleftharpoons \sigma S + \tau T + \ldots \qquad\qquad\qquad (19)$$

Forward reaction rate $= k_{+}[A]^{\alpha}[B]^{\beta}\ldots$ $\qquad\qquad\qquad\qquad$ (20)

Backward reaction rate $= k_{-}[S]^{\sigma}[T]^{\tau} + \ldots,$ $\qquad\qquad\qquad$ (21)

where $[A], [B], [S]$ and $[T]$ are active masses and k_{+} and k_{-} are called affinity constants or rate constants. Since at equilibrium the affinities and reaction rates for the forward and backward reactions are equal, so

$$K = \frac{k_{+}}{k_{-}} = \frac{[S]^{\sigma}[T]^{\tau} + \ldots,}{[A]^{\alpha}[B]^{\beta}\ldots} \qquad\qquad\qquad (22)$$

The equilibrium constant K was obtained by setting the rates of forward and backward reactions to be equal. Today the expression for the equilibrium constant is derived by setting the chemical potential of forward and backward reactions to be equal. The units of K depend on the units used for concentrations. If M is used fo all concentrations then K has the units " $M^{(\sigma+\tau)-(\alpha+\beta)}$ " if the system is not at equilibrium, the ratio is different from equlibrium constants. In such cases the tatio is called "reaction quotient" denoted by Q,

$$\frac{[c]^{\sigma}[D]^{\tau}}{[A]^{\alpha}[B]^{\beta}} = Q. \qquad\qquad\qquad (23)$$

A system which is not at equilibrium tends to reach at equilibrium and any changes in the system will cause changes in Q so that the value of the reaction quotient approaches the value of the equilibrium constant K, i.e., $Q \to K$.

For a list of components $A_1, A_2, \ldots, A_n$, where each comoponent has a real variables of particles, and for concentrations $c_1, c_2, \ldots, c_n$, we have algebra of reactions,

$$\alpha_1 A_1 +, \ldots, \alpha_n A_n, \rightleftharpoons \beta_1 A_1 +, \ldots, \beta_n A_n, \qquad\qquad (24)$$

where $\alpha, \beta \geq 0$. we can write the stoichiometric equation for this algebra of reactions as,

$$R_r = k_r \prod_i c_i^{\alpha i}, \qquad\qquad\qquad (25)$$

With stoichiometric vector, $\gamma_{ri} = \beta_{ri} - \alpha_{ri}$,

$$I \;\Big\| \; II$$

$$A^I, C^I, N^I \Big\| A^{II}, C^{II}, N^{II}$$

Fig. 1 Cell jump model.

11 A Simple Cell Jump Model

The standard form of reaction diffusion equation is,

$$\frac{\partial c_i}{\partial t} = f(c) + d_i \Delta c, \tag{26}$$

where $d_i > 0,$ is diffusion coefficient and c should be literally small for very dialuted media (Gorban et al., 2011).

The thermodynamic ideality is the general requirement of a system to apply the law of mass action. Let us consider our space divided on cells, a system represented as a chain of cells each with homogeneous composition and elementary acts of transfers on the boundary (for us there are only two cells). On each cell we have some concentration for these processes. Let $c^I,$ is the vector concentration in the first cell I and $c^{II},$ is the vector of concentration in the second cell II .using the general secheme of the formal kinetics which is considered to be complete if we work with list of macroscopic variables with balanced equations, a mechanism of transformation (elementary process) and the functions of the rates of these elementary processes. The mechanism of diffusion is defined as a list of elementary transitions between cells described by their stoichiometric equation. Since diffusion is a sort of jumping reaction on the border, so for these jumps the stoichiometric equation is written as,

$$\sum_i \alpha_{ri}^I A_i^I + \sum_i \alpha_{ri}^{II} A_i^{II} \rightarrow \sum_i \beta_{ri}^I A_i^I + \sum_i \beta_{ri}^{II} A_i^{II}, \tag{27}$$

where r is the number of processes, $\alpha_{ri}^{I,II},$ and $\beta_{ri}^{I,II}$ are the stoichiometric coefficients which indicate the number of particles in cells involved in the process, where each process is is compared with the components of the stoichiometric vector defined by $\gamma_{ri}^I = \beta_{ri}^I - \alpha_{ri}^I,$ also there should be sysymmetry between these cells, i.e. $A_i^I \rightleftharpoons A_i^{II}$, and diffusion coefficient d_i depends on the cell. Each elementary process is characterized by an intensive quantity, termed as velocity function $\omega_r(c^I, c^{II})$, (the number of elementary acts in a unit time per unit surface of the cell). In the absence of convection and any chemical transformation the equation of kinetics for the vector quantities, in isotropic and isothermal enviroment, is written as the equation of rate of reaction,

$$\frac{dN^I}{dt} = \frac{dN^{II}}{dt} = S \sum_r \gamma_r \omega_r, \tag{28}$$

where S is the area of the boundary between two cells. The i-flux density of a substance through a unit surface of section is,

$$J_{ri}(I, II) = -\gamma_{ri} \omega_r(c^I, c^{II}). \tag{29}$$

As a result of elementary acts of diffusion (27), the density of the total flux of the i-substance will be,

$$J_i = \sum_r J_{ri}(I, II) = -\sum_r \gamma_{ri} \omega_r, \tag{30}$$

And the velocity function in the form of mass action law can be written as,

$$\omega(c^I, c^{II}) = k_r \prod_i (c_i^I)^{\alpha_{ri}^I} (c_i^{II})^{\alpha_{ri}^{II}}, \tag{31}$$

where $k_r \geq 0$ is the rate constant depending upon the kinetic factors of the investigated system and its thermodynamic properties. It is necessary to fulfil the no advection condition:

$$\sum_r \gamma_r \omega_r(c,c) = 0, \tag{32}$$

which means the absence of flux at $c^I = c^{II}$, in a homogeneous composition enviroment. Introducing flux

J_{ri} as the first approximation of the Taylor series expansion in ω Subject to the condition (0.2.47),

$$\overrightarrow{J_{ri}} = -\gamma_{ri} \sum_j \frac{\partial \omega_r(c^I, c^{II})}{\partial c_j^{II}} \bigg|_{c^I = c^{II} = c(x)} . \vec{\nabla} c_j. \tag{33}$$

This expression represents the transition from flux density between the cells to the vector of the flux density. It should be noted that model will work well, especially in the selection of the suitable relationship between size of unit cell and diffusion zone (Gorban et al., 2011).

Gorban et al. (2011), Kuttler (2011), and Murray (2003) discussed a few diffusion reactions mechanisms for cell jump models and determined their reactions rates, vectors of total flux density, and their diffusion equations by using the law of mass action along with principle of detailed balance. Some of the mechanisms are presented here.

12 Models of Non Linear Diffusion

The simplest mechanism of diffusion between any two cells is the process of jumping from one cell to another. This type of mutually inverted and mutually inverse process can be written as (Gorban et al., 2011),

$$\begin{aligned} A^I &\to A^{II} \\ A^{II} &\to A^I \end{aligned} \tag{34}$$

12.1 Mechanism of sharing place

The simplest mechanism of diffusion with interaction of n different substances of a multi component system is given by stoichiometric equation of the form (Gorban et al., 2011):

$$\begin{aligned} A_i^I + A_j^{II} &\to A_j^I + A_i^{II} \\ A_i^{II} + A_j^I &\to A_i^I + A_j^{II} \end{aligned} \tag{35}$$

12.2 Diffusion of mechanisms of attraction and repulsion

The diffusion mechanisms leading to the inhomogeneous structures can be described by the multi solution immersed into the external conditions under which non linear mechanisms are possible with the stoichiometric equations, describing the mechanism of attraction (Gorban et al., 2011):

Repulsion:

$$\begin{aligned} 2A^I &\to A^I + A^{II} \\ 2A^{II} &\to A^I + A^{II} \end{aligned} \tag{36}$$

Attraction:

$$\begin{aligned} A^I + A^{II} &\to 2A^I \\ A^I + A^{II} &\to 2A^{II} \end{aligned} \tag{37}$$

12.3 Autocatalysis mechanisms

A process where a chemical is involved in its own production is called autocatalysis. Feedbacks controls exist into many biological systems. One must be familiar with how to model them because they have enormous importance. In 1978, Tyson and Othmer introduced the dynamics of metabolic feedback control systems and a theoretical models review. A process when the product of one step in a reaction sequence has an effect on other reaction steps in the sequence. The effect is generally nonlinear and may be to activate or inhibit these reactions. A very simple pedagogical example is,

$$A + Y \underset{k_2}{\overset{k_1}{\rightleftharpoons}} 2Y, \qquad (38)$$

where a molecule of Y combines with one of A to form two molecules of Y (Murray, 2003).

13 Foxes and Rabbits Interaction- A Non-Linear System

The humans are and they make open systems. Open systems interactions with the other open systems in an immeasurable set of connections of open systems. Open systems become accustomed to these connections such that their behaviour is non linear in general. It is difficult to predict the behavior of our artificial systems when they are introduced into some operational atmosphere.

It is a constant battle among the animals on a daily basis to survive. They have to find their food and avoid becoming a food. These species can be divide up on the basis that how they get their food, if they provide it they are producers (prey), if they need to find it they are consumers (predator) and if they breakdown dead material they are decomposers.

This correlation between predator and prey intertwines into a complex food chain. In the ecosystem there are various species and every one play a significant role in maintaining populations near the carrying capacity and in keeping the system in balance. In the ecosystems Predator and prey assist them through particular adaptations to compete for food resources. Predator and prey populations are directly related and they cannot survive without each other. Here this relationship is illustrated by using foxes and rabbits.

In artificial world and in the social systems non-linear behaviors are the order of the day. Rabbits and foxes share some behaviors but vary in others and their interaction is non-linear in nature.

The rabbits are proverbially very good in reproduction. In case when more rabbits are there, the more foxes will prey on them, so that the number of foxes will increase, when more foxes will predate a lot such that the number of rabbits start to decrease, swiftly it follows by falling fox numbers as there are a small amount of rabbits to sustain them therefore it is observed that the populations of rabbits and foxes are locked together in an interactive population 'dance.' This system is non-linear in its behavior. In mathematics non-linear simultaneous equations are used to represent these interactions for which there exist an infinite number of solutions.

Between the fox and rabbit populations here is a complicated and natural predator-prey relationship, since rabbits thrive in the absence of foxes and foxes thrive in the presence of rabbits. Foxes like to occupy a combination of forest and open fields. Foxes usually define their territory zones and use the transition zone or "edge" between these habitats as hunting areas. Foxes do not interfere in the defined areas of each others.

13.1 Predator prey interactions

In this idealized ecological system, two populations are considered of which one may be hunted by the second.

- Each prey gives rise to a constant number of off-spring preys each year.
- A constant proportion of the prey population is hunted by each predator each year.
- Predator reproduction is directly proportional to the constant proportion of the prey population hunted by each predator.

- A constant proportion of predator population dies each year.

If R is the number of prey population (rabbits) and F is the number of predator population (foxes), then conventionally this model has a mathematical representation as,

$$\frac{dR}{dt} = AR - BRF,$$

$$\frac{dF}{dt} = CRF - DF. \tag{39}$$

The above equations are called Lotka-Volterra equations or the predator-prey equations, which are used to describe the dynamics of biological systems in which two species interact, one is a Predator-the hunter, and other is its prey-the hunted.

Here the derivatives $\frac{dR}{dt}$ and $\frac{dF}{dt}$ represent the growth of the two populations with respect to time and

A, B, C and D are parameters representing the interaction of the two species in the following manner:

- A - The number of offspring per prey per year.
- B - The proportion of the prey population hunted by one predator per year.
- C - Conversion of one prey hunted and consumed by the predator into new predators.
- D - The proportion of predator population dying per year.

Physically, the Lotka-Volterra model requires a number of assumptions about the environment and evolution of the predator and prey populations such as food availability at all times to predator populations depending entirely upon the prey populations, rate of change of population which is proportional to the size of the population and an unchanged environment in favour of one species during the process.

13.2 Rabbits on the island

Taking an island inhabited by a small population of rabbits (prey) with no other animals around. The prey are assumed to have an unlimited food supply and the number of rabbits born and die in a certain period is proportional to the total number of rabbits at that time, then the change in their population can be written as:

$$\frac{dR}{dt} = rR \tag{40}$$

where R is the number of rabbits at time t and r is the growth rate of the rabbit population defined as the difference between birth and death rates. The equation has a solution as:

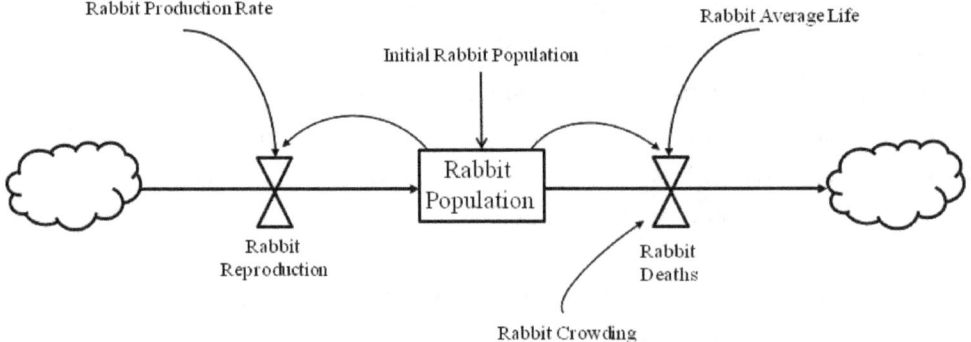

Fig. 2 Single population model-1.

$$R = R_0 e^{rt} \tag{41}$$

where R_0 is the initial population of the rabbits (prey) at time $t = 0$. This model is also called exponential growth model due exponential factor involved. The exponential growth rate will be positive if births are more than deaths and negative if deaths are dominant.

The exponential model is not very realistic because with a negative exponential growth rate, there will not be any life left on the island. On the other hand, if the growth rate is positive, the rabbits will cover the entire island with enough time. This will not happen actually because rabbits will run out of space and food due to maximum carrying capacity of island. This kind of problem was handled through the law of population growth derived by Alfred J. Lotka in 1925 by adding the concept of carrying capacity K to the exponential model. This model is called logistic model formalized by the differential equation:

$$\frac{dx}{dt} = rx\left(1 - \frac{x}{k}\right) \tag{42}$$

Clearly when $R < K$, and $R \to k$ the equation (42) is close to the equation (40) and rabbit population stops growing. If $R > K$, then there will be a crowding of rabbit population on the island and from equation (42) a negative sign will reduce the population growth to a manageable level. The behaviour of the model can be predicted through the growth rate r.

13.3 Fox as a predator

In the presence of hundreds of different animals on the island, the modeling of the interaction among these animals is not an easy task. However, the interaction between two species is possible taking one as a prey and the other as a predator. The same single population growth model can be used for predators (foxes), if prey (rabbits) is considered as an unlimited food supply to foxes. But Foxes are territorial animals, and in general each fox claims its own territory by marking their territories with signals that other foxes will recognize e. g., by leaving their droppings in prominent positions. They pair up only in winter. So it will be difficult for the foxes to cover the whole island without defining their own territories.

In order to describe the interaction between a predators (foxes) and its prey (rabbits), introduce a small number of fox population on the island with rabbits having unlimited food supply and following the equation (42). If R represents the rabbit's population and F is the fox's population, then population rates for each species will be affected and described by the following equations as:

$$\frac{dR}{dt} = aR - bRF,$$
$$\frac{dF}{dt} = cRF - dF. \tag{43}$$

where a, b, c and d are parameters as defined for predator (fox) and its prey (rabbits). The additional terms will now be defined as:

- $-bRF$ Rabbits hunted by the foxes, reducing the rabbit population
- $+cRF$ Fox population growth by eating the rabbits

Since some of the rabbits will be hunted by the foxes (as a food), so the birth rate r is not the only factor describing the rabbit population. Without the presence of foxes, i.e., $F = 0$, the equation (43) will reduce to a simple exponential model (40).

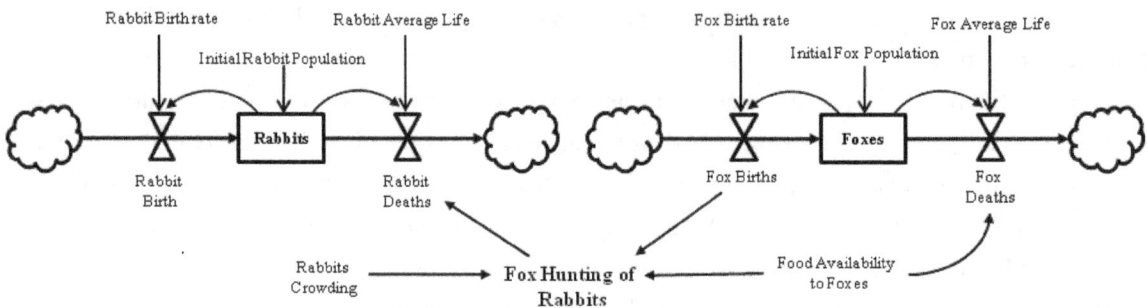

Fig. 3 Model structure.

On the other hand, without food supply $(R = 0)$ to foxes, the equation (43) will reduce to:

$$\frac{dF}{dt} = -dF \qquad (44)$$

Which means foxes will die in turn if they do not get enough food for their survival. In this case, the rabbit population will begin to increase in the absence of prey. The populations of foxes will again starts increasing with the increase in rabbit population, and this cycle will continue.

13.4 Basic reactions of foxes and rabbits

Let us suppose that our space is divided on two territories, a system represented as a chain of territories each with homogeneous composition and elementary acts of transfers on the boundary (for us there are only two territories). On each territory we have some concentrations (number of foxes and rabbits) for these processes. Let F^I (fox/ foxes in first territory) and R^I (rabbit/ rabbits in first territory) is the vector concentration in the first territory I and F^{II} (fox/ foxes in second territory) and R^{II} (rabbit/ rabbits in second territory) is the vector of concentration in the second territory II.

In any territory when foxes and rabbits both are present, the following basic mechanisms are possible in that situation;

1.   When in any territory a fox (F) and total n number of rabbits (R) are present, let us

suppose that r is the number of rabbits which fox (F) prey in that territory. When fox will prey rabbits (R),

rabbit's population will start to decrease. The rabbits are proverbially very good in reproduction so their population will increase again; in this case the possible mechanism will be,

$$F + (n)R \rightarrow F + (n-r)R, r < n \qquad (45)$$

2.   Now in the case of foxes, when two foxes, a male fox $F_{(m)}$ and a female fox $F_{(f)}$ interact, let

n be the number of foxes which born as the result of that interaction, they all can be males, females or a combination of both $F_{(m+f)}$, this circle repeats itself; in this case the possible mechanism will be,

$$F_{(m)} + F_{(f)} \rightarrow (n)F_{(f+m)} \qquad (46)$$

3.   Now in the case of rabbits, when two rabbits, one male $R_{(m)}$ and the other female rabbit $R_{(f)}$ interact, let n be the number of rabbits which born as the result of that interaction, they can be males, females or a combination of both sexes $R_{(m+f)}$, this circle repeats itself; in this case the possible mechanism will be,

$$R_{(m)} + R_{(f)} \rightarrow (n)R_{(m+f)} \tag{47}$$

13.5 Mechanism of circulation

The simplest mechanism of circulation between any two foxes or rabbits is the process of moving one fox or rabbit from area defined for it, to the area defined for another fox or rabbit and vise versa. This type of mutually inverted and mutually inverse processes for the case of foxes can be written as,

$$
\begin{aligned}
F^{I} &\rightarrow F^{II} \\
F^{II} &\rightarrow F^{I}
\end{aligned} \tag{48}
$$

This mechanism for the case of rabbits will be written as,

$$
\begin{aligned}
R^{I} &\rightarrow R^{II} \\
R^{II} &\rightarrow R^{I}
\end{aligned} \tag{49}
$$

As it is stated in the early lines that foxes are territorial and stay in their own territorial zones, the above mechanism may not be possible in practical in the case of foxes, but rabbits move freely in nature without any restrictions so this mechanism may or may not be possible in that case.

13.6 Mechanism of sharing place

1.   The mechanism of sharing place with interaction of n different foxes or rabbits i.e. when a fox F_i^{I} is in territory one and another fox F_j^{II} is in territory two and then fox F_i^{I} moves from territory one to territory two and vise versa, in a multi component system is given by stoichiometric equations of the form:

$$
\begin{aligned}
F_i^{I} + F_j^{II} &\rightarrow F_j^{I} + F_i^{II} \\
F_i^{II} + F_j^{I} &\rightarrow F_i^{I} + F_j^{II}
\end{aligned} \tag{50}
$$

Practically in the case of foxes the above mechanism seems to be almost impossible because foxes usually define their territory zones and use the transition zone or "edge" between these habitats as hunting areas. Foxes do not interfere in the defined areas of each other's and foxes fight other foxes if they find them in their territory.

2.   In the case of rabbits it may or may not be possible because rabbit's movement is free naturally without any restrictions. In that case the possible mechanisms will be,

$$R_i^{I} + R_j^{II} \rightarrow R_j^{I} + R_i^{II} \tag{51}$$

$$R_i^{II} + R_j^{I} \rightarrow R_i^{I} + R_j^{II} \tag{52}$$

3.   Now in the case of mix situation i.e. by taking foxes and rabbits case together,

$$F_i^{I} + R_j^{II} \rightarrow F_i^{I} + R_j^{I} \rightarrow F_i^{I} \tag{53}$$

$$F_i^{II} + R_j^{I} \rightarrow F_i^{II} + R_j^{II} \rightarrow F_i^{II} \qquad\qquad (54)$$

As the foxes are territorial while the rabbits move freely, in the above mechanism (53) it is described that when a fox is in territory one and a rabbit is in territory two, if rabbit move from territory two to territory one, the fox in that area will prey that rabbit and a similar case happens in mechanism (54) for second territory.

13.7 Mechanism of attraction and repulsion

1.   Foxes are solitary and necessitate quite huge hunting areas. A fox constantly patrols its territory looking for food, using its urine to mark places it has completed searching. Foxes are territorial and fight other foxes that they find on their territory. Because they wander over such a wide area, foxes maintain several burrows and dens across their territory. The possible mechanisms of attraction and repulsion for the foxes and rabbits can be;

Repulsion:

$$2F^{I} \rightarrow F^{I} + F^{II}$$
$$2F^{II} \rightarrow F^{I} + F^{II} \qquad\qquad (55)$$

Attraction:

$$F^{I} + F^{II} \rightarrow 2F^{I}$$
$$F^{I} + F^{II} \rightarrow 2F^{II} \qquad\qquad (56)$$

Both the mechanisms of attraction are unnatural in the case of foxes because foxes are territorial and fight other foxes that they find on their territory so two or more foxes cannot stay in the same territory thus we will have to move towards mechanisms of repulsion consequently there will be no attraction in the case of foxes.

2.   Now taking the mechanisms of attraction and repulsion for the case of rabbits the following mechanisms are possible,

Attraction:

$$R^{I} + R^{II} \rightarrow 2R^{I}$$
$$R^{I} + R^{II} \rightarrow 2R^{II} \qquad\qquad (57)$$

Repulsion:

$$2R^{I} \rightarrow R^{I} + R^{II}$$
$$2R^{II} \rightarrow R^{I} + R^{II} \qquad\qquad (58)$$

In this case the above stated both the mechanisms of attraction and repulsion may or may not be possible because rabbits move freely in nature without any restrictions. If two rabbits are in two different territories and they both move in the same territory then it is described by mechanisms of attraction and if two rabbits are in the same territory and one of them leaves that territory and moves to another territory in that case, it is described by mechanisms of repulsion.

3.   Now for the mix situation i.e. when in any territory the mechanisms of attraction and repulsion are considered for both, the foxes and the rabbits at a time, the following mechanisms for attraction and repulsion are possible,

Attraction:

$$R^{I} + F^{I} \rightarrow F^{I}$$
$$R^{II} + F^{II} \rightarrow F^{II} \qquad\qquad (59)$$

In attraction mechanism when both the rabbit and the fox are in the same territory, then fox will prey the rabbit and the result will be the fox.

Repulsion:

$$R^I + F^I \rightarrow F^I + R^{II}$$
$$R^{II} + F^{II} \rightarrow F^{II} + R^I$$

(60)

In repulsion mechanism when both the rabbit and the fox will be in the same territory then it is possible that a rabbit may change its territory but fox does not do so because foxes are territorial. When rabbit in the first territory will change its territory then the fox of that territory will not be in a position to prey that rabbit so it will survive that fox but fox present in that territory will be in a position to prey that rabbit and vise versa.

13.8 Pair wise interaction

In pair wise interaction the following mechanisms are possible because foxes are territorial while rabbits move freely in nature and rabbits change their territories regularly while foxes does not do so. In this situation the possible mechanisms will be;

$$(R^I + F^I) + (R^{II} + F^{II}) \rightarrow (R^{II} + F^I) + (R^I + F^{II})$$

(61)

$$(R^I + F^I) + (R^{II} + F^{II}) \rightarrow (R^I + R^I + F^I) + F^{II}$$

(62)

$$(R^I + F^I) + (R^{II} + F^{II}) \rightarrow F^I + (R^{II} + R^{II} + F^{II})$$

(63)

In the above mechanism (61) a rabbit and a fox are present in territory one and a similar situation is in the territory two. In mechanism one, rabbit from territory one moves to territory two and vise versa. Other possibilities are that both the rabbits move to territory one or both of them move to territory two. These possibilities are described in mechanism (62) and (63) respectively.

13.9 Autocatalysis *mechanisms*

Autocatalysis is the procedure whereby a chemical is involved in its own production. In the following mechanisms, in the case of foxes, when two foxes, a male fox $F_{(m)}$ and a female fox $F_{(f)}$ interact, let n be the number of foxes which born as the result of that interaction, they all can be males, females or a combination of both $F_{(m+f)}$, this circle repeats itself; in this case the possible mechanism will be,

$$F_{(m)} + F_{(f)} \rightarrow (n)F_{(f+m)}$$

(64)

Now in the case of rabbits, when two rabbits, one male $R_{(m)}$ and the other female rabbit $R_{(f)}$ interact, let n be the number of rabbits which born as the result of that interaction, they can be males, females or a combination of both s $R_{(m+f)}$, this circle repeats itself; in this case the possible mechanism will be,

$$R_{(m)} + R_{(f)} \rightarrow (n)R_{(m+f)}$$

(65)

14 The Observations

The sense in which the discrete equations for cells converge to the partial differential equations of diffusion is that the cell models give the semi-discrete approximation of the partial differential equations for diffusion.

They result in a system of ordinary differential equations in cells. Such approximation appears often in finite element methods and cells are discontinuous finite elements. There are jumps of concentrations on the boundary of cells. The Taylor expansion of the right hand sides of the discrete system of ordinary differential equations for cells produces the second order in the cell size approximation to the continuous diffusion equation (the standard result for the central differences).

Significant difference from the classical finite elements is in construction of right hand sides of the ordinary differential equations for concentrations in cells: there are flows in both directions: from cell 1 to cell II and from cell II to cell I. These flows have a simple mass action law construction and the resulting diffusion flow is the difference between them. The kinetic constants should be scaled with the cell size to keep $kl = d$ (k is the kinetic constant, l is the cell size, d is the diffusion coefficient for the particular mechanism).

This is the approximation of the right hand sides. The approximation of solutions is a more difficult problem and depends upon the properties of solutions of PDE. It seems a good hypothesis that for obtained diffusion systems with convex Lyapunov functional and "no-flux" boundary conditions in bounded areas with smooth boundaries the cell model gives the uniform approximation to the solution of the correspondent PDE.

The system of semi-discrete models for cells with "no-flux" boundary conditions has all the nice properties of the chemical kinetic equations for closed systems. Under the proper relations between coefficients, like complex balance or detailed balance, this system demonstrated globally stable dynamics. This global stability property can help with the study of the related PDE.

Finally, the cell-jump models may be considered as the proper diffusion models by themselves, for the finite physically reasonable cell size, without limit. This size may be quite large for the coarse-grained models (it depends on the medium microstructure and on the smoothness of the concentration fields).

14.1 For mechanisms of circulation

For this mechanism,

$$F^I \to F^{II}$$
$$F^{II} \to F^I$$
$$R^I \to R^{II}$$
$$R^{II} \to R^I$$

It is stated in the early lines that the above stated mechanism is not possible in the case of foxes because the foxes are territorial and does not change their territory while it may or may not be possible in the case of rabbits because their movement is free naturally and if this mechanism occurs in their case, then the equation of kinetics for the above mechanism the condition of absence of flux in a homogenous environment with the rate of constant of direct and inverted processes as $k_i^+ = k_i^- = k_i$ give the following form of the diffusion equation,

$$\frac{\partial c_i}{\partial t} = k_i \Delta c_i.$$

14.2 For sharing place mechanism

Sharing place mechanism is also not possible in the case of foxes due to the fact that the foxes are territorial and does not change their territory while it may or may not be possible in the case of rabbits because their movement is free naturally and if this mechanism occurs in their case, then for the mechanisms of sharing place the diffusion equations for the above mechanism using stoichiometric vectors $\vec{\gamma}_{ri}$, fluxes of the substances F_i and F_j and the functions w_1 and w_2 for the first and second elementary process are calculated as,

For process:
$$F_i^I + F_j^{II} \rightarrow F_j^I + F_i^{II}$$

$$\gamma_{1i} = -1 \qquad \gamma_{1j} = 1 \qquad w_1(c^I, c^{II}) = k_1 c_i^I c_j^{II}$$

$$\vec{J}_{1i} = k_1 c_i \vec{\nabla} c_j \qquad \vec{J}_{1i} = -k_1 c_i \vec{\nabla} c_j$$

$$w_1 = div\left(k_1 c_i \vec{\nabla} c_j \right) = k_1 \left(c_i \Delta c_j + (\vec{\nabla} c_i, \vec{\nabla} c_j) \right)$$

For process:
$$F_i^{II} + F_j^I \rightarrow F_i^I + F_j^{II}$$

$$\gamma_{2i} = 1 \qquad \gamma_{2j} = -1 \qquad w_1(c^I, c^{II}) = k_k c_j^I c_i^{II}$$

$$\vec{J}_{2i} = -k_2 c_j \vec{\nabla} c_i \qquad \vec{J}_{1i} = k_2 c_j \vec{\nabla} c_i$$

$$w_2 = div\left(k_2 c_j \vec{\nabla} c_i \right) = k_2 \left(c_j \Delta c_i + (\vec{\nabla} c_i, \vec{\nabla} c_j) \right)$$

The equation for mutual diffusion for ith and jth substances was determined as,

$$\frac{\partial c_i}{\partial t} = k(c_j \Delta c_i - c_i \Delta c_j)$$

$$\frac{\partial c_j}{\partial t} = k(c_i \Delta c_j - c_j \Delta c_i)$$

The above equation describes the diffusion on the surface of the catalyst for the mechanism of jumping to a neighboring free space (Gorban et al., 2011).

15 Our Objectives

Our goals are:

(a)   To build up a brief study of complex biological systems by taking a study case of foxes and rabbits.

(b)   To tackle key research questions about our case study by proposing new techniques and algorithms that are inspired by those complex biological systems.

Further for our case study we want to study, and aim to extend these ideas to all other possible mechanisms complete in all aspects between foxes and rabbits.   We aim a comparative study for all the possible mechanisms like, mechanism of attraction and repulsion, pair wise attraction, and for autocatalysis mechanisms between foxes and rabbits and want to determine their reactions rates, vectors of total flux density, and their diffusion equations by using the law of mass action along with the principle of detailed balance and aim to check their stability for these interactions.

References

Abrams PA, Ginzburg LR. 2000. The nature of predation: prey dependent, ratio-dependent or neither? Tree, 15: 337-341

Alabdullatif M, Abdusalam HA, Fahmy ES. 2007. Adomian Decomposition Method for Nonlinear Reaction Diffusion System of Lotka-Voltera Type. International Mathematical Forum, 2: 87-96

Berezovskaya F, Karev G, Arditi R. 2001. Parametric analysis of the ratio-dependent predator-prey model. Journal of Mathematical Biology, 43(3): 221-246

Beretta E, Kuang Y. 1998. Global analyses in some delayed ratio-dependent predator-prey systems. Nonlinear Analysis, 32: 381-408

Kuttler C. 2011. Reaction-Diffusion Equations with Applications. Sommersemester.

Douglas 1t. Boucher. 1982. The Ecology of Mutualism. 3: 315-347

Gorban AN, Sargsyan HP, Wahab HA. 2011. Quasichemical models of multicomponent nonlinear diffusion. Mathematical Modelling of Natural Phenomena 6(5): 184-262. E-print: arXiv: 1012.2908 [cond-mat.mtrl-sci].

Hsu SB, Hwang TW, Kuang Y. 2001. Global analysis of the michaelis-menten type ratio-dependent predator-prey system. Journal of Mathematical Biology, 42: 489-506

Hsu SB, Hwang TW, Kuang Y, 2001 Rich dynamics of a ratio-dependent one-prey two-predators model. Journal of Mathematical Biology, 43: 377-396

Lotka AJ. 19200. Undamped oscillations derived from the law of mass action. Journal of American Chemical Society, 42: 1595

Murray JD. 2003. Mathematical Biology I: An Introduction. Springer-Verlag

Voltera V. 1926.Variations and fluctuations of the number of the individuals in animal species living together. In: Animal Ecology (Chapman RN, ed). McGraw-Hill, New York, USA

Wahab HA, Shakil M, Khan T, et al. 2013.A comparative study of a system of Lotka-Voltera type of PDEs through perturbation methods, Computational Ecology and Software, 3(4): 110-125

14

Regression modeling of different proteins using linear and multiple analysis

Shruti Jain

Department of Electronics and Communication Engineering, Jaypee University of Information Technology, Solan-173234, India

E-mail: jain.shruti15@gmail.com

Abstract

There are different types of regression analysis. Out of which simple regression and multiple regressions was considered in this paper. For calculation purpose we have used PLS analysis which calculates squared r values. This paper considers eleven different proteins and one output. We have validated our results by calculating adjusted regression coefficient, predicted regression coefficient regression coefficient cross validation, rm^2 and F-test values. Later multiple regressions were used as we have different independent variable (proteins). For that analysis we have calculated the coefficient, standard error, standard coefficient, tolerance, t value and p value, variation explanation of predictors and estimators which gives percentage and cumulative percentage. Correlation matrixes were also shown at the end for eleven proteins and one output.

Keywords linear regression analysis; multiple regression analysis; marker proteins; PLS.

1 Introduction

Regression analysis (RA) is a statistical method for investigating relationship between one variable and other variables (Farahani, 2010; Ringle, 2010). A statistical model is a simple description of state. There are three types of RA: linear regression (LR), multiple linear regression (MLR) and non linear regression (NLR). If we have to model the linear relationship between dependent and independent variables than LR was used but if there are more than one independent variable and one dependent variable than MLR is used. The MLR considers co linearity, variance inflation, graphical display of regression diagnosis, and detection of regression outlier and influential observation. In NLR the variables (dependent and independent) are not linear. NLR can be written as

$$y = \frac{\alpha}{1 + e^{\beta t}} + \varepsilon$$

(1)

where y is the growth of a particular organism as a function of time t, α and β are model parameters, and ε is the random error. NLR model is more complicated than LR model in terms of estimation of model parameters, model selection, model diagnosis, variable selection, outlier detection, or influential observation identification. In this paper we have calculated regression coefficient or regression coefficient cross validation (r^2 or q^2_{cv}), adjusted regression coefficient (r^2_{adj}), predicted regression coefficient (r^2_{pre}), regression coefficient without intercept (r^2_0) for ten different concentration (Jain, 2012a; Suzzane, 2005; Weiss, 2001) of three input proteins TNF (Thoma, 1990; Jain, 2009a, 2009b, 2010a, 2011a), EGF (Janes, 2005; Normano, 2006; Jain, 2014, 2015a, 2016a, 2017) and Insulin (Jain, 2010b, 2010c, 2011b, 2012b, 2012c; Morris, 2003). For validation of our results we have calculated rm^2 and F-test values. Different plots were plotted which are showing r^2 values. Later in paper we have shown the results of multiple regression. We have different marker proteins: AkT (Coffer, 1998; Hemmings, 1997; Bruent, 1999; Jain, 2010d, 2012d, 2015b, 2017b), MK2 (Jain, 2011c, 2016b, 2016c), JNK (Jain, 2010e, 2015b, 2015c), FKHR (Jain, 2015c, 2011d), MEK, ERK, IRS, IKK, pAkT, ptAkT and EGFR for the HT carcinoma cells which helps in cell survival/ apoptosis. If these proteins are present in the pathway than it leads to cell survival otherwise cell death (Jain, 2009c).

2 Material and Methods

There are different types of regression analysis. In this paper we are working on Linear Regression/ Simple regression (LR) and Multiple regression analysis (MR). For calculation of LR there are two techniques: Ordinary least square method (OLS) and Partial least square method (PLS). PLS is a technique which is helpful in predictive models when the factors are many and highly collinear. PLS approach is beneficial for relating dependent variables to many independent variables.

LR as the name suggests the shape of regression line is linear whose intercept is a and slope of line is b. For LR the dependent variable (Y) is continuous while independent variable (X) can be continuous or discrete. We can consider the error term 'e or ε' . LR is expressed as:

$$\underbrace{Y}_{actual\,(observed)} = \underbrace{aX+b}_{explained\,(predicted)} + \underbrace{\varepsilon}_{error} \tag{2}$$

Equation 1 is also known as linear population regression model, or linear population regression. For LR error ε is normally distributed with $E(\varepsilon) = 0$ and a constant variance $Var(\varepsilon) = \sigma^2$. LR can be also be represented as :

$$\underbrace{Y}_{observed} = \underbrace{\hat{Y}}_{predicted\,/\,estimator} + \underbrace{\hat{\varepsilon}}_{error\,/\,residual} \tag{3}$$

Predicted value is also known as conditional mean. For predicted values equation 2 can be written as:

$$\hat{Y} = \hat{a}X + \hat{b} \tag{4}$$

Equation 4 is known as sample regression function (SRF) where intercept is represented by equation 5.

$$\hat{b} = \bar{Y} - \hat{a}\bar{X} \tag{5}$$

$\bar{X}$ and $\bar{Y}$ are the sample means of X and Y. Slope is represented as:

$$\hat{a}=\frac{SS_{XY}}{SS_{XX}}=\frac{\sum(X-\bar{X})(Y-\bar{Y})}{\sum(X-\bar{X})^2}=\frac{\sum xy}{\sum x^2}=\frac{Cov(X,Y)}{Var(X)} \tag{6}$$

where

$$Cov(X,Y)=Var(X,Y)=\frac{\sum(X-\bar{X})(Y-\bar{Y})}{n-1}$$

and

$$Var(X)=\frac{\sum(X-\bar{X})^2}{n-1} \quad Var(Y)=\frac{\sum(Y-\bar{Y})^2}{n-1},$$

equation 6 can be written as

$$\hat{a}=\frac{\sum xY}{\sum X^2-n\bar{X}^2}=\frac{\sum Xy}{\sum X^2-n\bar{X}^2} \tag{7}$$

We can define $x=(X-\bar{X})$ and $y=(Y-\bar{Y})$ where lower case letters denote deviations from mean values.

OLS method and PLS is used for calculation of LR. OLS minimizes the SS of the vertical deviations from each data point to the line while in PLS, initially the values are squared, then added up so as there is no cancellation of positive and negative terms. Finally, the minimum (least) square value is considered.

We have three different types of sum of squares (SS):

a. regression sum of squares (SSreg) / explained SS which is a measure of explained variation,

$$SS_{reg}=\sum(\hat{y}-\bar{y})^2 \tag{8}$$

b. residual sum of squares or error sum of squares (SSerr) / unexplained SS which is a measure of unexplained variation and

$$SS_{err}=SS_{residual}=\sum(y-\hat{y})^2=\sum\hat{\varepsilon}^2 \tag{9}$$

c. total sum of squares (SS_{total}) which is a measure of total variation.

$$SS_{total}=SS_{reg}+SS_{err} \tag{10}$$

$$SS_{total}=\sum(y-\bar{y})^2 \tag{11}$$

The ratio of SS_{reg} to SS_{total} is known as coefficient of determination (r^2) is expressed as:

$$r^2=\frac{SS_{reg}}{SS_{total}}=\frac{SS_{reg}}{SS_{reg}+SS_{err}}=1-\frac{SS_{err}}{SS_{total}} \tag{12}$$

For deviation form, the SRF can be written as:

$$= \frac{\sum \hat{y}^2}{\sum y^2} \tag{13}$$

$$= \frac{\hat{a}^2 \sum x^2}{\sum y^2} \tag{14}$$

$$= \hat{a}^2 \left(\frac{\sum x^2}{\sum y^2} \right) \tag{15}$$

If numerator and denominator are divided by sample size 'n' than r^2 is expressed as:

$$r^2 = \hat{a}^2 \left(\frac{S_x^2}{S_y^2} \right) = a^2 \left(\frac{Var\ X}{Var\ Y} \right) \tag{16}$$

where S_x^2 and S_y^2 is a sample variance of X and Y respectively.

Replace the value of $\hat{a}$ in equation 15 by equation 6; we get

$$r^2 = \frac{\left(\sum xy \right)^2}{\sum x^2 \sum y^2} \tag{17}$$

Or
$$r = \frac{\left(\sum xy \right)}{\sqrt{\sum x^2 \sum y^2}} = \frac{Cov(X,Y)}{\sqrt{Var(X).Var(Y)}} \tag{18}$$

In above equation; r is known as sample/linear correlation coefficient. Equation 12 can be written as

$SS_{reg} = r^2 SS_{total}$ where

$$SS_{total} = \sum y^2 \tag{19}$$

Equation 10 can be written as $$SS_{reg} = r^2 \sum y^2 \tag{20}$$

$$SS_{err} = SS_{total} - SS_{reg}$$

Placing values from eq 19 and eq 20 to equation 10 we get:

$$SS_{err} = \sum y^2 \left(1 - r^2 \right) \tag{21}$$

Finally, placing all the values i.e. from eq 19 , eq 20 and eq 21 in equation 120 we get

$$\sum y^2 = r^2 \sum y^2 + \sum y^2 \left(1 - r^2 \right) \tag{22}$$

The r^2 lies in the range of 0 and 1, greater the value of r^2 more accurate the model. If the value of r^2 is 1 means a perfect fit on the other hand if r^2 value is zero it means that there is no relationship between regress and regressor. r^2 is a measure of goodness of fit which tells how close the estimate values are to their actual values. r^2 can also be calculated as the squared coefficient of correlation between actual Y, estimated Y i.e. $\hat{Y}$ and is expressed as

$$r^2 = \frac{\sum\left(Y-\overline{Y}\right)^2\left(\hat{Y}-\overline{Y}\right)^2}{\sum\left(Y-\overline{Y}\right)^2\sum\left(\hat{Y}-\overline{Y}\right)^2}$$

(23)

$$= \frac{\left(\sum y\,\hat{y}\right)^2}{\sum y^2\sum\hat{y}^2}$$

(24)

Multiple Regression (MR): If the x parameters are more than one i.e. $x_1, x_2....$ than the regression analysis is known as multiple regression (MR) but if the x parameter is one than it is LR. MR equation can be represented as:

$$y = a + b_1x_1 + b_2x_2 + b_3x_3 + ... b_nx_n + e$$

(25)

For MR it is usually assumed that the error term ε follows the normal distribution with $E(\varepsilon) = 0$ and a constant variance $Var(\varepsilon) = \sigma^2$. Forward selection (FS), backward elimination (BE) and step wise approximation (SWA) is used for analysis of MR.

3 Validation Tests

In this paper we are considering LR and MR methods for which we have discussed how to calculate r^2 values. To validate the results we have different approaches. In this paper we are using r^2_{adj}, r^2_{pre}, q^2_{cv}, rm^2 and F-test values. The predictive capability of the equation is determined using the leave-one-out cross validation method. q^2_{cv} was calculated by the following equation:

$$q^2_{cv} = 1 - \frac{PRESS}{TOTAL} \approx r^2$$

(26)

For a perfect model value of q^2_{cv} should be close to one and its value is approximately equal to r^2.

The evaluation of the predictive ability of the model for the external test set compounds was done by determining the value of rm^2 which was determine by :

$$rm^2 = r^2\left(1-\left|\sqrt{r^2-r_0^2}\right|\right)$$

(27)

r^2_0 is the squared correlation coefficient for regression without using intercept and the regression equation becomes $y = ax$. For a perfect model value of rm^2 should be close to one.

4 Discussion

In this paper we have considered eleven different proteins MK2, JNK, FKHR, MEK, ERK, IRS, AkT, IKK, pAkT, ptAkT and EGFR for the HT carcinoma cells which occurs due to the combination of TNF, EGF and Insulin. These proteins yield four different output: phosphatidylserine exposure (PE), membrane permeability (MP), nuclear fragmentation (NF) and caspase substrate cleavage (CCK). We have first taken average of all outputs and then normalized the output to maximum that's why in results we are showing only one output. Table 1 shows the minimum, maximum, median, mean, standard deviation, variance, and coefficient of variance for eleven different proteins and output.

Table 1 Different values for eleven different proteins and output.

	MK2	JNK	FKHR	MEK	ERK	IRS	AKT	IKK	pAKT	pTAKT	EGFR	OUTPUT
N of Cases	300	300	300	300	300	300	300	300	300	300	300	300
Minimum	221.466	221.526	191.019	110.233	140.711	191.123	164.909	59.976	110.359	90.321	49.437	0.407
Maximum	262.76	255.651	245.391	189.092	189.29	244.545	188.985	90.748	201.321	133.322	78.153	0.593
Median	242.156	235.749	210.593	136.447	172.221	216.314	177.376	76.804	157.058	108.792	68.604	0.48
Arithmetic Mean	242.798	236.224	214.208	144.297	167.113	216.768	177.868	75.085	154.684	110.499	64.849	0.483
Standard Deviation	10.541	8.302	14.94	26.971	14.425	14.304	6.149	9.22	30.922	13.876	9.505	0.051
Variance	111.115	68.923	223.217	727.422	208.083	204.606	37.808	85.006	956.141	192.552	90.348	0.003
Coefficient of Variation	0.043	0.035	0.07	0.187	0.086	0.066	0.035	0.123	0.2	0.126	0.147	0.106

In this section we have shown all the results r^2, r^2_{adj}, q^2_{cv}, rm^2 and F-test values for our 10 data sets of TNF/ EGF and Insulin in ng/ml.

1. **For r^2**: We have observed values of our 10 data set. First we have predicted the values using STAT SOFT software. We have put all results in excel and get the r^2 value from their and then calculate the same r^2 value using formula as shown in Eq 1. We have also calculated the r^2 value using MINITAB software. All the results of r^2 are shown in Table 2 which proves that all the values are same.

Table 2 All possible values of r^2 using excel, formula and software.

S. No	Possible Values	r^2 from excel	r^2 from formula using equation 1	r^2 from software
1	0-0-0	0.984	0.984	0.985
2	5-0-0	0.991	0.991	0.991
3	100-0-0	0.966	0.966	0.966
4	0-100-0	0.765	0.8265	0.765
5	5-1-0	0.953	0.9589	0.953
6	100-100-0	0.981	0.9817	0.981
7	0-0-500	0.991	0.992	0.991
8	0.2-0-1	0.985	0.985	0.985
9	5-0-5	0.9916	0.992	0.992
10	100-0-500	0.9105	0.9116	0.911

2. **For r^2_{pred}, r^2_{adj}** : We have calculated the r^2_{pred}, r^2_{adj} using MINITAB software shown in Table 3 which is coming out to be OK. Table 4 shows the cumulative r^2 values for X and Y. this table also represents the Eigen values and Q^2 cumulative values.

Table 3 Values of r^2, r^2_{pred}, r^2_{adj}.

S. No	Possible Values	r^2 (%)	r^2_{pred} (%)	r^2_{adj} (%)
1	0-0-0	98.5	98.5	98.49
2	5-0-0	99.1	99.13	99.1
3	100-0-0	96.6	96.59	96.6
4	0-100-0	76.5	76.44	76.5
5	5-1-0	95.3	95.32	95.3
6	100-100-0	98.1	98.10	98.1
7	0-0-500	99.1	99.10	99.1
8	0.2-0-1	98.5	98.49	98.5
9	5-0-5	99.2	99.16	99.2
10	100-0-500	91.1	91.05	91.1

Table 4 Eigen values and cumulative Q2 values of ten different concentrations.

S. No	Possible Values	$r^2 X$ (Cumul.)	Eigen values	$r^2 Y$ (Cumul.)	Q^2 (Cumul.)	Iterations
1	0-0-0	0.661527	6.778547	0.642263	0.530556	3
2	5-0-0	0.706644	7.770976	0.664049	0.662104	4
3	100-0-0	0.716885	7.654734	0.70051	0.648019	6
4	0-100-0	0.643378	6.55393	0.589539	0.48238	5
5	05-01-2000	0.449319	4.906164	0.515036	0.460442	49
6	100-100-0	0.704479	7.743815	0.663912	0.661397	4
7	0-0-500	0.794276	8.711558	0.517066	0.512291	4
8	0.2-0-1	0.76338	8.297655	0.704509	0.685306	3
9	5-0-5	0.806147	8.854371	0.482898	0.480075	4
10	100-0-500	0.658964	7.232619	0.6151	0.609608	9

3. **For q^2_{cv}:** As we know for a perfect model value of q^2_{cv} should be close to one and its value should be equal to r^2. We have already calculated the value of r^2. q^2_{cv} is calculated from MINITAB software both the values are equal. So it proves that the value of q^2_{cv} is close to r^2.

4. **For rm^2:** Fig. 1 shows the r^2 and r^2_0 value for ten different concentrations. For r^2 and r^2_0 value we have plotted the observed value and predicted value in Excel and get the equations from the same. Similarly we have plotted the data in MINITAB software and get the same equation which verifies the result as shown in Table 5. Putting the values of r^2 and r^2_0 in Eq. 27 we get the rm^2 value which is coming close to one as shown in Table 6 which verifies our result.

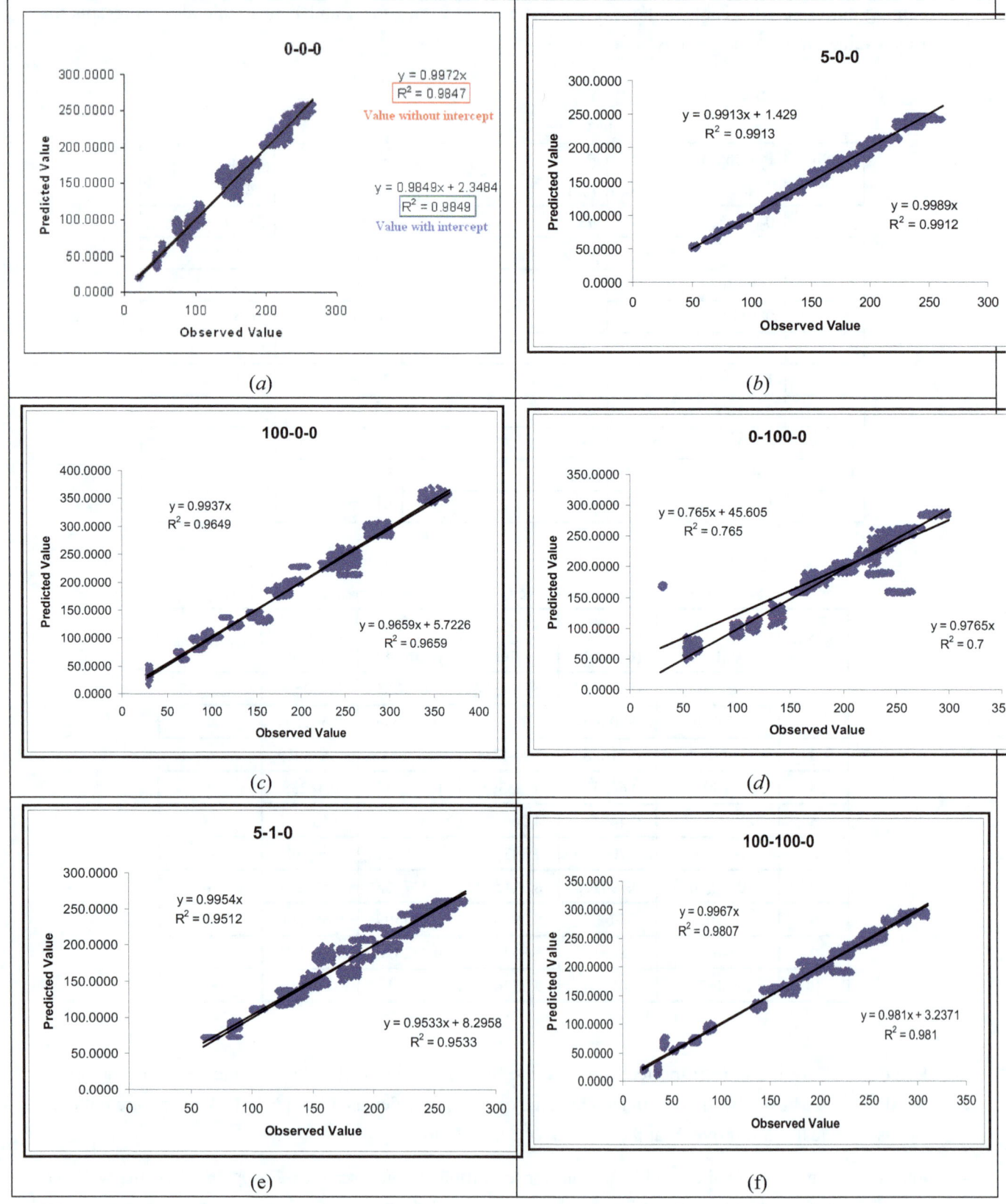

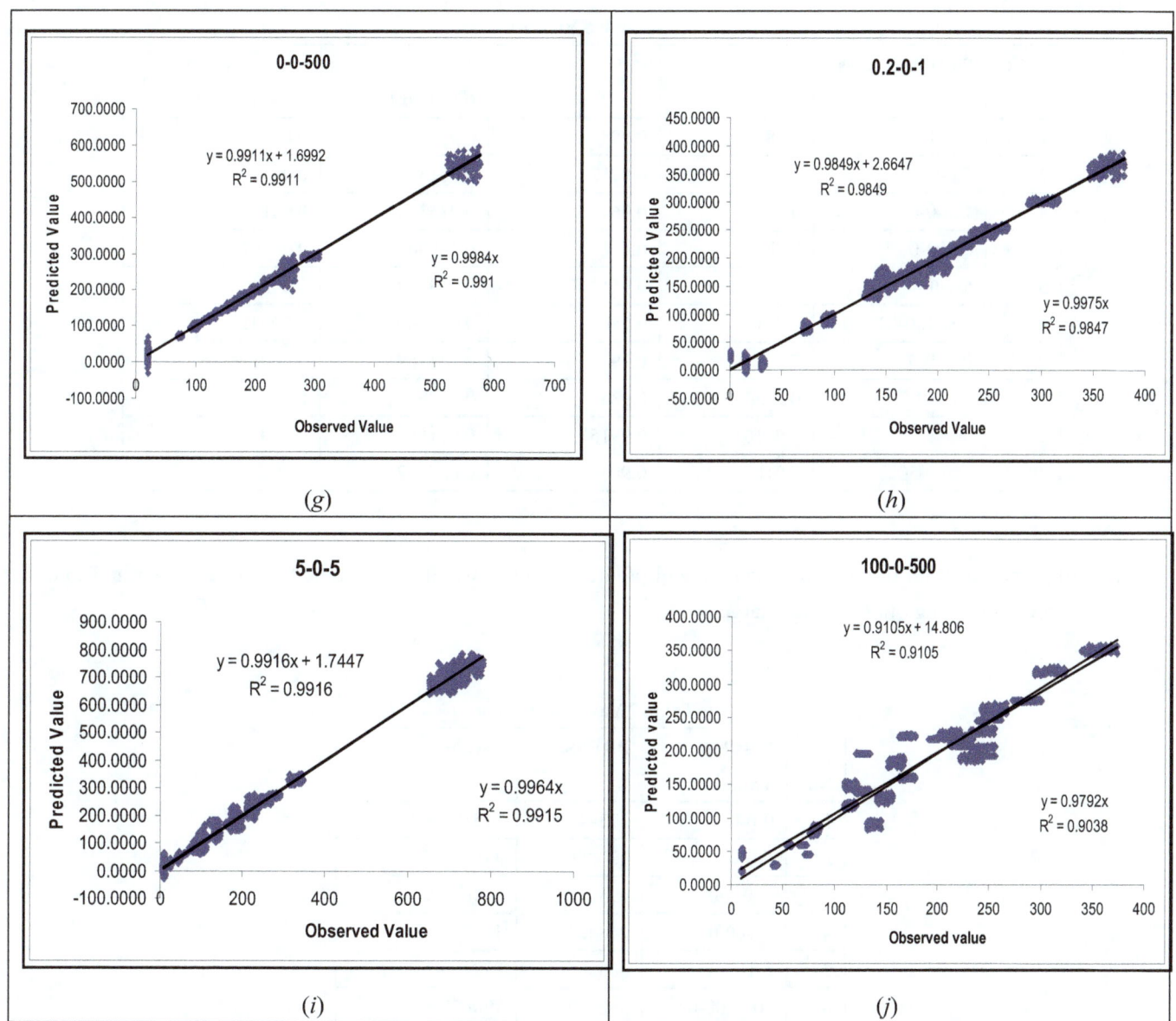

Fig. 1 Values for r^2 and r^2_0 for 10 data sets.

Table 5 Equations of r^2 and r^2_0.

S. No	Possible Values	r^2 with intercept From excel	r^2 with intercept From MINITAB	r^2_0 without intercept
1	0-0-0	$0.9849x + 2.3484$	$0.985x + 2.35$	$0.9972x$
2	5-0-0	$0.9913x + 1.429$	$0.991x + 1.43$	$0.9989x$
3	100-0-0	$0.9659x + 5.7226$	$0.966x + 5.72$	$0.9937x$
4	0-100-0	$0.765x + 45.605$	$0.765x + 45.6$	$0.9765x$
5	5-1-0	$0.9533x + 8.2958$	$0.953x + 8.30$	$0.9954x$
6	100-100-0	$0.981x + 3.2371$	$0.981x + 3.24$	$0.9967x$
7	0-0-500	$0.9911x + 1.6992$	$0.991x + 1.7$	$0.9984x$
8	0.2-0-1	$0.9849x + 2.6647$	$0.985x + 2.66$	$0.9975x$
9	5-0-5	$0.9916x + 1.7447$	$0.992x + 1.74$	$0.9964x$
10	100-0-500	$0.9105x + 14.806$	$0.911x + 14.8$	$0.9792x$

Table 6 Values of rm^2.

S. No	Possible Values	r^2	r^2_0	rm^2 (close to 1)	$(r^2 - r^2_0) / r^2$ (less than 0.1)
1	0-0-0	0.984	0.9847	0.970971	0.0002
2	5-0-0	0.991	0.9912	0.981387	0.0001
3	100-0-0	0.966	0.9649	0.935356	0.00104
4	0-100-0	0.765	0.7	0.569963	0.08497
5	5-1-0	0.953	0.9512	0.909614	0.0022
6	100-100-0	0.981	0.9807	0.964009	0.00031
7	0-0-500	0.991	0.991	0.981189	0.0001
8	0.2-0-1	0.985	0.9847	0.970971	0.0002
9	5-0-5	0.9916	0.9915	0.981684	0.0001
10	100-0-500	0.9105	0.9038	0.835972	0.00736

5.   **For F-value**: With the help of observed and predicted values we have calculated F-value shown in Table 7 which is coming out to be very large.

Table 7 Values of F-value.

S. No	Possible Values	PRESS	F-value
1	0-0-0	273446	215195.4
2	5-0-0	106089	375401.5
3	100-0-0	698648	93465.53
4	0-100-0	2484216	10735.86
5	5-1-0	505386	97249.82
6	100-100-0	382434	170650
7	0-0-500	235785	365494
8	0.2-0-1	296709	215009.8
9	5-0-5	896652	388299.1
10	100-0-500	2231417	33567.88

As the independent variables are many so MR can be applied. Table 8 shows the coefficient, standard coefficient, t value and p values of the independent variables. Table 9 shows the variation explanation for predictors and responses of eleven proteins and output. Table 10 shows the correlation matrixes for eleven different proteins and one output.

Table 8 Different parameters for MR.

Predictor	Coeff	SE coeff	t	p
Constant	0.2225	0.187	1.19	0.235
MK2	0.001157	0.000186	6.21	0
JNK	-0.00046	0.000222	-2.07	0.039

FKHR	-0.0007	0.000232	-3.03	0.003
MEK	-0.00045	0.000296	-1.5	0.133
ERK	0.001518	0.000306	4.97	0
IRS	0.000117	0.000249	0.47	0.639
AkT	-0.00124	0.00027	-4.6	0
IKK	-0.00033	0.000624	-0.53	0.597
pAkT	-0.00122	0.000266	-4.58	0
ptAkT	0.001206	0.000423	2.85	0.005
EGFR	0.005013	0.000674	7.44	0

Table 9 Variation explained for predictors and responses of eleven proteins.

Factors	Variation Explained for Predictor(s)		Variation Explained for Response(s)	
	Percentage	Cum. Percentage	Percentage	Cum. Percentage
MK2	68.938	68.938	45.166	45.166
JNK	14.369	83.307	22.41	67.576
FKHR	4.333	87.64	3.151	70.727
MEK	2.077	89.717	3.485	74.212
ERK	4.783	94.5	0.648	74.86
IRS	2.176	96.675	0.639	75.499
AkT	0.797	97.473	0.221	75.72
IKK	0.971	98.443	0.042	75.763
pAkT	0.738	99.181	0.017	75.779
ptAkT	0.287	99.468	0.002	75.781
Output	0.532	100	0	75.781

Table 10 Correlation matrixes.

	MK2	JNK	FKHR	MEK	ERK	IRS	AKT	IKK	PAKT	PTAKT	EGFR	OUTPUT
MK2	1											
JNK	-0.026	1										
FKHR	0.226	0.511	1									
MEK	0.263	0.538	0.895	1								
ERK	-0.223	-0.533	-0.85	-0.934	1							
IRS	0.366	0.442	0.8	0.876	-0.831	1						
AKT	-0.313	0.163	0.108	0.105	-0.135	0.029	1					
IKK	-0.295	-0.515	-0.875	-0.955	0.909	-0.888	-0.08	1				
PAKT	-0.377	-0.517	-0.882	-0.966	0.917	-0.899	-0.043	0.954	1			
PTAKT	0.364	0.507	0.865	0.947	-0.903	0.884	0.035	-0.938	-0.962	1		
EGFR	-0.218	-0.564	-0.889	-0.963	0.925	-0.865	-0.162	0.937	0.945	-0.929	1	
OUTPUT	0.306	-0.497	-0.596	-0.6	0.637	-0.444	-0.442	0.546	0.497	-0.487	0.666	1

5 Conclusion

We have made a best fit linear model using partial least square for ten cytokine combinations of TNF, EGF and Insulin. In this we have find all the results like regression coefficient, adjusted regression coefficient, regression coefficient cross validation, rm^2 and F-test values for our 10 data sets which comes out to be correct. Later multiple regressions were applied as we have eleven different input independent variables (proteins). We have calculated coefficient, standard error, standard coefficient, tolerance, t value and p value, variation explanation of predictors and estimators which gives percentage and cumulative percentage of all eleven proteins and one output. Later, Correlation matrixes were also for eleven proteins and one output.

References

Brunet A, Bonni A, Zigmond MJ, Lin MZ, Juo P, Hu LS. 1999. Akt promotes cell survival by phosphorylating and inhibiting a Forkhead transcription factor. Cell, 96: 857-868

Coffer PJ, Jin J, Woodgett JR. 1998. Protein kinase B (c-Akt): a multifunctional mediator of phosphatidylinositol 3-kinase activation. Biochemical Journal, 335: 1-13

Farahani H.A., Rahiminezhad A, Samec L, Immannezhad K. 2010. A comparison of Partial Least Squares (PLS) and Ordinary Least Squares (OLS) regressions in predicting of couples mental health based on their communicational patterns. Procedia Social and Behavioral Sciences, 5: 1459-1463

Hemmings BA. 1997. Akt signaling: linking membrane events to life and death decisions. Science, 275: 628-630

Jain S. 2012a. Communication of Signals and Responses Leading To Cell Survival/Cell Death Using Engineered Regulatory Networks. PhD Thesis. Jaypee University of Information Technology, Solan, Himachal Pradesh, India

Jain S. 2014. Implementation of fuzzy system using operational transconductance amplifier for ERK pathway of EGF/ Insulin leading to cell survival/ death. Journal of Pharmaceutical and Biomedical Sciences, 4(8): 701-707

Jain S. 2015d. Mathematical Analysis and Probability Density Function of FKHR pathway for Cell Survival /Death. Control System and Power Electronics – CSPE 2015. 84-93, Banglore, India

Jain S. 2016a. Compedium model using frequency / cumulative distribution function for receptors of survival proteins: Epidermal growth factor and insulin, Network Biology, 6(4): 101-110

Jain S. 2016b, Regression analysis on different mitogenic pathways. Network Biology, 6(2): 40-46

Jain S. 2016c. Mathematical analysis using frequency and cumulative distribution functions for mitogenic pathway. Research Journal of Pharmaceutical, Biological and Chemical Sciences, 7(3): 262-272

Jain S. 2017a. Implementation of marker proteins using standardised effect. Journal of Global Pharma Technology, 9(5): 22-27

Jain S. 2017b. Parametric and non parametric distribution analysis of AkT for cell survival/death. International Journal of Artificial Intelligence and Soft Computing, 6(1): 43-55

Jain S, Bhooshan SV, Naik PK. 2010a. A system model for cell death/ survival using SPICE and ladder logic. Digest Journal of Nanomaterials and Biostructures, 5(1): 57-66

Jain S, Bhooshan SV, Naik PK. 2010e. Model of mitogen activated protein kinases for cell survival/death and its equivalent bio-circuit. Current Research Journal of Biological Sciences, 2(1): 59-71

Jain S, Bhooshan SV, Naik PK. 2011a. Mathematical modeling deciphering balance between cell survival and cell death using Tumor Necrosis Factor α. Research Journal of Pharmaceutical, Biological and Chemical Sciences, 2(3): 574-583

Jain S, Bhooshan SV, Naik PK. 2011b. Mathematical modeling deciphering balance between cell survival and cell death using insulin. Network Biology, 1(1): 46-58

Jain S, Bhooshan SV, Naik PK. 2011c. Model of Protein Kinase B for Cell Survival/Death and its Equivalent Bio Circuit. In: Proceedings of the 2nd International Conference on Methods and Models in Science and Technology (ICM2ST-11). 69-73, Institution of Engineers, Technocrats and Academician Network (IETAN), Jaipur, Rajasthan, India

Jain S, Bhooshan SV, Naik PK. 2012d. Compendium model of AkT for cell survival/death and its equivalent BioCircuit. International Journal of Soft Computing and Engineering (IJSCE), 2(3): 91-97

Jain S, Chauhan DS. 2015a. Mathematical analysis of receptors for survival proteins. International Journal of Pharma and Bio Sciences, 6(3): 164-176

Jain S, Chauhan DS. 2015b. Linear and Non Linear Modeling of Protein Kinase B/ AkT. International Conference on Information and Communication Technology for Sustainable Development (ICT4SD - 2015). 81-88, Ahmedabad, India

Jain S, Chauhan DS. 2015c. Implementation of fuzzy system using different voltages of OTA for JNK pathway leading to cell survival/ death. Network Biology, 5(2): 62-70

Jain S, Naik PK, Sharma R 2009a. A computational modeling of cell survival/death using VHDL and MATLAB simulator. Digest Journal of Nanomaterials and Biostructures (DJNB), 4(4): 863-879

Jain S, Naik PK, Sharma R. 2009b. A Computational modeling of apoptosis signaling using VHDL and MATLAB simulator. International Journal of Information and Communication Technology, 2(1-2): 7-12

Jain S, Naik PK, Chauhan DS, Sharma R. 2009c. Computational Modeling of Cell Survival/ Death Using MATLAB Simulator. 6th International Conference on Information Technology: New Generations ITNG. 64-68, Las Vegas, Nevada, USA

Jain S, Naik PK. 2010b. Computational modeling of cell survival using VHDL, BVICAM's. International Journal of Information Technology (BIJIT), 2(1): 47-51

Jain S, Naik PK, Bhooshan SV. 2010c. Computational modeling of cell survival/death using BiCMOS. International Journal of Computer Theory and Engineering, 2(4): 478-481

Jain S, Naik PK, Bhooshan SV, 2010d. Petri net implementation of cell signaling for cell death. International Journal of Pharma and Bio Sciences, 1(2): 1-18

Jain S, Naik PK, Bhooshan SV. 2011d. Nonlinear Modeling of cell survival/ death using artificial neural network. In: The Proceedings of International Conference on Computational Intelligence and Communication Networks. 565-568, Gwalior, India

Jain S, Naik PK. 2012b. System modeling of cell survival and cell death: A deterministic model using Fuzzy System. International Journal of Pharma and BioSciences, 3(4): 358-373

Jain S, Naik PK. 2012c. Communication of signals and responses leading to cell death using Engineered Regulatory Networks. Research Journal of Pharmaceutical, Biological and Chemical Sciences, 3(3): 492-508

Janes KA, John AG, Suzanne G, Peter SK, Douglas LA, Michael YB. 2005 A systems model of signaling identifies a molecular basis set for cytokine-induced apoptosis. Science, 310: 1646-1653

Morris F. White 2003 Insulin signaling in health and disease science, 302: 1710-1711

Normanno N, De Luca A, Bianco C, Strizzi L, Mancino M, Maiello MR, et al. 2006. Epidermal growth factor receptor (EGFR) signaling in cancer. Gene, 366: 2-16

Ringle C, Wende S, Will A. 2010. Finite mixture partial least squares analysis: Methodology and numerical examples. In: Handbook Partial Least Squares (Vinzi VE, Chin W, Henseler J, Wang H, eds). Springer, Heidelberg, Germany

Suzanne G, Janes KA, John AG, Emily PA, Douglas LA, Peter SK. 2005. A compendium of signals and responses trigerred by prodeath and prosurvival cytokines. Manuscript M500158-MCP200.

Thoma B, Grell M, Pfizenmaier K, Scheurich P. 1990. Identification of a 60-kD tumor necrosis factor (TNF) receptor as the major signal transducing component in TNF responses. The Journal of Experimental Medicine, 172: 1019-1023

Weiss R., 2001 Cellular Computation and Communications Using Engineered Genetic Regulatory Networks. PhD Thesis. MIT, USA

15

Regulatory switches for hierarchical use of carbon sources in *E. coli*

Ruth S. Pérez-Alfaro[1], **Moisés Santillán**[2], **Edgardo Galán-Vásquez**[1], **Agustino Martínez-Antonio**[1]

[1]Departamento de Ingeniería Genética, Centro de Investigación y de Estudios Avanzados del IPN, Unidad Irapuato, Km. 9.6 Libramiento Norte Carr. Irapuato-León 36821 Irapuato, Guanajuato, México

[2]Centro de Investigación y de Estudios Avanzados del IPN, Unidad Monterrey, Vía del Conocimiento 201, 66600 Apodaca NL, México

E-mail: amartinez@ira.cinvestav.mx

Abstract

In this work we study the preferential use of carbon sources in the bacterium *Escherichia coli*. To that end we engineered transcriptional fusions of the reporter gene *gfpmut2,* downstream of transcription-factor promoters, and analyzed their activity under several conditions. The chosen transcription factors are known to regulate catabolic operons associated to the consumption of alternative sugars. The obtained results indicate the following hierarchical order of sugar preference in this bacterium: glucose > arabinose > sorbitol > galactose. Further dynamical results allowed us to conjecture that this hierarchical behavior might be operated by at least the following three regulatory strategies: 1) the coordinated activation of the corresponding operons by the global regulator catabolic repressor protein (CRP), 2) their asymmetrical responses to specific and unspecific sugars and, 3) the architecture of the associated gene regulatory networks.

Keywords *Escherichia coli*; transcription factors; carbon sources; hierarchical use.

1 Introduction

The way bacteria use different carbon sources (Monod, 1942) has been studied for a long time, in which *Escherichia coli* has been the favorite model organism. We learned from the very beginning that glucose is the carbon source supporting the fastest growth on this bacterium (Walker et al., 1934). This sugar is also the preferred one if bacteria are exposed to a mixture of carbon sources. It seems that *E. coli* uses carbon sources on the basis of "best food served first". The molecular mechanisms behind this operating principle are various, the best-known ones are: inducer exclusion, local or dedicated transcriptional regulation, global transcriptional regulation, small RNAs, and catabolite repression.

Inducer exclusion (Jones-Mortimer and Kornberg, 1974; Chen et al., 2013) takes place when, in the presence of glucose or other PTS sugars, the unphosphorylated EIIA^{Glc} (part of the PTS system) binds to and stabilizes the resting state of non PTS-sugar transporters, inhibiting the transport (and use) of alternative carbon sources.

Local or dedicated transcriptional regulation operates at the initiation of gene transcription of sugars catabolic operons. These operons are normally subject to repression by at least one specific regulator, whose derepression occurs when the corresponding specific sugar is available and binds to it. This binding causes the effector-repressor complex to unbind from the operator zone, which is a necessary condition for the corresponding operon to become active (Jacob and Monod, 1961; Sellitti et al., 1987).

Global transcriptional regulation. The complementary condition for the transcription of sugar catabolic genes is given by the activity of the global regulator CRP (catabolic repressor protein or cAMP regulatory protein). CRP becomes active when bound by cyclic adenosine monophosphate (cAMP). The cAMP-CRP complex is then capable of recruiting RNA polymerase to promoter zones of catabolic operons so their transcription is started if no repressor is present. Hence, a condition for the transcription of catabolic operons is that high cAMP levels are present. High cAMP levels are in general achieved in the absence of glucose although, as mentioned below, it could be the result or a wider physiological status (Gottesman, 1984; Martínez-Antonio and Collado-Vides, 2003).

Small RNAs (sRNA). Arguably, the best known sRNA is the multi-target Spot42, which inhibits the translation of at least 14 genes, mostly related to the use of non-PTS sugars. Spot42 is activated by cAMP-CRP and together form a coherent feed-forward loop to avoid use of non-PTS sugars when the preferred sugars are available (Beisel and Storz, 2001; Wright et al., 2013).

Catabolite repression (Magasanik, 1961; Görke and Stülke, 2008), a physiological concept so-named by Boris Magasanik as a generalization of the "glucose effect" described many years earlier (Cohn, 1957). It was derived after observing the repression, when glucose is present, of catabolic enzymes specific for carbon and nitrogen metabolism. This phenomenon was related to cAMP levels that increase when poor carbon sources are present in the milieu (Epstein et al., 1975). cAMP is synthesized by the CyaA enzyme, which is activated by phosphorylated EIIA^{glu} but requires an additional unidentified factor (Park et al., 2006). It was postulated that a derived catabolite of carbon sources (the repressor catabolite) is the responsible to trigger cAMP syntesis. Only recently, a high-throughput proteome analysis (studying carbon, nitrogen and sulfur sources metabolism) in *E. coli* revealed that cAMP levels are diminished by α-ketoacids (mainly by oxaloacetate) through the inhibition of adenylate cyclase, the enzyme responsible for cAMP synthesis (You et al., 2013). This explains how this central metabolite is balancing the overall bacterial physiology throughout the nitrogen/carbon metabolism (Rabinowitz and Silhavy, 2013).

Here we present a study that copes with the activities of the promoters of specific catabolic regulators, which in addition to self-regulation, respond to the global regulator CRP (points 2 and 3 above). These locals and the global regulator operate together to regulate transcriptional initiation in *E. coli* catabolic operons for the transport and use of carbon sources other than glucose.

We tackle the question of how bacteria decide to consume alternative carbon sources, focusing in L-arabinose, D-sorbitol and D-galactose. The regulation, transport and first catabolic steps in the metabolism of these sugars are depicted in Fig. 1. As we can see, not only the corresponding genes are activated by CRP, but they are also repressed by specific transcription factors. We investigate in this work the promoter activities of these specific regulators. Importantly, all of them use the transported sugars as signal effectors to modulate their activities. The signal sugar binds to the repressor and unbinds it from the operator zone, thus allowing transcription of the corresponding genes. Finally, all the promoters here analyzed require of the housekeeping σ^{70} to be transcribed so this is not a variable to consider in this study.

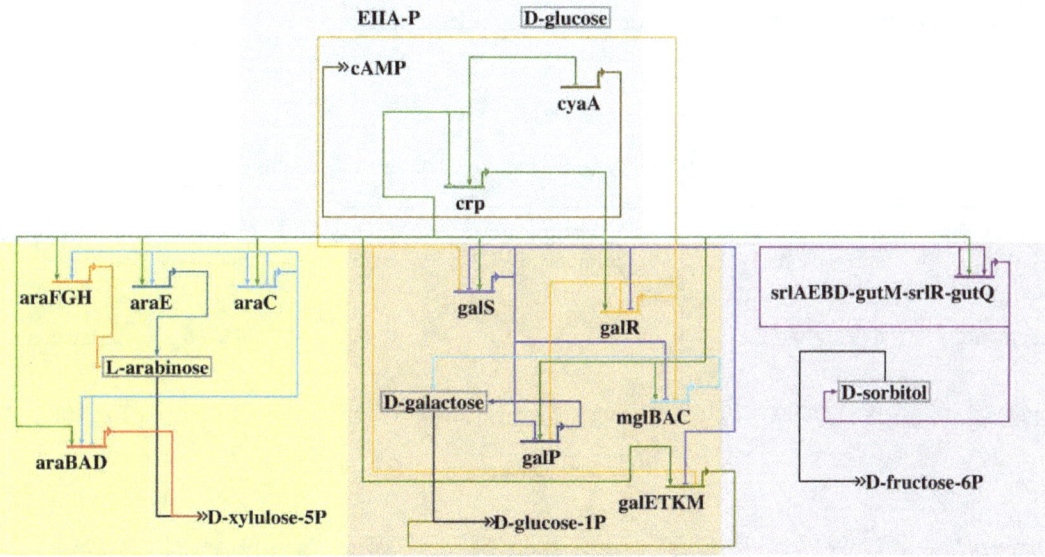

Fig. 1 Regulatory network controlling the use of the carbon sources employed in this study. It is show the module corresponding to the global regulator CRP. cAMP, a co-activator of CRP, is synthesized via CyaA when glucose is absent. The arabinose module includes the dual regulator AraC, which transcriptionally regulates the arabinose transporter genes AraE (low affinity) and AraFGH (high affinity). AraC also regulates the genes of enzymes isomerase (AraA), ribolukinase (AraB) and epimerase (AraD), which metabolize arabinose to D-xylulose 5-P. The galactose module has two repressors, GalR and GalS, in different transcription units. In absence of galactose they repress the genes for galactose transporters GalP (low affinity) and MglBAC (high affinity), as well as those for the enzymes GalK (galactokinase), GalT (uridiltransferase), and galM (epimerase), which metabolize galactose to glucose 1-P. The sorbitol module is also regulated by two transcription factors, SrlR and GutM, encoded in the same operon, which also includes genes for high affinity transporter (SrlAEB) and for the enzymes SrlD (dehydrogenase) and GutQ (isomerase), which transform sorbitol to fructose 6-P. This figure was created using the BioTrapestry software (Longabaugh et al, 2009).

2 Material and Methods

2.1 Strains

In all our experiments we employ *Escherichia coli* K-12 MG1655 strain and derivatives harboring the different transcriptional fusions show in Table 1. Most of the used transcriptional fusions were taken from a collection reported previously (Zaslaver et al., 2006). However, we rebuilt the transcriptional fusions for *gutM* and *crp* promoters in order to include regulatory sites for transcription factors not comprised in fusions from the collection. We realized the necessity of such regulatory sites by inspecting the transcription-factor binding sites reported in RegulonDB (Salgado et al., 2013). These last fusions were engineered by amplifying (through PCR and specific primers) the corresponding regulatory regions, cloning the resulting DNA fragments on pUA66 with the aid of the BamH1 and XhoI restriction sites, and verifying the construction by means of DNA sequencing.

2.2 Bacterial growth

For strain maintenance we routinely used LB medium and for experimental tests we used M9 medium, supplemented with sugars as indicated. Also when indicated, we added kanamycin (Km) 50 µg ml-1. Pre-inoculates were grown overnight in 5 ml of LB medium at 37 ºC with agitation (200 rpm). Next, the cultures were diluted 1 : 100 in 150 μl of fresh M9 media in micro-titer plates of 96 wells and incubated for 12 h with agitation (250 rpm) at 37 °C. We supplemented M9 with 0.4% or 0.03% of glucose, and 0.2% of one or two alternative sugars as specified. We followed bacterial growth, by measuring OD595nm, and fluorescence (535 nm) every hour in a Perkin Elmer Victor X3 plate multi-lector.

Table 1 Regulatory regions employed on the transcriptional fusions.

Promoter fusions	E. coli chromosome coordinates	Designed primers* 5′-3′	Region size	Cloning vector	Reference
araCp::gfpmut2	69973-70452		479bp	pUA66	Zaslaver et al., 2006
crpp::gfpmut2	3483776-3484200	F:tgatgactcgaggcggattc R:tggcaatgagacaggatcca	424bp	pUA66	This study
galSp::gfpmut2	2239619-2239844		225bp	pUA139	Zaslaver et al., 2006
galRp::gfpmut2	2973960-2974698		738bp	pUA66	Zaslaver et al., 2006
gutMp::gfpmut2	2823533-2823932	F:cttgctgctcgaggcggcaa R:ccatccggatccacacctctccgc	399bp	pUA66	This study
srlRp::gfpmut2	2826905-2827074		169bp	pUA66	Zaslaver et al., 2006

*Underlined nucleotides define restriction sites for XhoI and BamH1 endonucleases on forward and reverse primers.

2.3 Data acquisition and processing

The raw numerical data obtained from the Victor X3 plate multi-lector consisted of discrete measurements of optical density (OD) and fluorescence (GFP) versus time along the growth curves, with a sampling frequency of 1 hr^{-1}. Although enough to provide an overview of the time evolution of variables OD and GFP, such sampling frequency is too low to perform more refined quantitative analyses. For that, it is necessary to find a function that fits the experimental data. Since the generalized logistic function is a widely used sigmoid function for growth modeling we decided to employ it. In all cases we found that it fits both the growth curves and the GFP profiles with correlation factors higher than 0.99. The functions used to fit the OD and GFP profiles are:

$$OD(t) = a_1 + \frac{k_1 - a_1}{\left(1 + q_1 e^{-b_2 t}\right)^{1/v_1}}, \tag{1}$$

$$GFP(t) = a_2 + \frac{k_2 - a_2}{\left(1 + q_2 e^{-b_2 t}\right)^{1/v_2}}, \tag{2}$$

in which a_i, b_i, k_i, q_i, and v_i ($i = 1, 2$) are fitting parameters. Zaslaver et al. (2006) and Martínez-Antonio et al.

(2012) have argued that promoter activity is proportional to $(dGFP(t) / dt) / OD(t)$. Thus, after finding the best fitting parameters we differentiated function (2) and divided the result by eq. (1) to compute the promoter activity in each case.

For every experimental condition and for every transcriptional function we periodically measured the values of optical density and green fluorescence in triplicate, computed the corresponding average values, and respectively fitted to Eqs. (1) and (2), and computed the promoter activity level as explained above.

In our experiments we could observe that the *crp*, *galR* and *srlR* promoters were unresponsive under all the tested conditions (data not shown). The invariable of *crp* promoter activity might be explained because it is the most global regulator in *E. coli*. Not only *crp* regulates itself, but it is also subject to dual regulation by another global regulator: FIS (factor for inversion stimulation). In the case of *galR* and *srlR*, the reason why they present constant low expression levels may be that they are constitutively expressed; up to date no regulator is known for these genes. Due this unresponsiveness and for the sake of clarity we excluded the results corresponding to these promoters in the fore coming sections.

3 Results

3.1 Different carbon sources support the grown of *E. coli* differentially

Our first objective was to analyze how the different carbon sources under study support the growth of *E. coli*. For this, we followed the progression of *E. coli* cultures growing in M9 minimal medium added with L-arabinose, D-sorbitol and D-galactose, both separately and in dual combinations. The growth profiles show in Fig. 2A confirm that glucose is by far the sugar that best supports *E. coli* growth. The hierarchical order of sugars in terms of their capacity to sustain cell growth is as follows: glucose > arabinose > sorbitol > galactose. When combinations of two alternative sugars were used, the bacterial growth rate almost equated that of glucose during the exponential growth phase. On the other hand, with all the sugar combinations, the maximal bacterial population density surpassed that of glucose alone. The decreasing order of alternative sugar combinations in terms of the exponential growth rate they are capable of sustaining is: arabinose+sorbitol > arabinose+galactose > galactose+sorbitol. In these experiments, glucose was set at a limiting amount (0.03%) from the very beginning to clearly distinguish the time at which *E. coli* starts using alternative carbon sources (Fig. 2A). We observed a differential growth of cultures with not limitation as compared with those limited on glucose as early as 3.5 hours after the start of the experiment. However, a careful observation on the alternative-sugar catabolic-operon promoter activity reveals that they become active after 2 hours of the experiment beginning (see below).

3.2 Glucose limitation triggers foraging alternatives

Our second objective was to study the dynamics of the alternative-sugar catabolic-operon promoters under glucose exhaustion conditions. Specifically, we were interested in the following scenarios: 1) when glucose is limiting from the culture at the very beginning and, 2) when glucose is exhausted after a normal period of bacterial growth. For that purpose we engineered specific reporters for relevant transcription factors (Table 1). These reporters were built by transcriptionally fusing each promoter to gene *gfpmut2*, and promoter activity was estimated by measuring fluorescence along the bacterial growth curves (Zaslaver et al., 2006). We made sure that the presence of the vector and genetic constructions were not detrimental for *E. coli* growth before the assays.

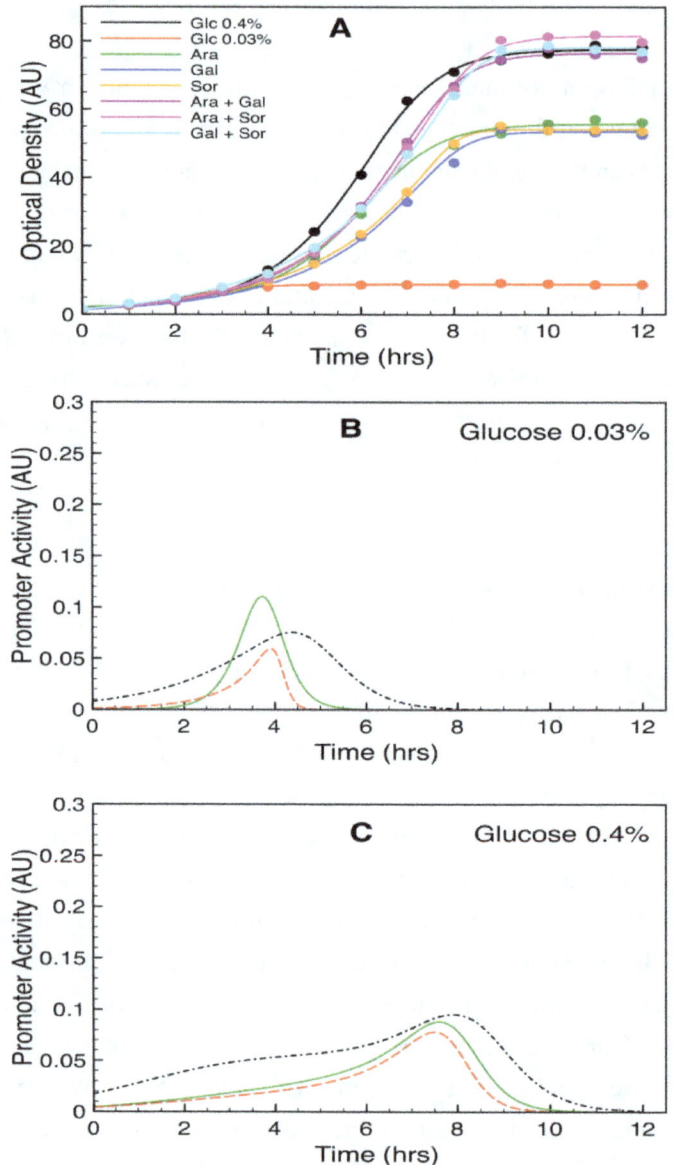

Fig. 2 Carbon source consumption by *Escherichia coli*. **A)** Wild-type *E. coli* growth curves (experimental data and best fitting generalized logistic functions) while cultured with different carbon-source concentrations and combinations. **B)** Transcription-factor promoter activities under glucose limitation conditions (M9 + 0.03% glucose). **C)** Transcription-factor promoter activities during a normal course of *E. coli* growth (M9 + 0.4% glucose). The color code in all graphs is as follows: *araC*, black; *gutM*, green; *galS*, red.

The results of these experiments can be summarized as follows. The time when alternative promoters are maximally active depends on the time at which glucose is exhausted. When glucose is limiting from the beginning of the experiment, the alternative-sugar catabolic-operon promoters start activating as soon as 2 hours after the experiment start, they reach their maximal of activity around the hour 4, and their activity starts declining thereafter (Fig. 2B). Contrarily, if there is a considerable amount of glucose at the culture beginning, the alternative-sugar catabolic-operon promoters become active only after glucose has been presumably exhausted, with the exemption of *araC* promoter that shows some activity during all the experiment. The three examined promoters reach their maximal activity about 8 hours after the beginning of the experiment (Fig. 2C).

The promoter behaviors reported in Figs. 2B and 2C, in absence of specific sugars in the milieu, can be explained by the activity of the master regulator CRP which only becomes active and turns up its target genes (among others the ones corresponding to the here studied transcription factors) when glucose is exhausted, and supposedly, when cAMP production is increased. Interestingly, no matter how fast glucose is exhausted, the studied promoters start showing some sign of activity about 2 hours after the cultures' start. Finally, under conditions of high initial glucose levels, not only all promoters reach their maximal activity at roughly the same time, but also their maximal activity levels are quite similar. When the initial glucose concentration is low, the maximal levels are dissimilar, although they are reached at similar times. However, it is important to emphasize that promoter activity is inversely proportional to bacterial density and that bacterial density (estimated by means of optical density measurements) is very low when glucose levels are initially low. All this implies that the obtained maximal promoter activity levels are not as reliable as those corresponding to high initial glucose concentration. Having this in mind it is remarkable that the maximal promoter activity levels have the same order of magnitude in all cases, see Figs. 2B and 2C.

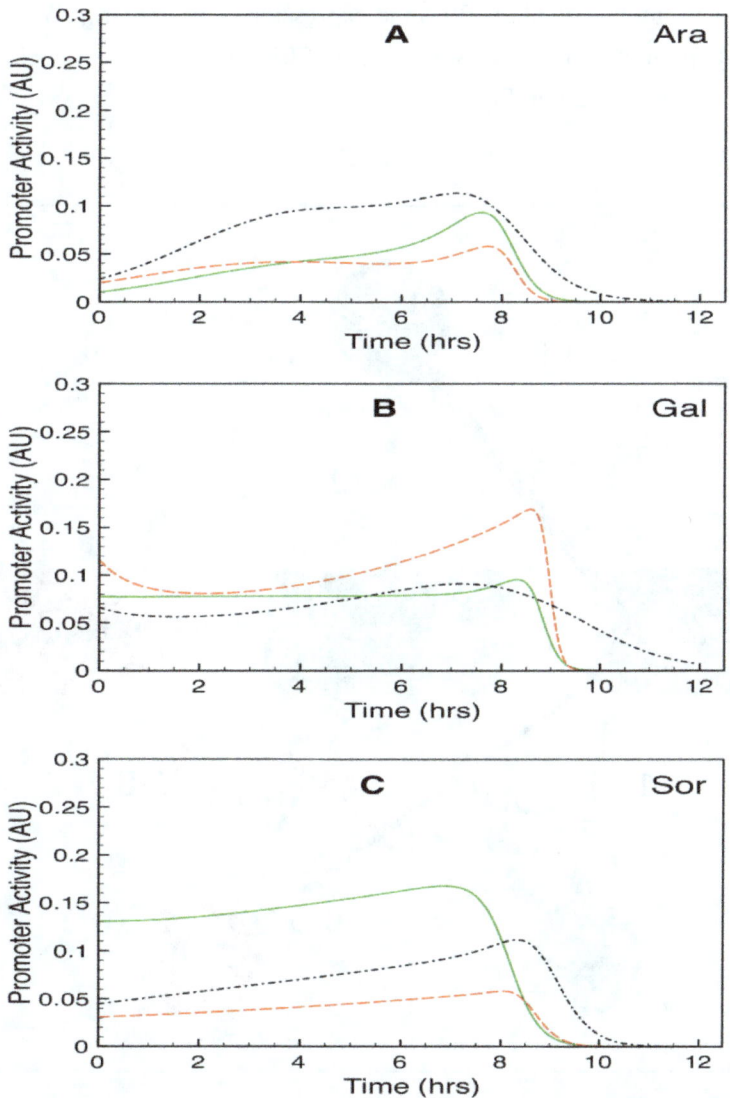

Fig. 3 Transcription-factor promoter activities as response to the presence of alternative carbon sources. **A)** Promoter activities in the presence of L-arabinose. **B)** Promoter activities in the presence of D-galactose. **C)** Promoter activities in the presence of D-sorbitol. The alternative sugars were supplemented at 0.2% (M9 + 0.03% glucose + 0.2% each alternative sugar). The color code in all panels is as follows: *araC*, black; *gutM*, green; *galS*, red.

3.3 Promoter specificity for sugar effectors

We further investigated the specificity of promoter response to different sugars. To that end we grew *E. coli* in M9 medium with a limiting amount of glucose plus different carbon sources, and measured the activity of the previously mentioned transcription-factor promoters. The results are reported in Fig. 3 in which the promoter activities are plotted vs. time under different conditions.

We found that each transcription-factor promoter primarily responds to the sugar whose catabolic operon it controls. However the individual effects vary in amplitude as follows. The largest effect is that of sorbitol on *gutM*, whose maximal activity is about 2.4 times larger the one caused by glucose exhaustion solely; then we have the effect of galactose on *galS* that shows a 1.8 fold increased activity; and the smallest response is that of *araC* to arabinose, with an increase of 1.3 in the promoter activity.

We could also observe a cross-regulation effect between signals (sugars) and the expression of transcription factors. This may allow transcription factors to display asymmetrical responses to specific and unspecific sugars. We can see that arabinose has a slight positive effect on *gutM* promoter, while it inhibits the expression level of *galS* promoter by about 40%. Regarding sorbitol, it increases by about 40% the expression level of promoter *araC*, and inhibits by about 30% the expression of promoter *galS*. Finally, galactose has no noticeable effect on *araC* promoter, but increases by about 20% the expression level of *gutM* promoter. The above-discussed results are summarized in Fig. 4.

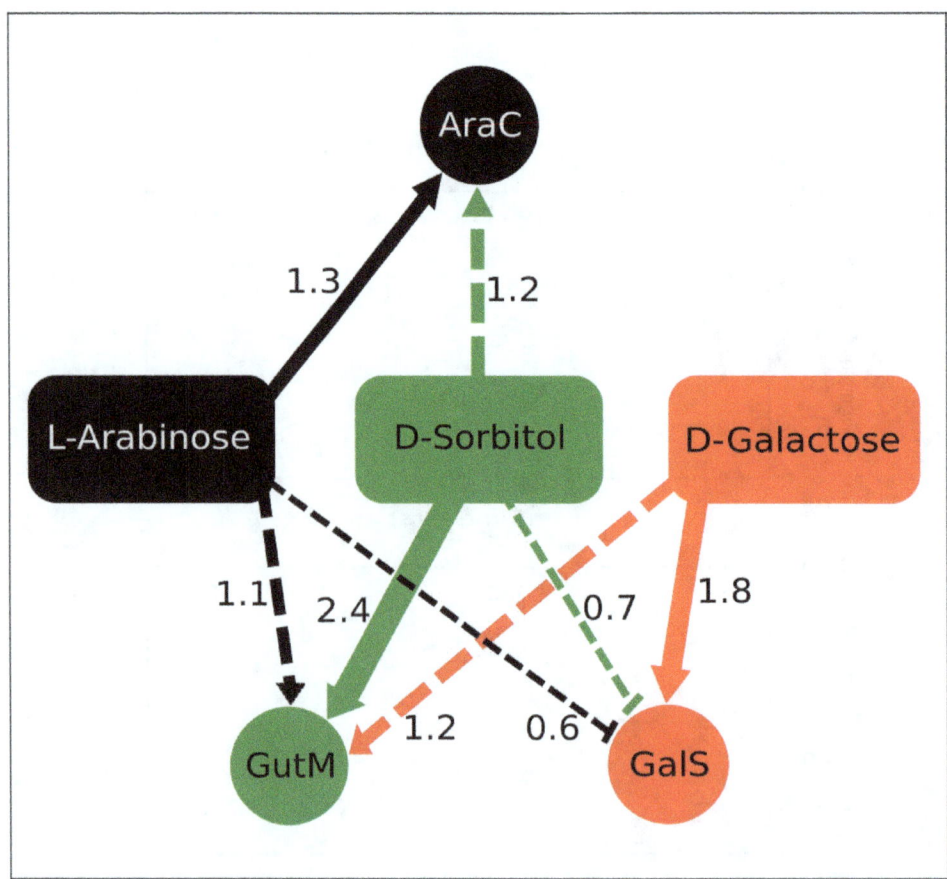

Fig. 4 Asymmetrical response of transcription factor promoters to sugar signals. The values next to the arrows indicate the percent of response of each promoter to every one of the three alternative sugars, as compared to the effect caused by glucose exhaustion solely. Continuous lines indicate the response of transcription factor to their effector sugar, while dashed lines indicate cross-response to the presence of other sugars.

Note from Fig. 4 that the promoter that is most responsive to sugars other than its own one is *araC*, followed by *gutM* and *galS* promoters. This behavior is interesting because a partial turn on of promoters due to non specific stimulus is an indicator of a putative conditioned behavior already observed in sugars consumption in *E. coli* (Tagkopoulos et al., 2008). Our results suggest that the promoter most prone to conditional behavior is that of *araC*.

Finally, the sugar that most enhances the activity of transcription-factor promoters other than the one specific to it is galactose. This suggests that, in agreement with (Liu et al., 2005), the worse a carbon source that sustains bacterial growth, the more it positively affects the activity of alternative-sugar catabolic operons.

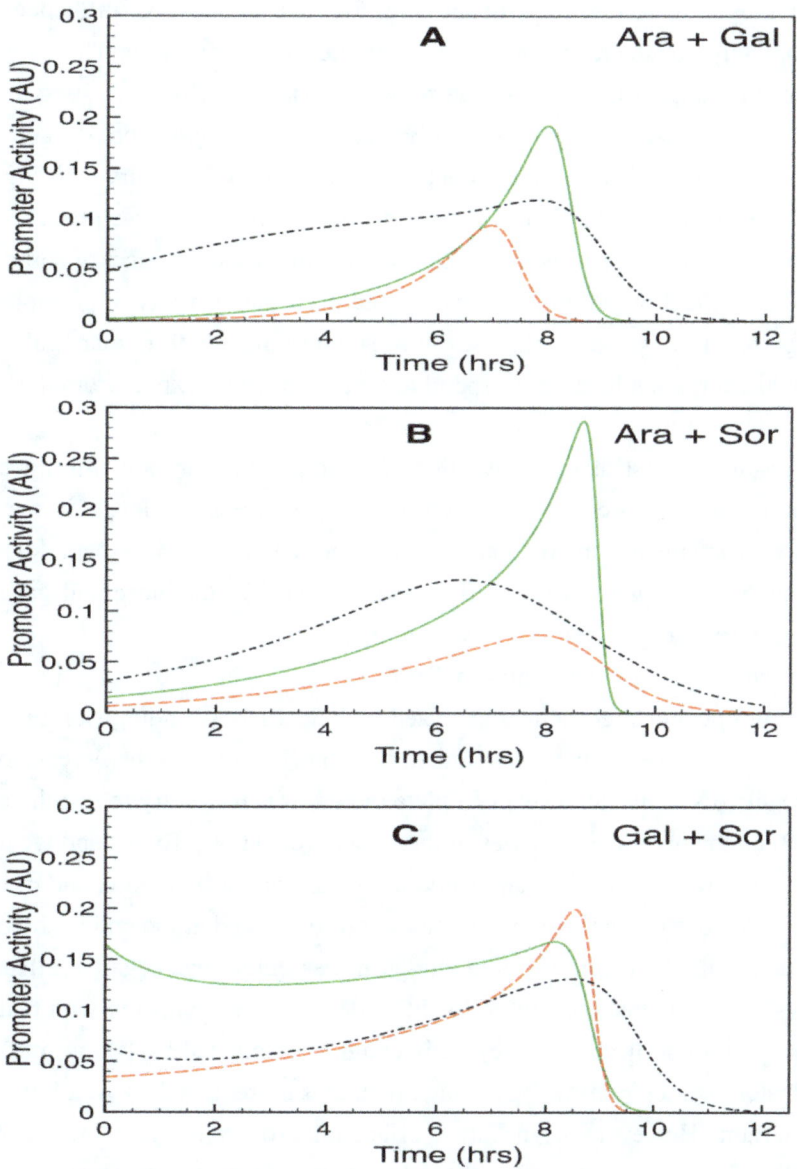

Fig. 5 Transcription-factor promoter activities in the presence of pairs of alternative sugars (M9 + 0.03% glucose + 0.2% sugar 1 + 0.2% sugar 2). **A)** Promoters activities in the presence of arabinose and galactose. **B)** Promoter activities in the presence of arabinose and sorbitol. **C)** Promoter activities in the presence of galactose and sobitol. The color code in all panels is as follows: *araC*, black; *gutM*, green; *galS*, red.

3.4 Promoter activities reflect carbon-use hierarchy

We finally performed experiments in which bacteria were grown in the presence of a limiting quantity of glucose (0.03%) and a mix of two alternative sugars (0.2% each). The rational behind this experiment was that once glucose is exhausted, *E. coli* should be forced to consume one of the two alternative sugars present in the milieu, and that this decision might be evidenced by the activity of the promoter associated to the transcription factor regulating the catabolic operon of the sugar of choice. The results of these experiments are reported in Fig. 5.

Notice that whenever bacteria are cultured in the presence of arabinose, promoter *araC* becomes active before the other two. The explanation of this observation is straightforward under the assumption that arabinose is preferred by bacteria over sorbitol and galactose.

Note from the arabinose + galactose experiment (Fig. 5A) that the positive influence of galactose upon *galS* is capable of completely counteracting the negative influence of arabinose. This is consistent with the supposition that arabinose is consumed before galactose by bacteria. It is also interesting that the rather small positive effects that both arabinose and galactose individually have on *gutM* boost each other to render a combined over-expression of more than 100%. A detailed observation of the curves in Fig. 5B (arabinose + sorbitol experiment), reveals that their amplitudes completely agree with the interaction scheme in Fig. 4.

In Fig. 5C (sorbitol + galactose experiment) we can see that the negative effect of sorbitol on *galS* is fully counteracted by the positive effect of galactose. However, the maximum activity of promoter *galS* is posterior to that of promoter *gutM*. This suggests that sorbitol is consumed before than galactose by bacteria. Furthermore, the maximal expression levels correspond to what one would expect from the interaction scheme in Fig. 4.

In summary, our results suggest that the investigated alternative sugars are consumed in the following order: arabinose, sorbitol and galactose. Moreover, the maximum expression levels in the experiments with two alternative sugars agree with the interaction scheme reported in Fig. 4, except for the expression of promoter *gutM* in the arabinose + galactose experiment. It seems that arabinose and galactose synergically make promoter *gutM* increase its expression level by more than 100%.

3.5 CRP as a global coordinator for carbon metabolism

All the transcription-factor operator regions here analyzed include a DNA-binding site for CRP, in addition to being self-regulated. In the previous sections we studied the contribution of specific catabolite signals to the activity of their local regulators. Thus, to have a complete picture it is necessary to test the effect of CRP (and indirectly that of cAMP) on the promoter activities of these local regulators. To this end we used a CRP mutant strain (Baba et al., 2006) as a receptor of the transcriptional fusions analyzed before, and measured growth and promoter activities when glucose is limiting at the beginning (0.03%) and at the end of the culture (0.4%), see Fig. 6. A first observation is that deletion of *crp*, although not essential, has negative effects on the bacterial growth rate (Figs. 6B vs 6A). Note that the culture final OD decreases as compared with that of the strain with an intact *crp* gene (comparable on 0.4% glucose). This could be explained by taking into consideration that CRP is a global coordinator of *E. coli* physiology, which regulates more, that 30% of all the genes with known regulation in this bacterium. However, the negative effect on growth is more pronounced in the strains harboring the transcriptional fusions of *gfp* with *galS* and *gutM* promoters that that with *araC*. We do not have a consistent explanation for this observation.

Regarding the promoter activities in the absence of *crp,* when glucose is depleted at the beginning of the culture (0.03%), the promoters activities changed as follows, as compared with the intact-*crp* strain: the *araCp* activity profile changed neither its amplitude nor the time at which the maximum value was achieved, yet the profile is now narrower; the maximal *gutMp* activity level was doubled, although it was retarded by more than one hour; the *galSp* activity profile became wider and its maximal level decreased by about 50% (Figs. 6D vs 6C).

On the other hand, when glucose is exhausted at the end of the culture (0.4%) all the promoter activities are diminished and retarded except that of *araCp* (Figs. 6F vs 6E). Together, these observations indicate that CRP contributes to the fitness and performance of bacterial growth and coordinates the response of alternative regulatory machinery for the use of alternative carbon sources in *E. coli*, although it seems not to be essential for the transcriptional response of local repressors.

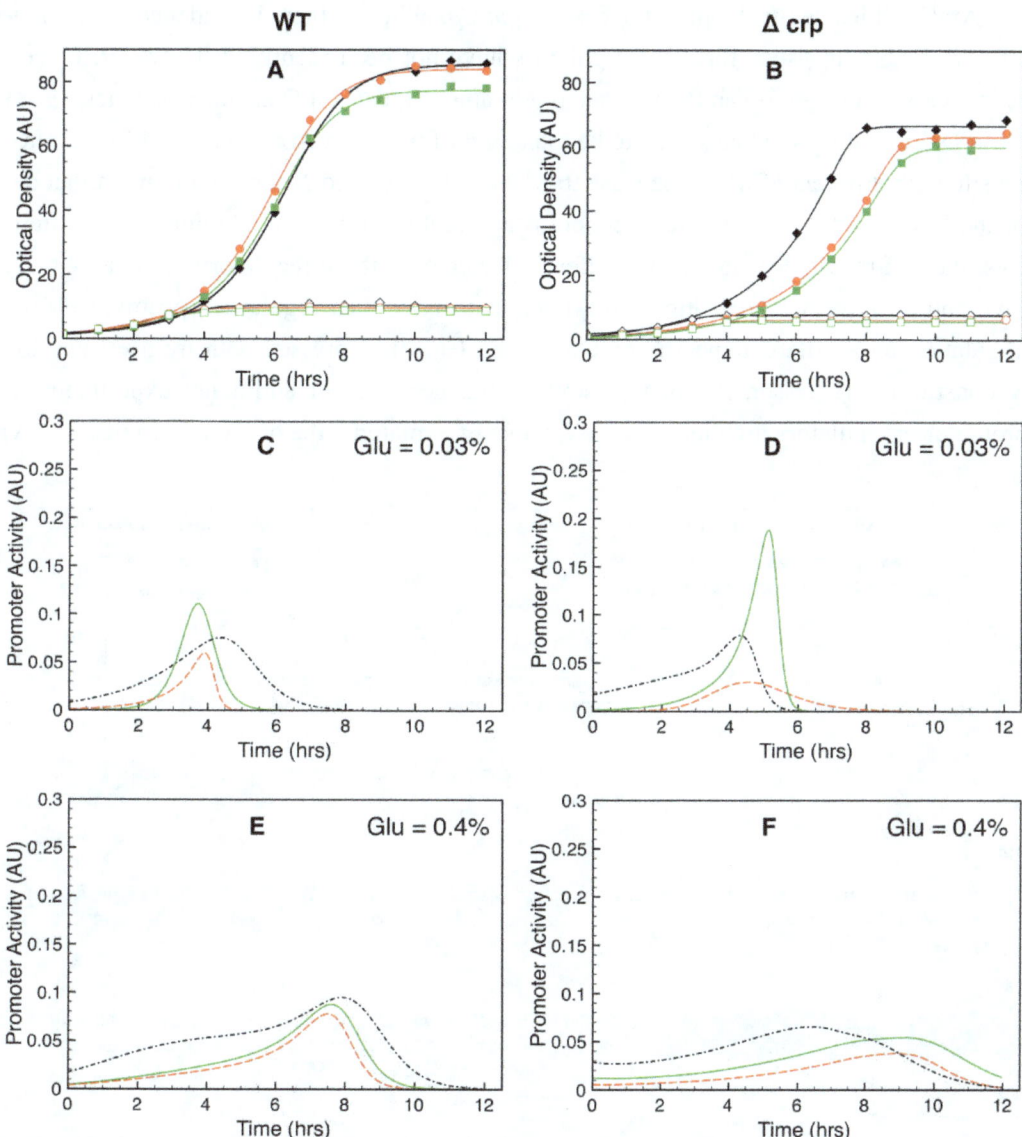

Fig. 6 Effect of CRP on the promoter activity of various catabolic repressors. Growth of bacteria harboring the transcriptional fusions on wild type background (A) and on *crp* deletion (B). Solid (empty) symbols correspond to a glucose concentration of 0.4% (0.03%). Promoter activities with glucose limited at the beginning of the culture on WT (C) and on Δ*crp* mutant (D). Promoter activities without glucose limitation at the beginning of the culture on WT (E) and on *crp* deletion (F). The color code in all panels is as follows: *araC*, black; *gutM*, green; *galS*, red. (C) and (E) are the same as Figs. 3B and 3C but are repeated here for the sake of clarity.

3.6 Could the architecture of regulatory circuits be responsible of this hierarchical behavior?

Given the displayed activities of specific regulators for alternative sugar consumption we decided to analyze the operator regions of the corresponding promoters, Fig. 7. It has been proposed that CRP recruits the *E. coli* RNA polymerase differentially to distinct promoters (Parkinson et al., 1996; Lee et al., 2012). As a result, CRP has different modes of activation. Briefly, in Class I activation, CRP is bound to an upstream site of the -35 element of promoters and contacts RNA polymerase throughout their αCTD subunit. In Class II activation, CRP binds to a site that overlaps the -35 element and contacts RNA polymerase throughout the domain 4 of σ⁷⁰subunit. Of the promoters here studied *araCp* correspond to the class I whereas *galS* and *gutMp* correspond to the class II (Fig. 7). In addition, there are some other differences on the promoters' architecture. For instance, *araCp* regulation involves a DNA loop by interaction of AraC dimmers (Gallegos, 1997) and the binding sites for CRP and AraC fall inside this loop. In the case of *gutMp*, although there is evidence of regulation by GutM and SrlR, the DNA binding sites for these regulators have not been identified. However, it seems that SrlR repress the transcription of *gutMp* and GutM positively auto-regulates it (Yamada and Saier, 1988). Regarding *galSp*, it has a DNA binding site that seems to be the target of two repressors: GalS and GalR. These regulators bind such a site with different affinities because they share a conserved N-terminal domain that determines the affinity for the DNA-binding region (Geanacopoulos and Adhya, 1997). In addition to the different promoter architectures, the self-regulatory circuits are different for each of these regulatory systems: AraC is subject to both positive and negative self-regulation; GutM seems to have dual regulation, positive self-regulation and negative regulation by a different specific regulator, and GalS is repressed both by itself and by and related apparently constitutive repressor. All of this stresses the necessity of additional experiments to determine whether promoter or regulatory architectures are capable of explaining the observed promoter activities.

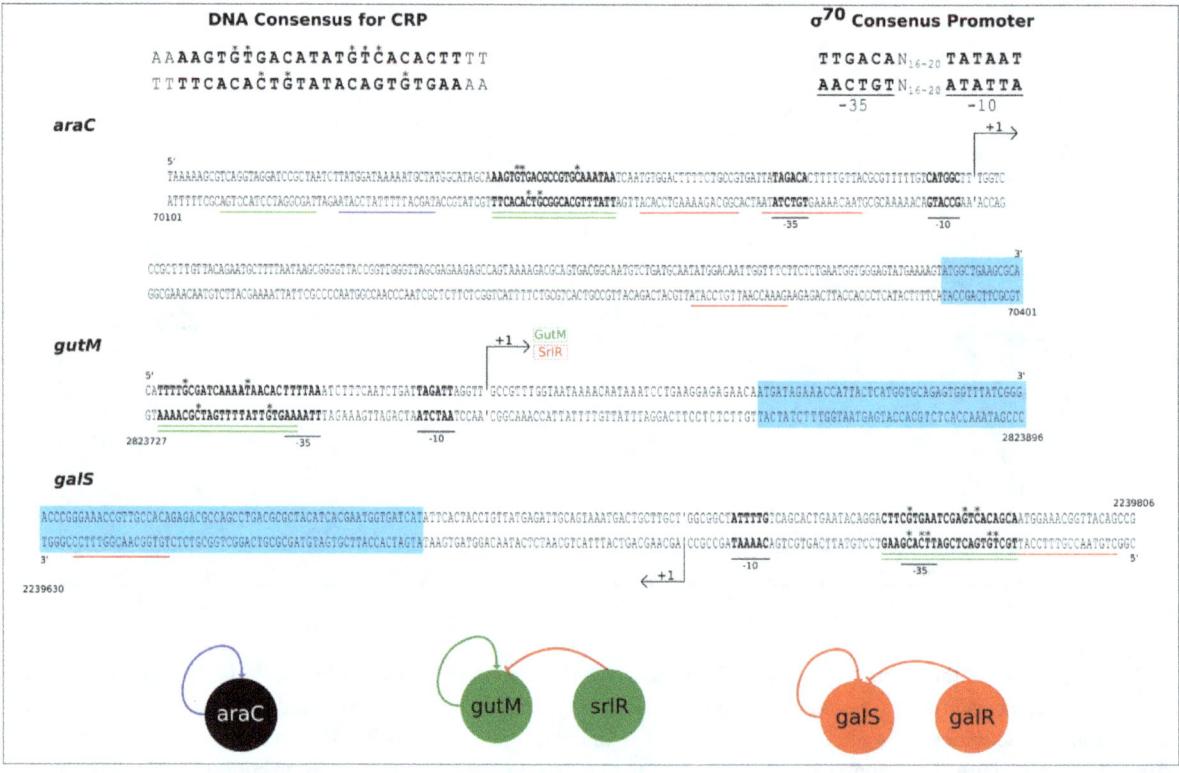

Fig. 7 Binding sites of CRP and σ70 in the studied promoters. The consensus binding sites for CRP and σ70 are show. Nucleotides marked with asterisks in the consensus for CRP are those that interact with the CRP protein (Parkinson et al., 1996). Numbers at the end of each sequence denotes genome positions of promoters. The initial translated amino acids of all regulatory proteins are shown in cyan. The DNA self-regulatory binding sites for each transcription factor are represented with single underlines. Double underlines are employed to denote the DNA binding sites for CRP. The -10 and -35 elements where the σ⁷⁰subunit of RNA polymerase binds are also marked. Colored underlines denote the type of regulation: green for activation, red for repression and blue for dual regulation. Finally, simplified schemes of regulatory switches for promoters are show.

4 Conclusions

It has been known for a long time that *E. coli* preferably consumes glucose over other carbon sources. However, we lack a complete knowledge of how this is achieved and regulated at the molecular level. In this work we present a proof of principle that permits to track the hierarchical use of carbon sources by following bacterial growth and promoter activities of the regulatory proteins that respond to specific sugars. We were able to identify the following order for the preferential use of carbon sources by *E. coli*: glucose > arabinose > sorbitol > galactose. A detailed analysis of regulator promoters for the corresponding catabolic operons indicates that this behavior can be due to at least three factors: 1) the coordinated activation of local regulators by the global regulator CRP, 2) the asymmetrical responses of transcription factors for specific and unspecific sugars and, 3) the architecture of promoters and operon-regulatory circuits. However, many questions remain open regarding the control mechanisms leading to this hierarchical behavior. Answering them will require a large amount of both experimental and mathematical modeling work. Finally, *E. coli* can consume more carbon sources than the ones here studied. It is still pending to test them to have a more complete scheme regarding the preferential use of carbon source in this bacterium.

Acknowledgments

We thank Alejandro Hernández-Morales and Susana Ruiz for their assistance in the construction of transcriptional fusions. This work was supported by CONACYT grant 103686 including an undergraduate fellow to R.S.P.-A. A.M.-A. conceived the study. R.S.P.-A. performed experiments. M.S., R.S.P.-A. and E. G.-V. analyzed data and prepared figures. A.M.-A. and M.S. drafted the manuscript. All the authors approved the manuscript.

References

Baba T, Ara T, Hasegawa M, et al. 2006. Construction of *Escherichia coli* K-12 in-frame, single-gene knockout mutants: the Keio collection. Molecular Systems Biology, 2: 0008

Beisel CL, Storz G. 2011. The base-pairing RNA spot 42 participates in a multi output feedforward loop to help enact catabolite repression in *Escherichia coli*. Molecular Cell, 41: 286-297

Chen S, Oldham ML, Davidson AL, et al. 2013. Carbon catabolite repression of the maltose transporter revealed by X-ray crystallography. Nature, 499: 364-368

Cohn M. 1957. Contributions of studies on the beta-galactosidase of *Escherichia coli* to our understanding of enzyme synthesis. Bacteriology Review, 21: 140-168

Epstein W, Rothman-Denes LB, Hesse J. 1975. Adenosine 3':5'-cyclic monophosphate as mediator of catabolite repression in *Escherichia coli*. Proceedings of the National Academy of Sciences of USA, 72: 2300-2304

Gallegos MT, Schleif R, Bairoch A, et al. 1997. AraC/XylS family of transcriptional regulators. Microbiology and Molecular Biology Reviews, 61: 393-410

Geanacopoulos M, Adhya S. 1997. Functional characterization of roles of GalR and GalS as regulators of the gal regulon. Journal of Bacteriology, 179: 228-234

Görke B, Stülke J. 2008. Carbon catabolite repression in bacteria: many ways to make the most out of nutrients. Nature Reviews Microbiology, 6: 613-624

Gottesman S. 1984. Bacterial regulation: global regulatory networks. Annual Review of Genetics, 18: 415-441

Jacob F, Monod J. 1961. Genetic regulatory mechanisms in the synthesis of proteins. Journal of Molecular Biology, 3: 318-356

Jones-Mortimer MC, Kornberg HL. 1974. Genetic control of inducer exclusion by *Escherichia coli*. FEBS Letters, 48: 93-95

Lee DJ, Minchin SD, Busby SJ. 2012. Activating transcription in bacteria. Annual Review of Microbiology, 66: 125-152

Liu M, Durfee T, Cabrera JE, et al. 2005. Global transcriptional programs reveal a carbon source foraging strategy by *Escherichia coli*. Journal of Biological Chemistry, 280: 15921-15927

Longabaugh WJR, Davidson EH, Bolouri H. 2009. Visualization, documentation, analysis, and communication of large-scale gene regulatory networks. Biochimica et Biophysica Acta, 1789: 363-374

Magasanik B. 1961. Catabolite repression. Cold Spring Harbor Symposium on Quantitative Biology, 26: 249-256

Martínez-Antonio A, Collado-Vides J. 2003. Identifying global regulators in transcriptional regulatory networks in bacteria. Current Opinion in Microbiology, 6: 482-489

Martínez-Antonio A, Velázquez-Ramírez DA, Mondragón-Sánchez J, et al. 2012. Hierarchical dynamics of a transcription factors network in *E. coli*. Molecular BioSystem, 8: 2932-2936

Monod J. 1942. Recherchessur la Croissance des Cultures Bactériennes. Hermann et Cie, Paris, France

Park YH, Lee BR, Seok YJ, et al. 2006. In vitro reconstitution of catabolite repression in *Escherichia coli*. Journal of Biological Chemistry, 281: 6448-6454

Parkinson G, Wilson C, Gunasekera A, et al. 1996. Structure of the CAP-DNA complex at 2.5 Å resolution: a complete picture of the protein-DNA interface. Journal of Molecular Biology, 260: 395-408

Rabinowitz JD, Silhavy TJ. 2013. Systems biology: metabolite turns master regulator. Nature, 500: 283-284

Salgado H, Peralta-Gil M, Gama-Castro S, et al. 2013. RegulonDB v8. 0: omics data sets, evolutionary conservation, regulatory phrases, cross-validated gold standards and more. Nucleic Acids Research, 41: D203–D213

Sellitti MA, Pavco PA, Steege DA. 1987. Lac repressor blocks in vivo transcription of lac control region DNA. Proceedings of the National Academy of Sciences of USA, 84: 3199-3203

Tagkopoulos I, Liu YC, Tavazoie S. 2008. Predictive behavior within microbial genetic networks. Science Signaling, 320: 1313

Walker HH, Winslow CE, Mooney MG. 1934. Bacterial cell metabolism under anaerobic conditions. Journal of General Physiology, 17: 349-357

Wright PR, Richter AS, Papenfort K, et al. 2013. Comparative genomics boosts target prediction for bacterial small RNAs. Proceedings of the National Academy of Sciences of USA, 110: E3487-96

Yamada M, Saier MH Jr. 1988. Positive and negative regulators for glucitol (gut) operon expression in *Escherichia coli*. Journal of Molecular Biology, 203: 569-583

You C, et al. 2013. Coordination of bacterial proteome with metabolism by cyclic AMP signalling. Nature, 500: 301-306

Zaslaver A, Bren A, Ronen M, et al. 2006. A comprehensive library of fluorescent transcriptional reporters for *Escherichia coli*. Nature Methods, 3: 623-628

16

Analysis of word occurrence frequency and word association in English text file: A big data analytics method

YanHong Qi[1], GuangHua Liu[2], WenJun Zhang[3]

[1]Sun Yat-sen University Libraries, Sun Yat-sen University, Guangzhou 510275, China

[2]Guangdong AIB Polytech College, Guangzhou 510507, China

[3]School of Life Sciences, Sun Yat-sen University, Guangzhou 510275, China

E-mail: qiyh@mail.sysu.edu.cn, zhwj@mail.sysu.edu.cn, ghliu@gdaib.edu.cn

Abstract

In present study, I presented an algorithm for analysis of word occurrence frequency and word association in English text file. Various delimiters were used for splitting words. In addition, common used grammatical words are ignored in word occurrence and association analysis. All different words were listed according to word occurrence frequency from the greater to the smaller. Word association was detected by using one-dimensional ordered cluster analysis. The words fallen in the same class may likely have strong association. Theoretically, various classes at distinct clustering hierarchical level may represent different hierarchical topics. Java software of the algorithm was provided.

Keywords big data analytics; word splitting; word occurrence frequency; word association; English text; algorithm; software.

1 Introduction

Big data analytics is a quickly growing technology, which has been using in various areas, including network biology. About the definition and explanation of big data, there are many different opinions (Zhang et al., 2013; Zhang, 2017a-d, 2017f, 2018a-b). In the general sense, big data is a collection of data that cannot be perceived, acquired, managed, processed, and serviced with traditional IT technologies and hardware and software tools for a limited period of time. In 2010, the Apache Hadoop organization defined big data as large-scale data sets that ordinary computer software cannot capture, manage, and manage in an acceptable time frame. International Data Corporation (IDC) held that big data technology describes a new generation of technology and architecture that, through high-speed acquisition, discovery or analysis, we can extract the economic value of a large amount of data. From this point of view, the characteristics of big data can be summarized as 4Vs, namely volume (massive size), variety (numerous modals), velocity (generated quickly) and value (the value is very remarkable but the density is very low) (Zhang et al., 2013). The 4Vs definition has been widely recognized. It stresses the significance and necessity of big data - that is, mining the huge value inside big data. Therefore, the most important topic of big data is how to extract valued information from large-scale, numerous, and rapidly growing data sets (Zhang, 2018).

In this study, I try to present an algorithm for analysis of word occurrence frequency and word association in English text file. Java is a powerful and widely used development tool (Zhang, 2011a-c; Zhang and Zhan, 2011; Zhang, 2012a; Zhang and Liu, 2012). Java software of the algorithm is provided.

2 Algorithm and Software
2.1 Algorithm
Read a text from a local English text file (*.txt), or from internet URL (http://www.../*.htm, or *.html). The following delimiters are used for splitting words:

Space, no-break space, soft hyphen, !, ", #, $, %, &, ', \r, \n, (,), * + , -, ., /, :, ;, <, =, >, ?, @, [,], ^, _, {, |, }, ~, ¯, °, ±, ·, ×, ÷, <, >, ∧, ∨, ×, \

As the default setting, the following common used grammatical words are ignored in word occurrence analysis:

the,and,of,is,are,has,have,was,were,had,etc,to,be,for,with,or,a,let,an,all,over,must,as,any,each,can,by,become,what,there,when,which,where,that,if,then,thus,than,we,you,our,your,yours,us,it,its,they,will,would,shall,should,may,from,their,while,at,in,on,how,once,again,ago,before,after,between,both,either,just,each,often,usually,without,use,also,high,low,large,small,great,only,most,go,going,went,but,some,such,as,who,whose,under,these,those,this

Users can enter their own new words to be ignored in word occurrence analysis.

All different words are listed according to word occurrence frequency from the greater to the smaller.

Closely associated words tend to occur simultaneously and are expected to have the similar occurrence frequencies. Therefore, by using one-dimensional ordered cluster analysis (Zhang and Fang, 1982; Zhang, 2017; Qi, 2005), the words with similar occurrence frequencies may be clustered into the same class. By doing so, we can try to find all closely associated words in the text file.

Suppose there are totally n different words, arranged with occurrence frequency rank from the greater to the smaller, $x_1, x_2, ..., x_n$. The distance between adjacent two words is the absolute of frequency difference between them

$$r_{i\,i+1} = |x_i - x_{i+1}| \quad i=1,2,...,n-1$$

Find the shortest distance $r_{i\,i+1}$, $i=1,2,...,n-1$, and combine the two words i and $i+1$ into the same cluster. Similarly, take the minimum between-word distance as the between-cluster distance. Find the two clusters with minimum between-cluster distance and combine into a single cluster. Finally all words are combined into a single cluster. The minimum between-cluster distance at a clustering hierarchical level is defined as the between-cluster distance of this hierarchical level. Theoretically, various classes at distinct clustering hierarchical level may represent different hierarchical topics.

Users can enter the required minimum difference of word occurrence frequency for cluster analysis.

2.2 Software

The software of the algorithm above, parseEnglish, is a Java applet (Fig. 1), developed using JDK (Java Development Kit). It includes the following major methods:

urlConnection(): Connect specified URL and obtain online HTML text.

parseArray(): Split words and store them in a vector. Delimiters above are used to separate words. Grammatical words above are ignored. The new words entered by the user are ignored also.

ranking(): Ranking words according to their occurrence frequency, from the greater to the smaller.

stat(): Calculate occurrence frequency of distinct words.

oneDimCluster(): Make one-dimensional ordered clustering for word series. Only the words with the occurrence frequency not less than specified minimum threshold frequency are clustered.

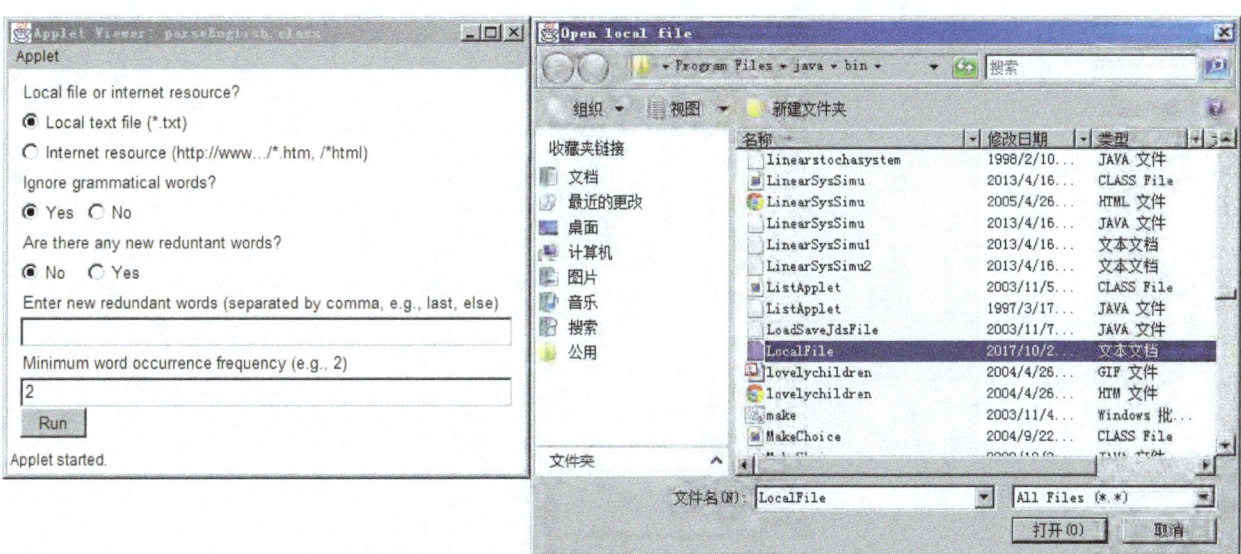

Fig. 1 Java applet of the algorithm.

The following are Java codes of three methods, parseArray(), ranking(), stat(), and oneDimCluster():

```java
public String[] parseArray(String s)
{
    Vector vector = new Vector();
    String s1 = s;
    int i = 0x186a0;
    String as1[] = {
        "the", "and", "of", "is", "are", "has", "have", "was", "were", "had",
        "etc", "to", "be", "for", "with", "or", "a", "let", "an", "all",
        "over", "must", "as", "any", "each", "can", "by", "become", "what", "there",
        "when", "which", "where", "that", "if", "then", "thus", "than", "we", "you",
        "our", "your", "yours", "us", "it", "its", "they", "will", "would", "shall",
        "should", "may", "from", "their", "while", "at", "in", "on", "how", "once",
        "again", "ago", "before", "after", "between", "both", "either", "just", "each", "often",
        "usually", "without", "use", "also", "high", "low", "large", "small", "great", "only",
        "most", "go", "going", "went", "but", "some", "such", "as", "who", "whose",
        "under", "these", "those", "this"
    };
    char ac[] = {
        '\u0020','\u00A0','\u00AD','\u0021','\u0022','\u0023','\u0024','\u0025','\u0026','\u0027','\r','\n','\t','\u0028','\u0029'
```

```
      ,'\u002A','\u002B','\u002C','\u002D','\u002E','\u002F','\u003A','\u003B','\u003C','\u003D','\u003E','\u003F','\u004
      0','\u005B','\u005D','\u005E','\u005F','\u007B','\u007C','\u007D','\u007E','\u00AF','\u00B0','\u00B1','\u00B7','\u0
      0D7','\u00F7','\u02C2','\u02C3','\u02C4','\u02C5','\u02DF','\\'
};
String as[];
if(choword == 1)
{
      int j2 = stword.length();
      int k2 = as1.length;
      String as2[] = new String[j2];
      String s3 = stword;
      int j = 0;
      int k1;
      do
      {
            k1 = s3.indexOf(',');
            if(k1 >= 0)
            {
                  as2[j] = s3.substring(0, k1).trim();
                  s3 = s3.substring(k1 + 1).trim();
                  j++;
            } else
            {
                  as2[j] = s3;
            }
      } while(k1 >= 0);
      as = new String[k2 + j + 1];
      for(int l = 0; l < j + k2 + 1; l++)
            if(l < k2)
                  as[l] = as1[l];
            else
                  as[l] = as2[l - k2];
} else
{
      as = new String[as1.length];
      as = as1;
}
int i2;
label0:
      do
      {
            i2 = i;
            for(int i1 = 0; i1 < ac.length; i1++)
                  if((s.indexOf(ac[i1]) < i2) & (s.indexOf(ac[i1]) >= 0))
                        i2 = s.indexOf(ac[i1]);
            if(i2 == i)
                  break;
            if(i2 >= 0)
            {
                  String s2 = s.substring(0, i2).trim();
                  s = s.substring(i2 + 1).trim();
                  if(cho == 1)
                  {
                        for(int j1 = 0; j1 < as.length; j1++)
                              if(s2.equalsIgnoreCase(as[j1]))
                                    continue label0;
                  }
                  vector.addElement(s2);
```

```
            } else
            {
                  vector.addElement(s);
            }
      } while(i2 >= 0);
      int l2 = vector.size();
      args1 = new String[l2];
      String as3[] = new String[l2];
      for(int k = 0; k < l2; k++)
            as3[k] = vector.elementAt(k).toString();
      return as3;
}


      public String[][] ranking(String as[][])
      {
            int i = as.length;
            int ai[] = new int[i];
            int ai1[] = new int[i];
            String as1[] = new String[i];
            String as2[][] = new String[i][2];
            for(int j = 0; j <= i - 1; j++)
            {
                  as1[j] = "";
                  for(int k = 0; k <= 1; k++)
                        as2[j][k] = "";
            }
            for(int l = 0; l <= i - 1; l++)
                  ai1[l] = Integer.parseInt(as[l][1]);
            for(int i1 = 0; i1 <= i - 1; i1++)
            {
                  ai[i1] = ai1[i1];
                  as1[i1] = as[i1][0];
            }
            for(int j1 = 0; j1 <= i - 2; j1++)
            {
                  int k1 = j1;
                  for(int j2 = j1; j2 <= i - 2; j2++)
                        if(ai[j2 + 1] >= ai[k1])
                              k1 = j2 + 1;
                  int k2 = ai[j1];
                  String s = as1[j1];
                  ai[j1] = ai[k1];
                  as1[j1] = as1[k1];
                  ai[k1] = k2;
                  as1[k1] = s;
            }
            for(int i2 = 0; i2 <= i - 1; i2++)
            {
                  as2[i2][0] = as1[i2];
                  as2[i2][1] = Integer.toString(ai[i2]);
            }
            return as2;
      }


      public String[][] stat(String as[])
      {
            int i = as.length;
            String as1[][] = new String[i][2];
```

```
        for(int j = 0; j <= i - 1; j++)
        {
            for(int k = 0; k <= 1; k++)
                as1[j][k] = "";
        }
        l1 = 0;
label0:
        for(int l = 0; l <= i - 1; l++)
        {
            for(int i1 = 0; i1 <= l1; i1++)
                if(as1[i1][0].equalsIgnoreCase(as[l]))
                    continue label0;
            as1[l1][0] = as[l];
            int j1 = 0;
            for(int i2 = 1; i2 <= i - 1; i2++)
                if(as[l].equalsIgnoreCase(as[i2]))
                    j1++;
            as1[l1][1] = Integer.toString(j1);
            l1++;
        }
        String as2[][] = new String[l1][2];
        for(int k1 = 0; k1 <= l1 - 1; k1++)
        {
            for(int j2 = 0; j2 <= 1; j2++)
                as2[k1][j2] = as1[k1][j2];
        }
        args2 = new String[l1][2];
        return as2;
    }

    public void oneDimCluster(int ai[], String as[])
    {
        int i = ai.length;
        String as1[] = new String[i + 2];
        int ai1[][] = new int[i + 2][i + 2];
        int ai4[] = new int[i + 2];
        int ai2[] = new int[i + 2];
        double ad[] = new double[i + 2];
        int ai3[] = new int[i + 2];
        double ad1[] = new double[i + 2];
        double ad2[] = new double[i + 2];
        for(int k = 1; k <= i; k++)
            ai4[k] = ai[k - 1];
        String s = "";
        for(int l = 1; l <= i - 1; l++)
            ad[l] = Math.abs(ai4[l] - ai4[l + 1]);
        for(int i1 = 1; i1 <= i - 1; i1++)
            ad1[i1] = ad[i1];
        for(int j1 = 1; j1 <= i - 2; j1++)
        {
            int j4 = j1;
            for(int j3 = j1; j3 <= i - 2; j3++)
                if(ad1[j3 + 1] <= ad1[j4])
                    j4 = j3 + 1;
            double d = ad1[j1];
            ad1[j1] = ad1[j4];
            ad1[j4] = d;
        }
```

```
                ad1[0] = 0.0D;
                int j = 1;
     label0:
                for(int k4 = 0; k4 <= i - 1; k4++)
                {
                    for(int k1 = 0; k1 <= k4 - 1; k1++)
                        if(Math.abs(ad1[k1] - ad1[k4]) <= 9.9999999999999995E-007D)
                            continue label0;
                    for(int i2 = 1; i2 <= i - 1; i2++)
                        if(ad[i2] - ad1[k4] <= 9.9999999999999995E-007D)
                            ai3[i2] = 1;
                        else
                            ai3[i2] = 0;
                    for(int j2 = 1; j2 <= i; j2++)
                        ai1[j][j2] = 0;
                    int k2 = 1;
                    ai2[j] = 1;
     label1:
                    while(k2 <= i - 1)
                    {
                        if(ai3[k2] == 0)
                        {
                            ai1[j][k2] = ai2[j];
                            ai1[j][k2 + 1] = ai2[j] + 1;
                        } else
                        if(ai3[k2] == 1)
                        {
                            for(int k3 = k2; k3 <= i - 1; k3++)
                            {
                                if(ai3[k3] == 0)
                                {
                                    k2 = k3;
                                    continue label1;
                                }
                                ai1[j][k3] = ai2[j];
                                ai1[j][k3 + 1] = ai2[j];
                                if(k3 == i - 1)
                                    break label1;
                            }
                        }
                        k2++;
                        ai2[j]++;
                    }
                    s = s + "\n";
                    s = s + "Cluster fineness (between-word frequency difference)=" + String.valueOf((double)(int)(ad1[k4] *
10000D) / 10000D) + "\n";
                    ad2[j] = ad1[k4];
                    String s1 = "";
                    for(int l2 = 1; l2 <= ai2[j]; l2++)
                    {
                        s1 = s1 + "(";
                        for(int l3 = 1; l3 <= i; l3++)
                            if(ai1[j][l3] == l2)
                                s1 = s1 + as[l3 - 1] + " ";
                        s1 = s1 + ")   ";
                    }
                    s = s + s1;
                    j++;
```

```
}
j--;
output.editt1.appendText(s + "\n");
String s2 = "One-dimensional Ordered Cluster";
int l4 = 1;
for(int i3 = 1; i3 <= ai2[1]; i3++)
{
    for(int i4 = 1; i4 <= i; i4++)
        if(ai1[1][i4] == i3)
        {
            as1[l4] = as[i4 - 1];
            l4++;
        }
}
(new GraphicsFrame(new ClusterGraphics(as1, j, i, i, ad2[j], ad2, ai2, ai1), s2)).resize(710, 560);
}
```

3 Application Example

Set the required minimum difference of word occurrence frequency for cluster analysis as 2. Table 1 and Fig. 2 show partial results of word occurrence frequencies of the mixed abstract text of Zhang (2012, 2015, 2016).

Table 1 Partial results of word occurrence frequencies of the mixed abstract text of Zhang (2012b, 2015, 2016)

Word	Frequency	%	Word	Frequency	%
network	16	4.507	degree	3	0.845
node	13	3.661	Various	3	0.845
Community	11	3.098	initial	3	0.845
species	9	2.535	Lamda	3	0.845
I	8	2.253	factor	3	0.845
evolution	8	2.253	dynamics	3	0.845
assembly	7	1.971	y	2	0.563
rule	7	1.971	x	2	0.563
present	6	1.69	emergency	2	0.563
connection	6	1.69	nodes	2	0.563
proposed	5	1.408	phenomena	2	0.563
model	5	1.408	disconnection	2	0.563
probability	5	1.408	composition	2	0.563
connections	5	1.408	think	2	0.563
organization	4	1.126	theory	2	0.563
self	4	1.126	rules	2	0.563
attraction	4	1.126	e	2	0.563
method	4	1.126	extinction	2	0.563
study	4	1.126	general	2	0.563
taxon	3	0.845	parameters	2	0.563
property	3	0.845	Effects	2	0.563
natural	3	0.845	simplified	2	0.563
growth	3	0.845	invasion	2	0.563
process	3	0.845	changes	2	0.563
random	3	0.845	number	2	0.563
Barabasi	3	0.845	different	2	0.563
Albert	3	0.845	1999	2	0.563
based	3	0.845	dependent	2	0.563
power	3	0.845	generate	2	0.563
addition	3	0.845	type	2	0.563
mechanism	3	0.845	Modeling	2	0.563

All words with frequency of 1 are omitted.

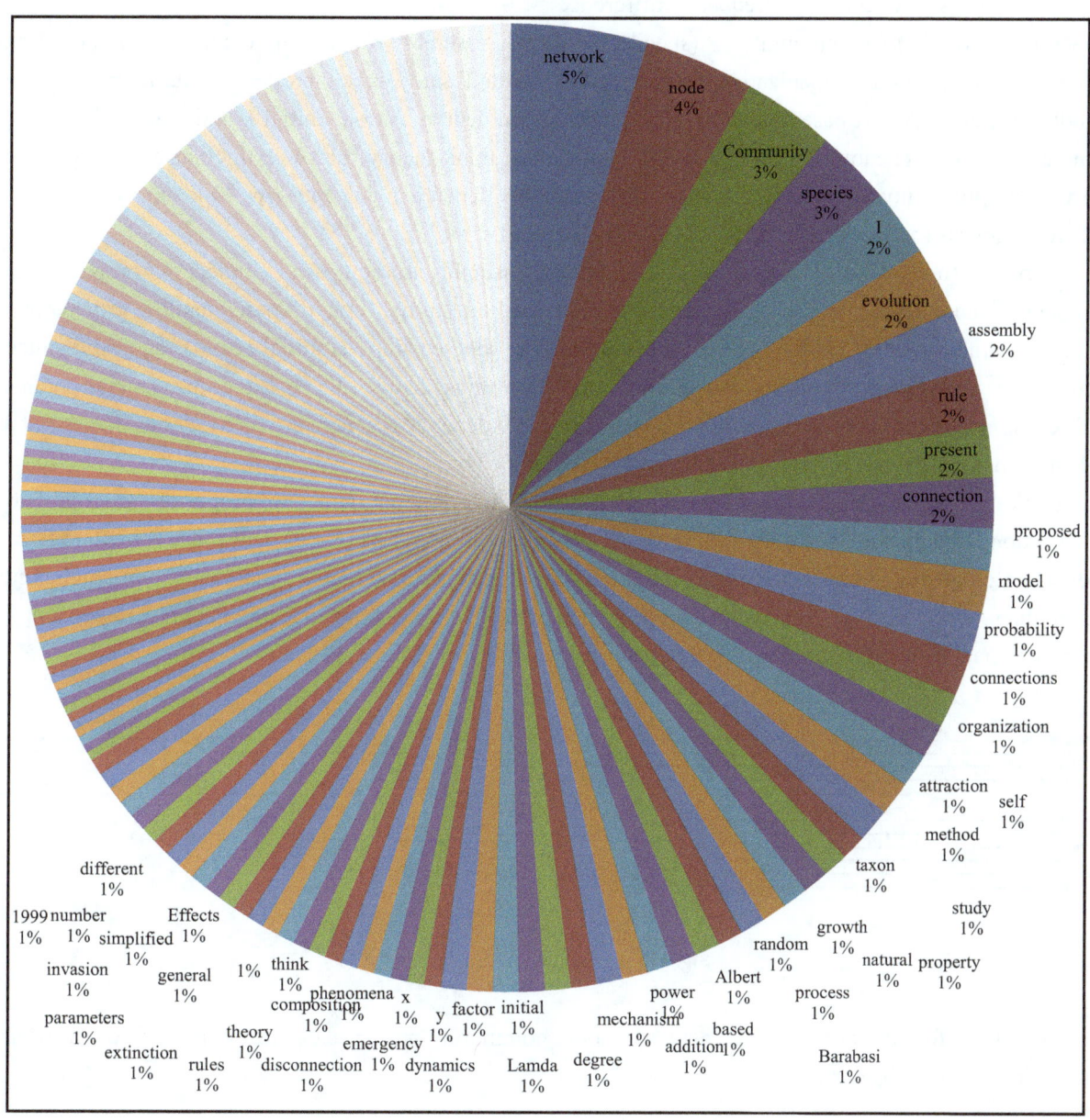

Fig. 2 Pie illustration of percentage of word occurrence.

The following and Fig. 3 show the results of one-dimensional ordered cluster analysis:

Cluster fineness (between-word frequency difference)=0.0

(network) (node) (Community) (species) (I evolution) (assembly rule) (present connection) (proposed model probability connections) (organization self attraction method study) (taxon property natural growth process random Barabasi Albert based power addition mechanism degree Various initial Lamda factor dynamics) (y x emergency nodes phenomena disconnection composition think theory rules e extinction general parameters Effects simplified invasion changes number different 1999 dependent generate type Modeling)

Cluster fineness (between-word frequency difference)=1.0

(network) (node) (Community) (species I evolution assembly rule present connection proposed model probability connections organization self attraction method study taxon property natural growth process random Barabasi Albert based power addition mechanism degree Various initial Lamda factor dynamics y x emergency nodes phenomena disconnection composition think theory rules e extinction general parameters Effects simplified invasion changes number different 1999 dependent generate type Modeling)

Cluster fineness (between-word frequency difference)=2.0

(network) (node Community species I evolution assembly rule present connection proposed model probability connections organization self attraction method study taxon property natural growth process random Barabasi Albert based power addition mechanism degree Various initial Lamda factor dynamics y x emergency nodes phenomena disconnection composition think theory rules e extinction general parameters Effects simplified invasion changes number different 1999 dependent generate type Modeling)

Cluster fineness (between-word frequency difference)=3.0

(network node Community species I evolution assembly rule present connection proposed model probability connections organization self attraction method study taxon property natural growth process random Barabasi Albert based power addition mechanism degree Various initial Lamda factor dynamics y x emergency nodes phenomena disconnection composition think theory rules e extinction general parameters Effects simplified invasion changes number different 1999 dependent generate type Modeling)

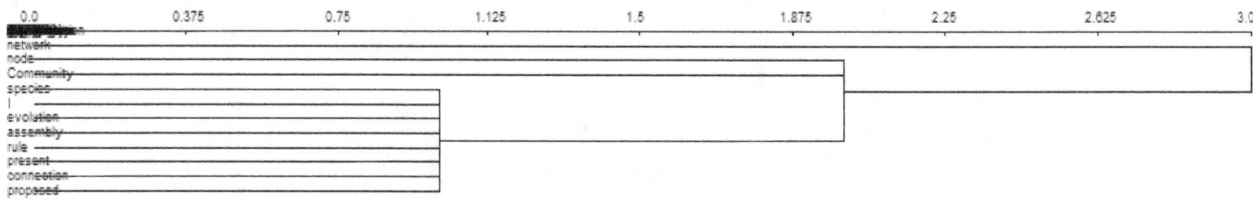

Fig. 3 One-dimensional ordered cluster.

It can be found that the words, network, node, community and species are the most frequent occurred words in the three abstracts.

Acknowledgment

We are thankful to the support of Discovery and Crucial Node Analysis of Important Biological and Social Networks (2015.6-2020.6), from Yangling Institute of Modern Agricultural Standardization, and High-Quality Textbook *Network Biology* Project for Engineering of Teaching Quality and Teaching Reform of Undergraduate Universities of Guangdong Province (2015.6-2018.6), from Department of Education of Guangdong Province, China.

References

Qi YH. 2005. The web software for ordered cluster analysis of sequential information. Information Science, 23(Suppl.): 99-101

Zhang WJ. 2011a. A Java algorithm for non-parametric statistic comparison of network structure. Network Biology, 1(2): 130-133

Zhang WJ. 2011b. A Java program for non-parametric statistic comparison of community structure. Computational Ecology and Software, 1(3): 183-185

Zhang WJ. 2011c. A Java program to test homogeneity of samples and examine sampling completeness. Network Biology, 1(2): 127-129

Zhang WJ. 2012a. A Java software for drawing graphs. Network Biology, 2(1): 38-44

Zhang WJ. 2012b. Modeling community succession and assembly: A novel method for network evolution. Network Biology, 2(2): 69-78

Zhang WJ. 2015. A generalized network evolution model and self-organization theory on community assembly. Selforganizology, 2(3): 55-64

Zhang WJ. 2016. A random network based, node attraction facilitated network evolution method. Selforganizology, 3(1): 1-9

Zhang WJ. 2017a. Network pharmacology of medicinal attributes and functions of Chinese herbal medicines: (I) Basic statistics of medicinal attributes and functions for more than 1100 Chinese herbal medicines. Network Pharmacology, 2(2): 17-37

Zhang WJ. 2017b. Network pharmacology of medicinal attributes and functions of Chinese herbal medicines: (II) Relational networks and pharmacological mechanisms of medicinal attributes and functions of Chinese herbal medicines. Network Pharmacology, 2(2): 38-66

Zhang WJ. 2017c. Network pharmacology of medicinal attributes and functions of Chinese herbal medicines: (III) Canonical correlation functions between attribute classes and linear eignmodels of Chinese herbal medicines. Network Pharmacology, 2(3): 67-81

Zhang WJ. 2017d. Network pharmacology of medicinal attributes and functions of Chinese herbal medicines: (IV) Classification and network analysis of medicinal functions of Chinese herbal medicines. Network Pharmacology, 2(3): 82-104

Zhang WJ. 2017e. Phase recognition in network evolution. Selforganizology, 4(3): 35-40

Zhang WJ. 2017f. Some correlations between eight types of malignant neoplasms: A hint from cancer dynamics of 31 European countries in 20 years. Network Biology, 7(3): 76-79

Zhang WJ. 2018a. Fundamentals of Network Biology. World Scientific, London, Singapore

Zhang WJ. 2018b. Global pesticide use: Profile, trend and cost / benefit analysis. Proceedings of the International Academy of Ecology and Environmental Sciences, 8(1): 1-26

Zhang WJ, Liu GH. 2012. Creating real network with expected degree distribution: A statistical simulation. Network Biology, 2(3): 110-117

Zhang WJ, Zhan CY. 2011. An algorithm for calculation of degree distribution and detection of network type: with application in food webs. Network Biology, 1(3-4): 159-170

Zhang Y, Chen M, Liao XF. 2013. Big Data Applications: A Survey. Journal of Computer Research and Development, 50(Suppl.): 216-233

Zhang YT, Fang KT. 1982. Introduction to Multivariate Statistics. Science Press, Beijing, China

17

A node-similarity based algorithm for tree generation and evolution

WenJun Zhang

School of Life Sciences, Sun Yat-sen University, Guangzhou 510275, China; International Academy of Ecology and Environmental Sciences, Hong Kong

E-mail: zhwj@mail.sysu.edu.cn, wjzhang@iaees.org

Abstract

In present study we proposed a node-similarity based algorithm for tree generation and evolution. In this algorithm, we assume that each isolated node is a node set at the beginning, two node sets with the greatest similarity tend to connect into a new node set firstly. Repeat this procedure, until all isolated nodes are combined into a tree. Pearson correlation measure, cosine measure, and (negative) Euclidean distance measure (the three measures are for interval attributes), contingency correlation measure (for nominal attributes), or Jaccard coefficient measure (for binary attributes) were used as the between-node similarity. In this way, all connections are sequentially generated and it thus forms the evolution process of a spanning tree of maximum likelihood. The similarity value of a connection can be considered as the weight of the connection. Matlab codes of the algorithm are provided.

Keywords tree; generation; evolution; node similarity; algorithm.

1 Introduction

Early in 1998, Watts and Strogatz developed a method for generating random graphs. Barabasi and Albert (1999) proposed a general mechanism for network evolution. Cancho and Sole (2001) algorithm can generate a variety of complex networks with diverse degree distributions. Fath et al. (2007) defined a step-by-step procedure for constructing an ecological network. Zhang (2011, 2012b, 2012c, 2012d, 2015a, 2016) proposed a series of methods and models for network generation and evolution. So far, research on network generation and evolution is still fewer. In present study, we will propose a node-similarity based algorithm for tree generation and evolution. Matlab codes of the algorithm are presented for further use.

2 Algorithm

Suppose there are m isolated nodes (or objects, etc.) and n attributes. The raw data matrix is $a=(a_{ij})_{m \times n}$. In the generation/evolution of a tree, I assume that in a set of isolated nodes (node sets), two nodes (node sets) with the greatest similarity tend to connect firstly. Pearson correlation measure, cosine measure, and (negative)

Euclidean distance measure (the three measures are for interval attributes), contingency correlation measure (for nominal (1, 2, 3…) attributes), or Jaccard coefficient measure (for binary (0, 1) attributes) can be used as the between-node similarity.

Pearson correlation measure is (Zhang, 2011, 2016; Zhang et al., 2014; Zhang, 2012b, c; Zhang and Li, 2015)

$$r_{ij} = \sum_{k=1}^{n} ((a_{ik} - a_{ib})(a_{jk} - a_{jb})) / (\sum_{k=1}^{n} (a_{ik} - a_{ib})^2 \sum_{k=1}^{n} (a_{jk} - a_{jb})^2)^{1/2}$$
$$i, j = 1, 2, \cdots, m$$

where $-1 \leq r_{ij} \leq 1$, $a_{ib} = \sum_{k=1}^{n} a_{ik}/n$, $a_{jb} = \sum_{k=1}^{n} a_{jk}/n$, $i, j = 1, 2, \cdots, m$.

Cosine measure is (Zhang, 2007; Zhang, 2012a)

$$r_{ij} = \sum_{k=1}^{n} a_{ik} a_{jk} / (\sum_{k=1}^{n} a_{ik}^2 \sum_{k=1}^{n} a_{jk}^2)^{1/2}$$
$$i, j = 1, 2, \cdots, m$$

Euclidean distance measure is (Zhang, 2007, 2012a)

$$d_{ij} = (\sum_{k=1}^{n} (a_{ik} - a_{jk})^2)^{1/2}$$

Its negative value is used as the similarity measure

$$r_{ij} = -d_{ij}$$

Contingency correlation measure is (Zhang, 2007, 2012b; Zhang et al., 2014):

$$r_{ij} = 2(h/(s(p-1)))^{1/2} - 1 \quad i, j = 1, 2, \cdots, m$$

where $-1 \leq r_{ij} \leq 1$, and

$$h = s_{..}(\sum_{i=1}^{p} \sum_{j=1}^{p} s_{ij}^2 / (s_{i.} s_{.j}) - 1)$$
$$s_{.} = \sum_{i=1}^{p} s_{i.}, \quad s_{i.} = \sum_{j=1}^{p} s_{ij}, \quad n_{.j} = \sum_{i=1}^{p} s_{ij}$$

where there are p available nominal values, i.e., $t_1, t_2, \ldots, t_p$, for attributes i and j, s_{kl} is the number of attributes of node i takes value t_k and node j takes value t_l, $k, l = 1, 2, \ldots, p$.

Jaccard coefficient measure is (Zhang, 2015b)

$$r_{ij} = (e - (c+b)) / (e+c+b) \quad i, j = 1, 2, \cdots, m$$

where $-1 \leq r_{ij} \leq 1$, c is the number of node pairs of 1 for attribute i but not for j; b is the number of node pairs of 1 for attribute j but not for i; e is the number of node pairs of 1 for both attribute i and attribute j.

Between-node similarity matrix, $r = (r_{ij})_{m \times m}$, is a symmetric matrix, i.e., $r = r'$.

Calculate the similarity between node sets. Suppose there are two node sets, A and B. The similarity between A and B is defined as the greatest similarity between A and B

$$r_{AB}=max\ r_{ij}\ ,\qquad i\in A, j\in B$$

At the start, m isolated nodes (or objects, etc.) are m node sets respectively. In all of node sets, choose two node sets with the maximal r_{AB} to combine into a new node set, and the corresponding nodes, i and j, with the maximal r_{ij}, are connected. Repeat this procedure, until m isolated nodes are eventually combined into a tree. In this way, all connections are sequentially generated, and thus form the evolution process of a spanning tree (a more general network can be further generated from the spanning tree by adding more connections).

The following are Matlab codes of the algorithm

```
%Reference: Zhang WJ, Li X. 2016. A node-similarity based algorithm for tree generation and evolution. Selforganizology, 3(2):
choice=input('Input a number to choose similarity measure (1: Pearson linear correlation; 2: Cosine measure; 3: (Negative)
Euclidean distance; 4: Contingency correlation; 5: Jaccard coefficient): ');
a=load(str);
m=size(a,1);
for i=1:m-1
for j=i+1:m
ix=a(i,:); jx=a(j,:);
if (choice==1)
str='Pearson correlation';
ixbar=mean(ix);
jxbar=mean(jx);
aa=sum((ix-ixbar).*(jx-jxbar));
bb=sum((ix-ixbar).^2);
cc=sum((jx-jxbar).^2);
r(i,j)=aa/sqrt(bb*cc);
end
if (choice==2)
str='Cosine measure';
aa=sum(ix.*jx);
bb=sum(ix.^2);
cc=sum(jx.^2);
r(i,j)=aa/sqrt(bb*cc);
end
if (choice==3)
str='(Negative) Euclidean distance';
r(i,j)=-sqrt(sum((ix-jx).^2));
end
if (choice==4)
str='Contingency correlation';
xx=[ix;jx];
pn=1;
tt(1)=xx(1);
for kk=1:max(size(xx))
jj=0;
for ii=1:pn
```

```
if (xx(kk)~=tt(ii)) jj=jj+1; end;
end
if (jj==pn) pn=pn+1;tt(pn)=xx(kk); end;
end
for kk=1:pn
for jj=1:pn
temp(kk,jj)=0;
for ii=1:max(size(ix))
if ((ix(ii)==tt(kk)) & (jx(ii)==tt(jj))) temp(kk,jj)=temp(kk,jj)+1; end; end
end; end
for kk=1:pn
pp=0;
for jj=1:pn pp=pp+temp(kk,jj); end
ni(kk)=pp;
end
for kk=1:pn
pp=0;
for jj=1:pn pp=pp+temp(jj,kk); end
nj(kk)=pp;
end
summ=0;
for kk=1:pn
summ=summ+ni(kk);
end;
xsquare=0;
for kk=1:pn
for jj=1:pn
if (ni(kk)==0 | nj(jj)==0) continue; end
xsquare=xsquare+temp(kk,jj)*temp(kk,jj)/(ni(kk)*nj(jj));
end; end
xsquare=summ*(xsquare-1);
r(i,j)=2*sqrt(xsquare/(summ*(pn-1)))-1;
end
if (choice==5)
str='Jaccard coefficient';
bb=sum((ix==0) & (jx~=0));
cc=sum((ix~=0) & (jx==0));
dd=sum((ix~=0) & (jx~=0));
r(i,j)=(dd-(cc+bb))/(dd+cc+bb);
end
r(j,i)=r(i,j);
end; end
adj=zeros(m);
r0=r;
classid=1;
```

```
u(classid)=0;
classnum(classid)=m;
for i=1:classnum(classid) x(classid,i)=i; end
tree=zeros(m);
disp(['Step   Node    Node    ' str]);
while (classnum(classid)>1)
aa=-1e+10;
for i=1:classnum(classid)-1
for j=i+1:classnum(classid)
if (r(i,j)>aa) aa=r(i,j); end
end; end
aa1=0;
for i=1:classnum(classid)-1
for j=i+1:classnum(classid)
if (abs(r(i,j)-aa)<=1e-06)
aa1=aa1+1; v(aa1)=i; w(aa1)=j;
temp=-1e+10;
for k=1:m
if (x(classid,k)==i)
for kk=1:m
if (x(classid,kk)==j)
if (r0(k,kk)>temp) temp=r0(k,kk); end
end; end; end; end;
for k=1:m
if (x(classid,k)==i)
for kk=1:m
if (x(classid,kk)==j)
if (abs(r0(k,kk)-temp)<1e-06) tree(k,kk)=classid;
disp([num2str(classid) '      ' num2str(k) '       ' num2str(kk) '       ' num2str(r0(k,kk))] );
adj(k,kk)=1; adj(kk,k)=1;
end;
end; end; end; end
end; end; end
for i=1:classnum(classid) s(i)=0; end
nn1=0;
for i=1:aa1
if ((v(i)~=0) & (w(i)~=0))
nn1=nn1+1;
for j=1:aa1
if ((v(j)==v(i)) | (v(j)==w(i)) | (w(j)==w(i)) | (w(j)==v(i)))
s(v(j))=nn1; s(w(j))=nn1;
if (j~=i) v(j)=0; w(j)=0; end; end
end
v(i)=0; w(i)=0;
end; end
```

```
for i=1:nn1
for j=1:classnum(classid)
if (s(j)==i)
for k=1:m
if (x(classid,k)==j) x(classid+1,k)=i; end
end
end; end; end
for i=1:classnum(classid)
if (s(i)==0)
nn1=nn1+1;
for k=1:m
if (x(classid,k)==i) x(classid+1,k)=nn1; end
end; end; end;
classid=classid+1;
u(classid)=aa;
classnum(classid)=nn1;
for i=1:classnum(classid)-1
for j=i+1:classnum(classid)
r(i,j)=-1e+10;
for k=1:m
if (x(classid,k)==i)
for kk=1:m
if (x(classid,kk)==j)
if (r0(k,kk)>r(i,j)) r(i,j)=r0(k,kk); end
end; end; end; end;
r(j,i)=r(i,j);
end; end;
end;
fprintf('\nMatrix for tree evolution (elements are step IDs)')
tree
fprintf('\n')
disp([str ' for each step'])
corr=u(2:classid)
fprintf('\nThe final adjacency matrix\n')
adj
```

3 Application Example

3.1 Analysis of 54 human races and populations and 14 common HLA-DRB1 alleles

Data of the world's 54 human races and populations (nodes) and 14 common HLA-DRB1 alleles (attributes) are from Jia (2001) (Zhang and Qi, 2014; HLA_DRB1.txt; supplementary material). Here I use Pearson correlation measure to generate the tree. The process of tree evolution is listed in Table 1.

The distribution of node degrees of final tree for 54 races and populations are indicated in Table 2, and the corresponding tree graph is drawn for convenient comparison using Java software (Zhang, 2012a), as indicated in Fig. 1.

Table 1 The process of tree evolution of world's 54 human races (nodes) and populations.

Step	Node	Node	Pearson corr.	Step	Node	Node	Pearson corr.	Step	Node	Node	Pearson corr.	Step	Node	Node	Pearson corr.
1	24	28	0.9817	15	16	15	0.9182	29	39	32	0.8791	43	53	54	0.7374
2	33	35	0.9619	16	20	21	0.9182	30	49	13	0.8753	44	33	37	0.7330
3	28	26	0.9579	17	45	47	0.9181	31	11	10	0.8750	45	2	7	0.7211
4	5	11	0.9545	18	16	14	0.9144	32	44	43	0.8731	46	2	11	0.7065
5	42	44	0.9527	19	49	50	0.9141	33	10	31	0.8707	47	18	32	0.6789
6	45	46	0.9458	20	33	39	0.9136	34	46	48	0.8606	48	2	8	0.6581
7	49	51	0.9413	21	42	40	0.9103	35	5	4	0.8512	49	30	27	0.6562
8	36	40	0.9371	22	15	18	0.9092	36	18	19	0.8458	50	26	31	0.5708
9	33	34	0.9333	23	38	43	0.9082	37	5	1	0.8376	51	7	6	0.5526
10	16	17	0.9328	24	14	12	0.9077	38	1	3	0.8173	52	6	50	0.5353
11	49	52	0.9259	25	25	27	0.8962	39	3	9	0.8062	53	19	21	0.5132
12	46	42	0.9252	26	34	22	0.8915	40	22	23	0.8060				
13	38	41	0.9199	27	12	11	0.8862	41	33	53	0.7828				
14	28	25	0.9185	28	35	40	0.8823	42	29	30	0.7639				

Node IDs from 1 to 54 represent Lahu-China, Dai-China, Yao-China, Guangdong Han-China, Dulong-China, Buyi-China, Thais, Yi-China, Hunan Han-China, Southern Han-China, Singapore Han-Singapore, Pumi-China, Shanghai Han-China, Liaoning Han-China, Shegyang Han-China, Northwest Han-China, Northern Han-China, Manchu-China, Japanese, Hokkaido-Japan, Uighur-China, Kazak-China, Siberian Nivkhs population, Siberian Udegeys population, Siberian Koryaks population, Siberian Eskimo, Siberian Chukchi population, South American Indians Ticuna, South American Indians Terena, Siberian Evenki population, Siberian Kets population, USA whites, Spanish, German, Romanians, Bulgarian, Greek, Polish, Turks, Macedonians, Israeli Arabs, Iranian Jews, Ashkenazi Jews-Germany, Libyan Jews, Moroccan Jews, Ethiopian Jews, Native population-Australia's central desert, Yuendumu Native population-Australia, Kimberley native population-Australia, Cape York native population-Australia, North American blacks, and   South African blacks.

Table 2 Node degrees for 54 races and populations.

ID	1	2	3	4	5	6	7	8	9
Degree	2	3	2	1	3	2	2	1	1
ID	10	11	12	13	14	15	16	17	18
Degree	2	4	2	1	2	2	3	1	3
ID	19	20	21	22	23	24	25	26	27
Degree	2	1	2	2	1	1	2	2	2
ID	28	29	30	31	32	33	34	35	36
Degree	3	1	2	2	2	5	2	2	1
ID	37	38	39	40	41	42	43	44	45
Degree	1	2	2	3	1	3	2	2	2
ID	46	47	48	49	50	51	52	53	54
Degree	3	1	1	4	2	1	1	2	1

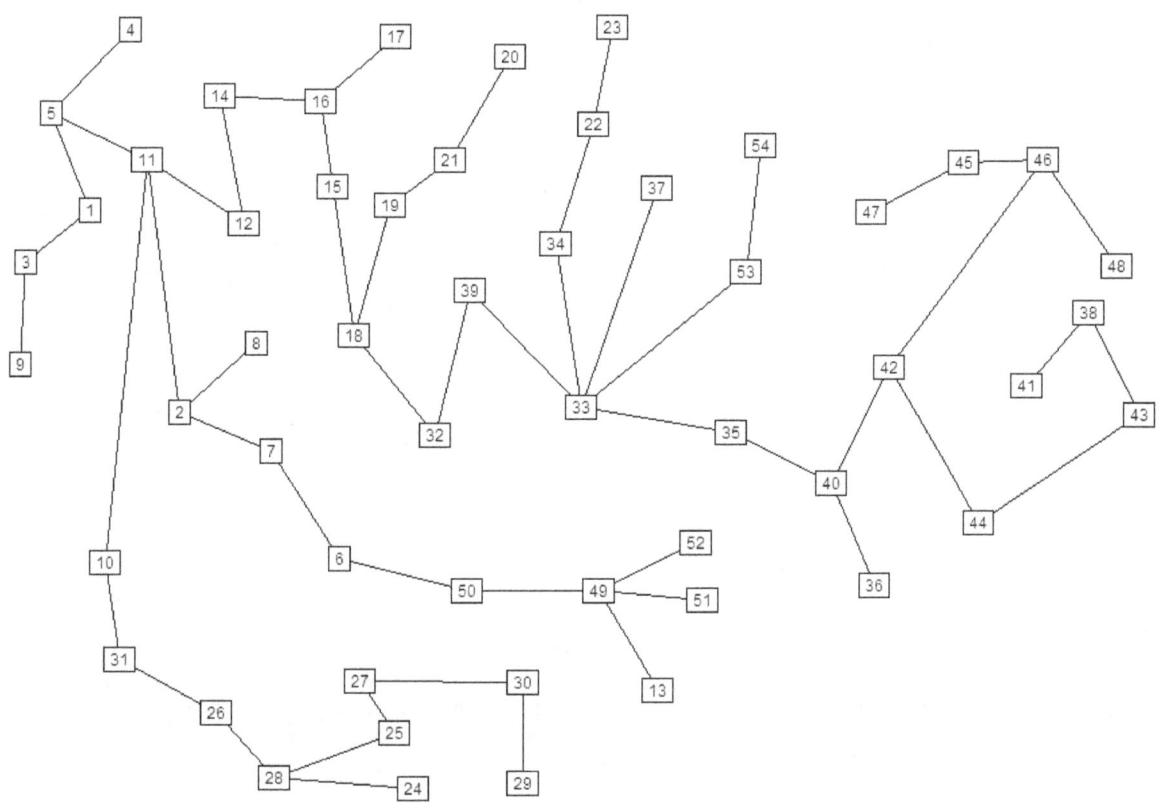

Fig. 1 The generated tree of 54 races and populations. Node IDs are explained in Table 1.

3.2 Analysis of 12 Chinese human populations and 17 HLA-DQB1 alleles

Data of the 12 Chinese human populations (nodes) and 17 common HLA-DQB1 alleles (attributes) (12×17; HLA_DQB1.txt; supplementary material) are from Geng et al. (1995), Chang et al. (1997), Mizuki et al. (1997, 1998), et al. Use Pearson cosine measure and results to generate the tree. The process of tree evolution is as follows

Step	Node	Node	Cosine measure
1	7	8	0.9764
2	6	10	0.97057
3	6	4	0.94927
4	10	5	0.94407
5	10	7	0.93692
6	4	3	0.92891
7	11	12	0.92311
8	3	2	0.90887
9	7	1	0.85977
10	6	11	0.78742
11	8	9	0.75927

where the node IDs from 1 to 12 represent Tibetan, Uighur, Kazak, Xinjiang Han, Taiwanese, Hong Kong,

Northern Han, Shanghai Han, Hunan Han, Manchu, Buyi, and Dai.

The corresponding tree is indicated in Fig. 2.

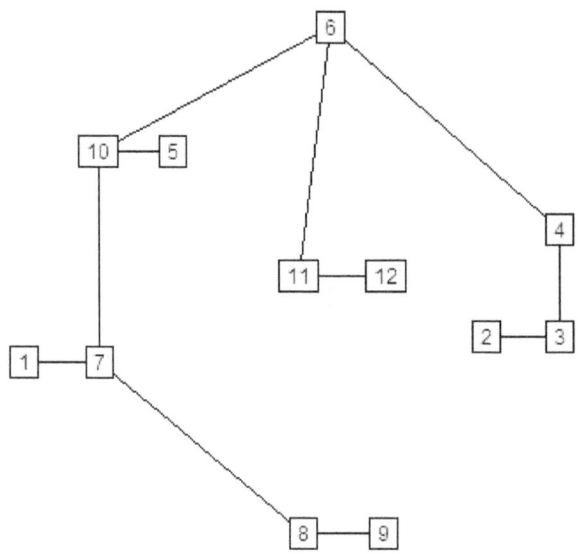

Fig. 2 The generated tree of 12 Chinese populations. Node IDs are explained in text.

4 Discussion

The present algorithm is based on similarity between nodes. The node sets with greater similarity tend to connect than that with less similarity. The final tree achieved is thus a spanning tree with maximum likelihood. The similarity value of a connection can be considered as the weight of the connection.

Considering the generality of such mechanism of tree generation in nature, it is expected to be a general method for tree generation. This algorithm can produce connections sequentially, thus can reflect the process of tree evolution. Further studies or applications on the algorithm are expected in the future.

Acknowledgment

We are thankful to the support of High-Quality Textbook *Network Biology* Project for Engineering of Teaching Quality and Teaching Reform of Undergraduate Universities of Guangdong Province (2015.6-2018.6), from Department of Education of Guangdong Province, Discovery and Crucial Node Analysis of Important Biological and Social Networks (2015.6-2020.6), from Yangling Institute of Modern Agricultural Standardization, and Project on Undergraduate Teaching Reform (2015.7-2017.7), from Sun Yat-sen University, China.

References

Barabasi AL, Albert R. 1999. Emergence of scaling in random networks. Science, 286(5439): 509

Cancho RF, Sole RV. 2001. Optimization in complex networks. Santafe Institute, USA

Chang YW, Hawkins BR. 1997. HLA Class I and Class II frequencies of a Hong Kong Chinese population based on bone marrow donor registry data. Human Immunology, 56: 125-135

Geng L, Imanishi T, Tokunaga K, et al. 1995. Determination of HLA class II alleles by genotyping in a

Manchu population in the northern part of China and its relationship with Han and Japanese populations. Tissue Antigens, 46:111-116

Fath BD, Scharler UM, Ulanowiczd RE, Hannone B. 2007. Ecological network analysis: network construction. Ecological Modeling, 208: 49–55

Jia ZJ. 2001. Polymorphism of HLA-DRB1 gene in southern Chinese populations. PhD Thesis. 46-47, Sun Yat-sen University, Guangzhou, China

Mizuki N, Ohno S, Ando H et al. 1998. Major histocompatibility complex class II alleles in an Uygur population in the Silk Route of Northwest China. Tissue Antigens, 51: 287-292

Mizuki N, Ohno S, Sato T et al. 1997. Major histocompatibility complex class II alleles in Kazak and Han populations in the Silk Route of Northwest China. Tissue Antigens, 50: 527-534

Watts D, Strogatz S. 1998. Collective dynamics of 'small world' networks. Nature, 393:440-442

Zhang WJ. 2007. Computer inference of network of ecological interactions from sampling data. Environmental Monitoring and Assessment, 124: 253-261

Zhang WJ. 2011. Constructing ecological interaction networks by correlation analysis: hints from community sampling. Network Biology, 1(2): 81-98

Zhang WJ. 2012a. A Java software for drawing graphs. Network Biology, 2(1): 38-44

Zhang WJ. 2012b. Computational Ecology: Graphs, Networks and Agent-based Modeling. World Scientific, Singapore

Zhang WJ. 2012c. How to construct the statistic network? An association network of herbaceous plants constructed from field sampling. Network Biology, 2(2): 57-68

Zhang WJ. 2012d. Modeling community succession and assembly: A novel method for network evolution. Network Biology, 2(2): 69-78

Zhang WJ, 2015a. A generalized network evolution model and self-organization theory on community assembly. Selforganizology, 2(3): 55-64

Zhang WJ. 2015b. Calculation and statistic test of partial correlation of general correlation measures. Selforganizology, 2(4): 55-67

Zhang WJ. 2016. Selforganizology: The Science of Self-Organization. World Scientific, Singapore

Zhang WJ, Li X. 2015. Linear correlation analysis in finding interactions: Half of predicted interactions are undeterministic and one-third of candidate direct interactions are missed. Selforganizology, 2(3): 39-45

Zhang WJ, Qi YH. 2014. Pattern classification of HLA-DRB1 alleles, human races and populations: Application of self-organizing competitive neural network. Selforganizology, 1(3-4): 138-142

Zhang WJ, Qi YH, Zhang ZG. 2014. Two-dimensional ordered cluster analysis of component groups in self-organization. Selforganizology, 1(2): 62-77

18

A hierarchical method for finding interactions: Jointly using linear correlation and rank correlation analysis

WenJun Zhang

School of Life Sciences, Sun Yat-sen University, Guangzhou 510275, China; International Academy of Ecology and Environmental Sciences, Hong Kong

E-mail: zhwj@mail.sysu.edu.cn, wjzhang@iaees.org

Abstract

In the earlier studies, I pointed out that a network changed in a local domain can be approximated as a linear network, i.e., all between-node (or -taxon, -component, etc) changes in the local domain are treated as linear ones and Pearson linear correlation measure can be used. For a little wider domain, the quasi-linear measure, Spearman rank correlation can be used also. In present study, I jointly use Pearson linear correlation measure and Spearman rank correlation measure and their partial correlations to find interactions. First, I define some hierarchical principles for finding interactions. Reliability levels are then defined using set operations. The full algorithm and Matlab codes for finding interactions are given.

Keywords partial correlation; correlation measure; Pearson linear correlation; Spearman rank correlation; algorithm; set operation; statistic test; interaction finding.

1 Introduction

In our earlier studies (Zhang, 2007, 2011, 2012a, 2012b, 2015, 2016; Zhang and Li, 2015a, b), we have proposed a series of methods for finding interactions by correlation analysis of sampling data. For the studies of ecological communities and ecosystems, there are generally two types of interactions, direct interactions and indirect interactions. Direct interactions refer to predation, parasitism, competition, amensalism, mutualism, protocooperation, commensalism, etc. Two taxa may interact by acting to the same resource, or by changing the environment of opposite sides, etc. An interaction means a dependency relationship in state changes of two taxa (direct interaction). Conversely, a seeming dependency relationship in state changes of two taxa does not necessarily mean an interaction (indirect interaction). Definition of interactions in other biological networks (metabolic regulatory networks, cancer networks, etc) can be found in related references.

In present study, I jointly use Pearson linear correlation measure and Spearman rank correlation measure and their partial correlations for finding interactions. Some hierarchical principles for finding interactions are defined. Reliability levels are defined using set operations. The full algorithm and Matlab codes for finding interactions are given.

2 Method

In the earlier studies (Zhang and Li, 2015a, b; Zhang, 2015), we pointed out that a network changed in a local domain (a short time, a small extent) can be approximated as a linear network, i.e., in the local domain, all between-node (or -taxon, -component, etc) changes are treated as linear ones, and Pearson linear correlation measure can be used. For a little wider domain, the quasi-linear measure, Spearman rank correlation (Spearman, 1904; Schoenly and Zhang, 1999; Zhang, 2011, 2012a, 2012b, 2015, 2016) can be used also. Therefore, we can jointly use Pearson linear correlation measure and Spearman rank correlation measure, and their partial correlations to find interactions.

Here I define the sets of interactions

A: Interactions detected by statistically significant partial correlations for Pearson linear correlation measure;

B: Interactions detected by statistically significant partial correlations for Spearman rank correlation measure;

C: Interactions detected by statistically significant Pearson linear correlations;

D: Interactions detected by statistically significant Spearman rank correlations.

When finding interactions, some principles are followed: (1) Pearson linear correlation measure is more reliable than Spearman rank correlation measure; (2) partial correlation measure is more reliable than correlation measure; (3) jointly use of Pearson linear correlation measure and Spearman rank correlation measure is more reliable than the single use of two measures, and (4) jointly use partial correlation measure and correlation measure is more reliable than the single use.

According to the above principles and the rules of set operation, I defined reliability levels (basically, from the set of most reliable interactions, L1, to the sets of most unreliable interactions, L61 and L62) as follows

L1: A∩B∩C∩D;	(candidate direct interactions)
L21: A∩C; L22: B∩D;	(candidate direct interactions)
L3: A∩B;	(candidate direct interactions)
L41: A; L42: B;	(candidate direct interactions)
L5: C∩D;	(indirect/direct interactions)
L61: C; L62: D.	(indirect/direct interactions)

Known the raw data is a matrix, X, with m rows, i.e., m attributes (taxa, proteins, genes, etc), and n columns, i.e., n samples. The Matlab codes of the full algorithm above are listed as follows:

```
% Reference: Zhang WJ. 2015. A hierarchical method for finding interactions: Jointly using linear correlation and
% rank correlation analysis. Network Biology, 5(4): 137-145
% X is the m*n raw data matrix. m: number of attributes (taxa, proteins, genes, etc); n: number of samples.
str=input('Input the file name of raw data matrix (e.g., raw.txt, raw.xls, etc. The file has m rows (taxa) and n columns
(samples)): ','s');
X=load(str);
sig=input('Input significance level(e.g., 0.01)');
dim=size(X);
m=dim(1); n=dim(2);
if (n<=m)
disp('The number of samples is not enough to support the required statistic test (DF=n-m) of partial correlations. Here
use the statistic test with DF=n-2 (not recommended). Please input the proportion of statistically significant pairs based
```

on DF=n-m vs. statistically significant pairs based on DF=n-2 (y, %) as the following. For Pearson linear correlation measure, the estimation formula, y1=88.748 exp(-0.045m), is suggested for use, and for Spearman rank correlation measure, y2=120.687exp(-0.045m), is suggested, where m is the number of attributes (taxa, proteins, genes, etc). If it is hard to be estimated, the full percent, 100, can be input. ')

```
y1=input('Input the proportion for Pearson linear correlation measure (a value between 0 and 100): ')
y2=input('Input the proportion for Spearman rank correlation measure (a value between 0 and 100): ')
end;
r=corr(X');
disp('Pearson linear correlation matrix')
r
for i=1:m-1; for j=i+1:m; spr(i,j)=spearman(X(i,:)',X(j,:)');end;end;
for i=1:m-1; for j=i+1:m; spr(j,i)=spr(i,j);end;end;
for i=1:m; spr(i,i)=1;end;
disp('Spearman rank correlation matrix')
spr
for k=1:2;
if (k==1) rr=r; else rr=spr; end;
inverse=inv(rr);
for i=1:m-1; for j=i+1:m; parr(i,j)=-inverse(i,j)/sqrt(inverse(i,i)*inverse(j,j));end;end;
for i=1:m-1; for j=i+1:m; parr(j,i)=parr(i,j);end;end;
for i=1:m; parr(i,i)=1;end;
if (k==1)
disp('Partial correlation matrix for Pearson linear correlation')
peparr=parr;
peparr
else
disp('Partial correlation matrix for Spearman rank correlation')
spparr=parr;
spparr
end;
end;
for k=1:2;
if (k==1)
tvalues=abs(r)./sqrt((1-r.^2)/(n-2));
alpha=(1-tcdf(tvalues,n-2))*2;
sigmat=alpha<sig;
sigmatr=sigmat.*r;
if (n>m)
partvalues=abs(peparr)./sqrt((1-peparr.^2)/(n-m));
paralpha=(1-tcdf(partvalues,n-m))*2;
else
partvalues=abs(peparr)./sqrt((1-peparr.^2)/(n-2));
paralpha=(1-tcdf(partvalues,n-2))*2;
end
parsigmat=paralpha<sig;
parsigmatr=parsigmat.*peparr;
if (n<=m) threshr=rrank(parsigmatr,y1); parsigmatr=parsigmatr>=threshr; parsigmatr=parsigmatr.*peparr; end;
else
tvalues=abs(spr)./sqrt((1-spr.^2)/(n-2));
alpha=(1-tcdf(tvalues,n-2))*2;
```

```
spsigmat=alpha<sig;
spsigmatr=spsigmat.*spr;
if (n>m)
partvalues=abs(spparr)./sqrt((1-spparr.^2)/(n-m));
paralpha=(1-tcdf(partvalues,n-m))*2;
else
partvalues=abs(spparr)./sqrt((1-spparr.^2)/(n-2));
paralpha=(1-tcdf(partvalues,n-2))*2;
end;
spparsigmat=paralpha<sig;
spparsigmatr=spparsigmat.*spparr;
if (n<=m) threshr=rrank(spparsigmatr,y2); spparsigmatr=spparsigmatr>=threshr; spparsigmatr=spparsigmatr.*spparr;
end;
end;
end;
L1=(parsigmatr & spparsigmatr & sigmatr & spsigmatr);          % candidate direct interactions
L21=(parsigmatr & sigmatr); L22=(spparsigmatr & spsigmatr);    % candidate direct interactions
L3=(parsigmatr & spparsigmatr);                                % candidate direct interactions
L41=parsigmatr; L42=spparsigmatr;                              % candidate direct interactions
L5=(sigmatr & spsigmatr);                                      % indirect/direct interactions
L61=sigmatr; L62=spsigmatr;                                    % indirect/direct interactions
for i=1:9;
switch (i)
    case 1
        mat=L1; s='L1';
    case 2
        mat=L21; s='L21';
    case 3
        mat=L22; s='L22';
    case 4
        mat=L3; s='L3';
    case 5
        mat=L41; s='L41';
    case 6
        mat=L42; s='L42';
    case 7
        mat=L5; s='L5';
    case 8
        mat=L61; s='L61';
    case 9
        mat=L62; s='L62';
end;
disp(['Interactions (the interactions found in higher levels are also listed) with statistically significance for reliability level:
' s])
[pairx,pairy]=find(mat);
temp1=pairx; temp2=pairy;
pairxs=pairx(temp1<temp2);
pairys=pairy(temp1<temp2);
InteractionPairs=[pairxs pairys]
end;
```

Two M function files, spearman.m, and rrank.m, are as follows:

```matlab
function spearm =spearman(x,y)          %x and y: two column vectors to be tested.
if (max(size(x))~=max(size(y)))
  error('Array sizes do not match.');
end
if ((min(size(x))~=1) | (min(size(y))~=1))
  error('Both x and y are vectors');
end
n=max(size(x));
for i=1:n
rx(i)=0;ry(i)=0;xx(i)=0;yy(i)=0;
end
for j=1:n
nx=1;ny=1;
for i=1:n
if (x(i)<x(j)) nx=nx+1; end
if (y(i)<y(j)) ny=ny+1; end
end
rx(j)=nx;
ry(j)=ny;
end
for j=1:n
if (rx(j)==(n+1)) continue; end
nx=rx(j);
ntie=-1;
for i=1:n
if (rx(i)~=nx) continue; end
ntie=ntie+1;
xx(i)=rx(i);
rx(i)=0;
end
for i=1:n
if (rx(i)~=0) continue; end
xx(i)=xx(i)+(ntie*0.5);
rx(i)=n+1;
end
end
for j=1:n
if (ry(j)==(n+1)) continue; end
ny=ry(j);
ntie=-1;
for i=1:n
if (ry(i)~=ny) continue; end
ntie=ntie+1;
yy(i)=ry(i);
ry(i)=0;
end
for i=1:n
```

```
if (ry(i)~=0) continue; end
yy(i)=yy(i)+ntie*0.5;
ry(i)=n+1;
end
end
rs=0;
rs=sum((xx-yy).^2);
spearm=1-((6*rs)/(n*(n^2-1)));

function threshr = rrank(mat,percent)
dim=size(mat); m=dim(1);
len=(m*m-m)/2;
vec=zeros(1,len);
n=0;
for i=1:m-1;
for j=i+1:m;
if (mat(i,j)~=0) n=n+1; vec(n)=mat(i,j); end;
end;
end;
num=round(percent/100*n);
vecc=sort(vec,'descend');
if (num~=0) threshr=vecc(num);
else threshr=1;
end;
```

3 Application

I use a dataset on arthropod ecosystem, PH-Apr-fg (20 functional groups, 60 samples; Schoenly and Zhang, 1999; Zhang, 2011; Table 1), to calculate interactions using the algorithm above. The IDs of 20 functional groups represent:

Herbivores
(1) Pollen feeder, (2) External plant feeder, (3) Leaf roller/webber, (4) Leaf miner, (5) Gall former, (6) Mixed (combination of two or more of above).

Predators
(7) Terrestrial flyer, (8) Terrestrial crawler, walker, jumper, or hunter, (9) Neustonic (water surface) swimmer (semiaquatic), (10) Planktonic (water column) swimmer and diver, (11) Terrestrial web-builder.

Parasitoids/parasites
(12) Terrestrial blood sucker, (13) Flying adult that is searching, ovipositing, or larvipositing, (14) Idiobiont (acarine ectoparasitoid).

Detritivores
(15) Collector (filterer, suspension feeder), (16) Collector (gatherer, deposit feeder), (17) Shredder, chewer of coarse particulate Matter.

Tourists
(18) Tourist (nonpredatory species with no known functional role other than as prey in ecosystem).

Omnivores
(19) Herbivore, predator, and detritivore.

Dual insectivores

(20) Predator and parasitoid.

Table 1 Mean number of individuals per sample for each functional group.

ID of functional group	1	2	3	4	5	6	7	8	9	10
Mean number of individuals per sample	0.07	51.47	0.02	0.18	0.03	4.67	2.80	27.60	25.15	1.25
ID of functional group	11	12	13	14	15	16	17	18	19	20
Mean number of individuals per sample	2.43	0.43	5.98	0.07	1.52	2.32	5.23	2.60	0.68	0.20

Choose the significance degree $p=0.01$, and some of the results are as follows:

Interactions (the interactions found in higher levels are also listed) with statistically significance for reliability level: L1
InteractionPairs =

    2   8

Interactions (the interactions found in higher levels are also listed) with statistically significance for reliability level: L21
InteractionPairs =

    1   2
    2   8
    2   9
    4   13
    14   16
    1   17
    8   17
    16   19

Interactions (the interactions found in higher levels are also listed) with statistically significance for reliability level: L3
InteractionPairs =

    2   8

Interactions (the interactions found in higher levels are also listed) with statistically significance for reliability level: L41
InteractionPairs =

    1   2
    2   8

2	9
1	11
4	12
4	13
12	16
14	16
1	17
8	17
11	19
16	19

Interactions (the interactions found in higher levels are also listed) with statistically significance for reliability level: L5

InteractionPairs =

1	2
2	8
2	9
8	9
2	13
4	13
2	15
2	16
8	16
14	16
1	17
2	17
8	17
9	17
16	17
13	18
16	19
1	20

It is obvious that the most reliable interaction is (External plant feeder -- Terrestrial crawler, walker, jumper, or hunter), which represents the interaction (herbivores, predators) - the most important ecological interaction between species. The two groups are also the most abundant and predominant arthropod groups in the rice field investigated (Table 1).

4 Discussion

Our knowledge of biological networks is so limited, e.g., 80% of themolecular interactions in cells of Yeast (Yu et al., 2008) and 99.7% of human (Amaral, 2008) are still unknown. Here I have given a set of rules for finding interactions. However, these rules can be revised and improved upon the requirement of researchers.

The final results for finding, candidate direct interactions, are true direct interactions only they are confirmed by experiments and observations. The present method provides a prompt and relatively reliable tool for primeval and batch screening of possible interactions. This will help to save time and cost for interaction finding by experiments or observations.

In a sense, the significance level is a reliability measure also. In present example, I use a significance level of $p=0.001$. To avoid missing candidate direct interactions as possible as, i.e., coarse screening of interactions, the significance level can be adjusted to a reasonable value, for example, $p=0.05$, or even $p=0.1$.

The validity of taxa by sample based interaction finding, as presented in this study, is dependent upon the representativeness of samples, reasonable distribution of samples over space or time, number of samples (sample size), etc. Thus how to take a set of statistically representative samples for correlation analysis is one of the keys for interaction finding.

Acknowledgment

I am thankful to the support of High-Quality Textbook *Network Biology* Project for Engineering of Teaching Quality and Teaching Reform of Undergraduate Universities of Guangdong Province (2015.6-2018.6), from Department of Education of Guangdong Province, and Project on Undergraduate Teaching Reform (2015.7-2017.7), from Sun Yat-sen University, China.

References

Amaral LA. 2008. A truer measure of our ignorance. PNAS, 105(19): 6795-6796

Schoenly KG, Zhang WJ. 1999. IRRI Biodiversity Software Series. V. RARE, SPPDISS, and SPPANK: programs for detecting between-sample difference in community structure. IRRI Technical Bulletin No.5. International Rice Research Institute, Manila, Philippines

Spearman C. 1904. The proof and measurement of association between two things. American Journal of Psychology, 15: 72-101

Yu H, Braun P, Yildirim MA, et al. 2008. High-quality binary protein interaction map of the yeast interactome network. Science, 322(5898): 104-110

Zhang WJ. 2007. Computer inference of network of ecological interactions from sampling data. Environmental Monitoring and Assessment, 124: 253-261

Zhang WJ. 2011. Constructing ecological interaction networks by correlation analysis: hints from community sampling. Network Biology, 1(2): 81-98

Zhang WJ. 2012a. Computational Ecology: Graphs, Networks and Agent-based Modeling. World Scientific, Singapore

Zhang WJ. 2012b. How to construct the statistic network? An association network of herbaceous plants constructed from field sampling. Network Biology, 2(2): 57-68

Zhang WJ. 2015. Calculation and statistic test of partial correlation of general correlation measures. Selforganizology, 2(4): 65-77

Zhang WJ, Li X. 2015a. General correlation and partial correlation analysis in finding interactions: with Spearman rank correlation and proportion correlation as correlation measures. Network Biology, 5(4): 163-168

Zhang WJ, Li X. 2015b. Linear correlation analysis in finding interactions: Half of predicted interactions are undeterministic and one-third of candidate direct interactions are missed. Selforganizology, 2(3): 39-45

19

An algorithm for calculation of degree distribution and detection of network type: With application in food webs

WenJun Zhang[1,2], ChenYuan Zhan[1]

[1]School of Life Sciences, Sun Yat-sen University, Guangzhou 510275, China; [2]International Academy of Ecology and Environmental Sciences, Hong Kong

E-mail: zhwj@mail.sysu.edu.cn, wjzhang@iaees.org

Abstract

In present study a Java algorithm to calculate degree distribution and detect network type was presented. Some indices, e.g., aggregation index, coefficient of variation, skewness, etc., were first suggested for detecting network type. Network types of some food webs reported in Interaction Web Database were determined using the algorithm. The results showed that the degree of most food webs was power law or exponentially distributed and they were complex networks. Different from classical distribution patterns (bionomial distribution, Poisson distribution, and power law distribution, etc.), both network type and network complexity can be calculated and compared using the indices above. We suggest that they should be used in the network analysis. In addition, we defined E, $E=s^2-\bar{u}$, where $\bar{u}$ and s^2 is mean and variance of degree respectively, as the entropy of network. A more complex network has the larger entropy. If $E \leq 0$, the network is a random network and, it is a complex network if $E>0$.

Keywords network; food web; type; degree distribution; aggregation indices; entropy; algorithm; Java.

1 Introduction

The food web is a set of species connected by trophic relations. It is an ecological network made of interactive species. Biodiversity, ecosystem structure and function, etc., can be represented by food webs. There are two kinds of food webs, i.e., the one with autotrophic species as base species and the one with scavenger animals as base species (Gonenc et al., 2007). The complexity and trophic levels of food web determine the stability, resilience and robustness of the community. An ecosystem resists the extinction of species if its food web is complex enough. The species loss in food web would at some extent detriment the stability of ecosystem.

Food webs have long been the center of ecological studies. They began with text and table expression and then linear and spatial expression. As the emerging of large numbers of algorithms and software, the studies of network structure are now becoming the focus of food webs. These algorithms and software have been used to explore the ecosystem stability and robustness. For example, they are used to study degree distribution, connectance and network size (Dunne et al., 2002). Odum (1983) pointed out that the community stability could be measured by energy path in the food web. MacArthur (1955) thought that stability may be increased by the increase of links in the food web. Pimm et al. (1991) have discussed effects of different models of food webs on ecosystem structure, stability and robustness. It was found that the mechanism of evolution and population size will affect food web topology (Rossberg et al., 2005). In addition, habitat destruction and climate change are likely to cause the extinction of key species. Once key species extinct, the robustness of food web will be profoundly affected (Allesina et al., 2009).

The networks, including food webs, met in last decade become more and more complex. There are usually large numbers of vertices and links in a complex network (Ibrahim et al., 2011; Goemann et al., 2011; Kuang and Zhang, 2011; Martínez-Antonio, 2011; Paris and Bazzoni, 2011; Rodriguez and Infante, 2011; Tacutu et al., 2011). It is hard to analyze such networks by using classical methods or algorithms. Graph theory, optimization, statistics, and stochastic processes, etc., are becoming the scientific basis and effective tools for studying complex networks (Ferrarini, 2011; Zhang, 2011a, b; Zhang, 2012). Degree distribution and network type is one of the research focuses based on those tools and methods. In this aspect some ecological networks have been proved to be scale-free networks (Zhang, 2011a).

The present study aimed to present a Java algorithm to calculate degree distribution and detect network type. Some indices were first suggested by us for detecting network type. Network types of food webs reported in Interaction Web Database were determined using the algorithm.

2 Materials and Methods

2.1 Methods

Suppose that the portion of nodes with k-degree is p_k, the degree will thus be a random variable and its distribution is degree distribution. It has found that in the random network, degree distribution is binomial distribution, and its limit model is Poisson distribution. In a random network, the majority of vertices have the same degree with the average. In the complex network, degree distribution is a power law distribution, and the network is called a scale-free network (Barabasi and Albert, 1999; Barabasi, 2009). A property of the scale-free network is that the structure and the evolution of network are inseparable. Scale-free networks constantly change because of the arrival of nodes and links (Barabasi and Albert, 1999).

In present algorithm, in addition to power law distribution, binomial distribution, Poisson distribution, and exponential distribution (Zhang, 2012), some other indices and methods were also used to detect network type:

(1) Skewness. This index is used to measure the degree of skewness of a degree distribution relative to the symmetric distribution, for example, the normal distribution (S=0) (Sokal and Rohlf, 1995):

$$S=v\sum(d_i-\bar{u})^2/[(v-1)(v-2)s^3].$$

where $\bar{u}$, s^2: mean and variance of degree; v: number of nodes; d_i: degree of node i, i=1,2,…,v. The smaller the skewness is, the more complex the network is.

(2) Coefficient of variation. In a random network, the majority of nodes have the same degree as the average. The coefficient of variation, H, can be used to describe the type of a network (Zhang, 2007):

$$H=s^2/\bar{u},$$
$$\bar{u}=\sum d_i/v,$$
$$s^2=\sum(d_i-\bar{u})^2/(v-1),$$

where $\bar{u}$, s^2: mean and variance of degree; v: number of nodes; d_i: the degree of node i, i=1,2,…,v. The network is a random network, if $H\leq1$. Calculate $\chi^2=(v-1)H$, and if $\chi_{1-\alpha}^2(v-1)<\chi^2<\chi_\alpha^2(v-1)$, the network is a complete random network. It is a complex network, if $H>1$, and to some extent, network complexity increases with H.

Here we define E, $E=s^2-\bar{u}$, as the entropy of network. A more complex network has the larger entropy. If $E\leq0$ the network is a random network and it is a complex network if $E>0$.

(3) Aggregation index. Network type can be determined by using the following aggregation index (Zhang, 2007):

$$H=v*\sum d_i(d_i-1)/[\sum d_i(\sum d_i-1)].$$

The network is a random network, if $H\leq 1$. Calculate $\chi^2=H(\sum d_i-1)+v-\sum d_i$, and if $\chi^2<\chi_a^2(v-1)$, the network is a complete random network. It is a complex network if $H>1$, and network complexity increases with H.

The following code is the Java algorithm, netType, to calculate degree distribution and detect network type:

```
/*v: number of vertice; d[1-v][1-v]: adjacency matrix to reflect the feature of edges, e.g., dij=dji=0 means no edge between
vertice i and j; dij=-dji, and |dij|=1, means there is an edge between vertice i and j; dij=dji=2, means there are parallel edges
between vertice i and j; dii=3 means there is a self-loop for vertex i; dii=4 means isolated vertex; dii=5 means isolated vertex i
with self-loop. */
public class netType {
public static void main(String[] args){
int i,j,v,n;
if (args.length!=1)
System.out.println("You must input the name of table in the database. For example, you may type the following in the
command window: java netType nettype, where nettype is the name of table. Graph is stored in the table using two arrays
listing and was transformed to adjacency matrix by method adjMatTwoArr.");
String tablename=args[0];
readDatabase readdata=new readDatabase("dataBase",tablename, 3);
n=readdata.m;
int a[]=new int[n+1];
int b[]=new int[n+1];
int c[]=new int[n+1];
int d[][]=new int[n+1][n+1];
for(i=1;i<=n;i++) {
a[i]=(Integer.valueOf(readdata.data[i][1])).intValue();
b[i]=(Integer.valueOf(readdata.data[i][2])).intValue();
c[i]=(Integer.valueOf(readdata.data[i][3])).intValue(); }
adjMatTwoArr adj=new adjMatTwoArr();
adj.dataTrans(a,b,c);
v=adj.v;
for(i=1;i<=v;i++)
for(j=1;j<=v;j++) d[i][j]=adj.d[i][j];
netType(v,d); }
public static void netType(int v, int d[][]) {
int i,j,k,l,m,rr,ty,r;
double it,pp,ss,qq,k1,k2,chi,mean,var,hr,h,skew;
int deg[]=new int[v+1];
int p[]=new int[v+1];
double fr[]=new double[v+1];
double pr[]=new double[v+1];
for(i=1;i<=v;i++) {
deg[i]=0;
for(j=1;j<=v;j++) {
if (Math.abs(d[i][j])==1) deg[i]++;
if ((d[i][j]==2) | (d[i][j]==3) | (d[i][j]==5)) deg[i]+=2; } }
for(i=1;i<=v;i++) p[i]=i;
for(i=1;i<=v-1;i++) {
k=i;
for(j=i;j<=v-1;j++) if (deg[j+1]>deg[k]) k=j+1;
l=p[i];
p[i]=p[k];
p[k]=l;
m=deg[i];
deg[i]=deg[k];
deg[k]=m; }
pp=qq=0;
System.out.println("Ranks        Vertice        Degrees\n");
for(i=1;i<=v;i++) {
System.out.println(i+"    "+p[i]+"    "+deg[i]);
pp+=deg[i];
```

```
qq+=deg[i]*(deg[i]-1); }
System.out.println();
rr=10;
it=(deg[1]-deg[v])/(double)rr;
for(i=1;i<=10;i++) {
fr[i]=0;
for(j=1;j<=v;j++)
if ((deg[j]>=(deg[v]+(i-1)*it)) & (deg[j]<(deg[v]+i*it))) fr[i]++; }
System.out.println("Frequency distribution of degrees:");
for(i=1;i<=10;i++)
System.out.print(deg[v]+it/2.0+(i-1)*it+"    ");
System.out.println();
for(i=1;i<=10;i++)
System.out.print(fr[i]/v+"    ");
System.out.println("\n");
mean=pp/v;
ss=0;
for(i=1;i<=v;i++)
ss+=Math.pow(deg[i]-mean,2);
var=ss/(v-1);
skew=v/((v-2)*Math.sqrt(var));
System.out.println("Skewness of degree distribution: "+skew+"\n");
h=v*qq/(pp*(pp-1));
System.out.println("Aggregation index of the network: "+h);
if (h<=1) System.out.println("It is a random network.\n");
if (h>1) System.out.println("It is a complex network.\n");
h=var/mean;
System.out.println("Variation coefficient H of the network: "+h);
System.out.println("Entropy E of the network: "+(var-mean));
if (h<=1) System.out.println("It is a random network.\n");
if (h>1) System.out.println("It is a complex network.\n");
ty=1;        //Binomial distri., pr= Crn pr qn-r, r=0,1,2,..., n;
ss=0;
for(i=0;i<=rr-1;i++) ss+=i*fr[i+1];
pp=ss/(v*(rr-1));
qq=1-pp;
pr[0]=Math.pow(qq,rr-1);
for(i=1;i<=rr-1;i++) pr[i]=(rr-i)*pp*pr[i-1]/(i*qq);
chi=xsquare(v, rr, pr, fr);
System.out.println("Binomial distribution Chi-square="+chi);
System.out.println("Binomial p="+pp);
k1=20.09;
coincidence(ty, k1, chi);
ty=2;
//Poisson distri., pr = e-λλr/r! , r=0,1,2,...
pr[0]=Math.exp(-mean);
for(r=1;r<=rr-1;r++) pr[r]=mean/r*pr[r-1];
chi=xsquare(v, rr, pr, fr);
System.out.println("Poisson distribution chi-square="+chi);
System.out.println("Poisson lamda="+mean);
k1=20.09;
coincidence(ty, k1, chi);
ty=3;        //Exponential distri., F(x) =1-e-λx, x≥0
chi=0;
for(i=1;i<=10;i++) {
k1=deg[v]+it/2.0+(i-1)*it;
k2=deg[v]+it/2.0+i*it;
pp=v*(Math.exp(-k1/mean)-Math.exp(-k2/mean));
chi+=Math.pow(fr[i]-pp,2)/pp; }
System.out.println("Exponential distribution lamda="+1.0/mean);
k1=20.09;
coincidence(ty, k1, chi);
powerDistr(v, deg); }
public static double xsquare(int v, int rr, double p[], double h[]) {
double hk,ss=0;
for(int i=0;i<=rr-1;i++) {
hk=p[i]*v;
if (p[i]==0) hk=h[i+1];
ss+=Math.pow(p[i]*v-h[i+1],2)/hk;}
return ss;
```

```
}
public static void coincidence(int ty, double k1, double ss) {
if (ss<=k1)
if (ss>=0) {
if (ty==1) System.out.println("Degrees are binomially distributed.\n");
if (ty==2) System.out.println("Degrees are Poisson distributed.\n");
if (ty==3) System.out.println("Degrees are exponentially distributed.\n");
if ((ty==1) | (ty==2)) System.out.println("It is a random network"); }
if ((ss>k1) & ((ty==1) | (ty==2))) System.out.println("It is likely not a random network\n");
if ((ss>k1) & (ty==3)) System.out.println("It is not an exponential network\n");
}
public static void powerDistr(int v, int x[]) {
//Power law distri., f(x)=x-α, x≥xmin
int i,j,k,n,r,xmin;
double xmax,a,alpha,dd,maa;
int xminn[]=new int[v+1];
int xmins[]=new int[v+1];
double z[]=new double[v+1];
double zz[]=new double[v+1];
double cx[]=new double[v+1];
double cf[]=new double[v+1];
double dat[]=new double[10000];
k=1;
xminn[1]=x[1];
for(i=1;i<=v;i++) {
n=0;
for(j=1;j<=k;j++)
if (x[i]!=xminn[j]) n++;
if (n==k) {
k++;
xminn[k]=x[i]; } }
for(i=1;i<=k-1;i++) xmins[i]=xminn[i];
for(i=1;i<=v-1;i++) {
k=i;
for(j=i;j<=v-1;j++)
if (x[j+1]>x[k]) k=j+1;
r=x[i];
x[i]=x[k];
x[k]=r; }
for(i=1;i<=v;i++) z[i]=x[v-i+1];
for(r=1;r<=v;r++) {
xmin=xmins[r];
n=0;
for(i=1;i<=v;i++)
if (z[i]>=xmin) {
n++;
zz[n]=z[i]; }
maa=0;
for(i=1;i<=n;i++) maa+=Math.log(zz[i]/xmin);
a=n/maa;
for(i=0;i<=n-1;i++) cx[i+1]=i*1.0/n;
for(i=1;i<=n;i++) cf[i]=1-Math.pow(xmin/zz[i],a);
dat[r]=0;
for(i=1;i<=n;i++) {
cf[i]=Math.abs(cf[i]-cx[i]);
if (cf[i]>dat[r]) dat[r]=cf[i]; } }
dd=1e+100;
for(i=1;i<=v;i++)
if (dat[i]<dd) dd=dat[i];
for(i=1;i<=v;i++)
if (dat[i]<=dd) {
k=i;
break; }
xmin=xmins[k];
n=0;
for(i=1;i<=v;i++)
if (x[i]>=xmin) {
n++;
zz[n]=x[i]; }
maa=0;
```

```
for(i=1;i<=n;i++) maa+=Math.log(zz[i]/xmin);
alpha=1+n/maa;
alpha=(n-1)*alpha/n+1.0/n;
System.out.println("Power law distribution KS D value="+dd);
if (dd<(1.63/Math.sqrt(n))) System.out.println("Degrees are power law distributed, it is a scale-free complex network");
System.out.println("Power law alpha="+alpha);
System.out.println("Power law xmin="+xmin); }
}
```

2.2 Data source

Interaction Web Database (National Center for Ecological Analysis and Synthesis, 2011; http://www.nceas.ucsb.edu/interactionweb/) was chosen as the data source of the present study. Interaction Web Database contains seven food webs, namely Anemone-Fish, Host-Parasite, Plant-Ant, Plant-Herbivore, Plant-Pollinator, Plant- Seed disperser, and Predator-Prey sub-webs. For each web, the species with corresponding inetrspecific relationship but not all species in the ecosystem or community, were included. Each of seven food webs was used to calculate degree distribution and detect network type.

For Anemone-Fish web, we used the data of Fautin and Allen (1997) and Ollerton et al. (2007), as indicated in Table 3. The data for other webs were chosen as follows:

Host-Parasite webs: we used the data for Canadian freshwater fish and their parasites (Arthur et al., 1976), which were from the investigation to seven water systems. Moreover, the data from Cold Lake (Leong et al., 1981; 10 hosts and 40 parasites) and Parsnip River (Arai et al.,1983; 17 hosts and 53 parasites) were also used.

Plant-Ant web: the data of Bluthgen (2004) from tropical rain forests, Australia, were used. There ware 51 plants and 41 ants in this web.

Plant-Pollinator webs: we used a set of data collected from KwaZulu-Natal, South Africa (Ollerton et al., 2003; 9 plants and 56 pollinators), and the data from Canada (Small, 1976; 13 plants and 34 pollinators)

Plant-Herbivore web: the data from Texas, USA (Joern,1979; 54 plants and 24 herbivores) were used.

Predator-Prey webs: four sets of data (Berwick, Catlins, Coweeta and Venlaw) were used. The major species included algae, fish, arthropods and amphibians.

Plant-Seed disperser webs: two sets of data were used. One from a forest in Papua New Guinea (Beehler, 1983; 31 plants and 9 birds), and one from a tropical forest in Panama (Poulin et al., 1999; 13 plants and 11 birds).

A typical raw data in Interaction Web Database is indicated in Table 1.

Table 1 An example of the data of Interaction Web Database.

Species	Unidentified detritus	Terrestrial invertebrates	Plant material	*Achnanthes lanceolata*
Unidentified detritus	0	0	0	0
Terrestrial invertebrates	0	0	1	0
Plant material	0	1	0	0
Achnanthes lanceolata	0	0	0	0

General information   In this paper, the authors examined the feeding patterns of grasshoppers from two arid grassland communities in Trans-Pecos, Texas. The studies took place from May until November in 1974 and 1975.

Date type   The authors recorded the identities of insect and plant species and their interactions. Data are presented as a binary interaction matrix, in which cells with a "1" indicate an interaction between a pair of species, and a "0" indicates no interaction.

In Table 1, the values 1 and 0 represent having or not having interspecific trophic relationship. The values neither 1 nor 0 represent frequencies and these values were transformed to 1 in present study. Table 1 should be transformed to the format needed by the Java algorithm above, as indicated in Table 2.

Table 2 The data transformed from Table 1.

ID of Taxon 1	ID of Taxon 2	Value
1	2	1
1	3	1
2	3	1
2	4	1
3	4	1

3 Results

The data of Anemone-Fish web is indicated in Table 3.

Table 3 Species and ID of Anemone-Fish web.

Genera	Species	ID	Genera	Species	ID
Amphiprion spp.	Akallopisos	1	Amphiprion spp.	percula	19
Amphiprion spp.	Akindynos	2	Amphiprion spp.	perideraion	20
Amphiprion spp.	Allardi	3	Amphiprion spp.	polymnus	21
Amphiprion spp.	Bicinctus	4	Amphiprion spp.	rubrocinctus	22
Amphiprion spp.	chrysogaster	5	Amphiprion spp.	sandaracinos	23
Amphiprion spp.	chrysopterus	6	Amphiprion spp.	sebae	24
Amphiprion spp.	clarkii	7	Amphiprion spp.	tricinctus	25
Amphiprion spp.	ephippium	8	Premnas	biaculeatus	26
Amphiprion spp.	frenatus	9	Heteractis	crispa	27
Amphiprion spp.	fuscocaudatus	10	Entacmaea	quadricolor	28
Amphiprion spp.	latezonatus	11	Heteractis	magnifica	29
Amphiprion spp.	latifasciatus	12	Stichodactyla	mertensii	30
Amphiprion spp.	leucokranos	13	Heteractis	aurora	31
Amphiprion spp.	mccullochi	14	Stichodactyla	gigantea	32
Amphiprion spp.	melanopus	15	Stichodactyla	haddoni	33
Amphiprion spp.	nigripes	16	Macrodactyla	doreensis	34
Amphiprion spp.	ocellaris	17	Heteractus	malu	35
Amphiprion spp.	omanensis	18	Cryptodendrum	adhaesivum	36

Table 3 is transformed to the data type needed by the Java algorithm, as indicated in Table 4.

Table 4 A data type of Anemone-Fish web.

ID of taxon 1	ID of taxon 2	Value	ID of taxon 1	ID of taxon 2	Value
1	27	1	7	32	1
1	28	1	7	34	1
1	29	1	8	28	1
1	30	1	8	30	1
1	31	1	8	31	1
1	32	1	9	27	1
1	33	1	9	28	1
1	34	1	9	29	1
1	35	1	9	32	1
1	36	1	10	29	1
2	27	1	10	30	1
2	28	1	10	32	1
2	29	1	11	27	1
2	30	1	11	29	1
2	31	1	11	32	1
2	32	1	12	27	1
2	33	1	12	29	1
3	27	1	12	30	1
3	28	1	13	27	1
3	29	1	13	33	1
3	30	1	13	34	1
3	31	1	14	27	1
3	33	1	14	28	1
3	34	1	14	33	1
4	29	1	15	28	1
4	30	1	15	32	1
4	31	1	16	27	1
4	33	1	16	30	1
5	27	1	17	29	1
5	28	1	17	30	1
5	29	1	18	27	1
5	30	1	18	28	1
5	31	1	19	28	1
5	32	1	20	30	1
6	27	1	21	27	1
6	28	1	22	30	1
6	30	1	23	28	1
6	31	1	24	29	1
7	27	1	25	33	1
7	29	1	26	28	1

Some results running the Java algorithm for Anemone-Fish web are as follows:

Skewness of degree distribution: 0.2739383901373063
Aggregation index of the network: 1.519811320754717
It is a complex network.
Variation coefficient H of the network: 3.361428571428571
It is a complex network.
Binomial distribution Chi-square=84779.33198479559
Binomial p=0.2222222222222222
It is likely not a random network
Poisson distribution chi-square=462.75941519476396
Poisson lamda=4.444444444444445
It is likely not a random network

Exponential distribution lamda=0.22499999999999998
Degrees are exponentially distributed.
Power law distribution KS D value=0.0
Degrees are power law distributed, it is a scale-free complex network
Power law alpha=NaN
Power law xmin=14

It is obvious that the food web is a complex network.
The results for all food webs are listed in Table 5 and 6.

Table 5 Summary of results for calculation of degree distribution and network type.

	Anemone-Fish web	Host-Parasite webs			Plant-Ant web	Plant-Pollinator webs	
Data source	Anemone fish	Aishihik Lake	Cold Lake	Parsnip River	Bluthgen, 2004	Ollerton et al,2003	Small, 1976
Skewness of degree distribution	0.2739	0.2524	0.2822	0.2404	0.1626	0.2065	0.2330
Aggregation index of the network	1.5198	1.6913	1.7425	1.6709	1.8597	3.1461	1.3843
Variation coefficient of the network	3.3614	4.0615	3.7425	4.0627	6.3754	7.8742	3.3478
Binomial distribution Chi-square	84779.33	316459.4	1500517	53631.9	346819.3	156.63	536.45
Binomial p	0.2222	0.2099	0.1556	0.1905	0.1715	0.0325	0.2577
Poisson distribution Chi-square	462.759415	538.79	498.22	1661.19	11352.2	1007.7	1197.5
Poisson lamda	4.4444	4.3333	3.6400	4.5143	6.1957	3.1692	6
Exponential distribution lamda	0.2249	0.2308	0.2747	0.2215	0.1614	0.3155	0.1667
Power law distribution KS D value	0	0	0	0	0.1586	0	0
Power law alpha	-	-	-	-	-	-	-
Power law Xmin	14	15	15	17	6	35	18
Type of degree distribution	Exponential, power law	Power law	Power law	Power law	Power law	Power law	Power law
Network type	Complex network	Complex network	Complex network	Complex network	Complex network	Complex network	Complex network

From variation coefficient and aggregation index in Table 5 and 6, we can find that all values are greater than 1 and all webs are thus complex networks. Plant-Pollinator web (Ollerton et al, 2003) is the most complex, seconded by Predator-prey web (Catlins) and Plat-Ant web (Bluthgen, 2004), the complexity of Plant-Seed disperser web (Poulin, 1999) is the lowest. It can be fond that the skewness of Plant-Pollinator web (Ollerton et al, 2003) is the smallest and its degree distribution is the most skewed, which reveals it is the most complex network.

The results show that the degree distribution of most of the food webs is power law and exponential distribution, and all of the food webs are complex networks.

Table 6 Summary of results for calculation of degree distribution and network type.

Data source	P-H web Joern, 1979	Plant-Seed disperser webs			Predator-Prey webs		
		Veehler 1983	Poulin, 1999	Berwick	Catlins	Coweeta1	Venlaw
Skewness of degree distribution	0.2261	0.1969	0.3612	0.1735	0.2069	0.2313	0.1918
Aggregation index of the network	1.8139	1.6254	1.2334	1.7743	2.0198	1.8211	1.7875
Variation coefficient H of the network	4.6468	4.8003	2.0656	5.7552	5.6528	4.6157	5.3199
Binomial distribution Chi-square	48066.1	1451.8	898.8	256.2	180662.6	382.9	3897.6
Binomial p	0.1182	0.2	0.3333	0.1252	0.1043	0.0958	0.1643
Poisson distribution Chi-square	2167.1	1306.7	64.4	8483.6	945.7	1286.7	2241.8
Poisson lamda	4.4359	5.95	4.4167	6.0759	4.4898	4.3448	5.4203
Exponential distribution lamda	0.2254	0.1681	0.2264	0.1646	0.2227	0.2302	0.1845
Power law distribution KS D value	0	0	0	0	0	0	0
Power law alpha	-	2.9967	-	-	-	-	-
Power law Xmin	23	6	11	35	27	26	26
Type of degree distribution	Power law	Exponential, power law	Exponential, power law	Exponential, power law	Exponential, power law	Power law	Exponential, power law
Network type	Complex network	Complex network	Complex network	Complex network	Complex network	Complex network	Complex network

Note: P-H web means Plant-Herbivore web.

4 Discussion

Different from classical distribution patterns (bionomial distri., Poisson distri., and power law distri., etc.), both network type and network complexity can be calculated and compared using the indices above, i.e., aggregation index, coefficient of variation, skewness, etc. We suggest they should be used in the network analysis.

Other indices to detect aggregation strength can also be used in network analysis. For example, the Lloyd index:

$$L=1+(s^2-\bar{u})/\bar{u}^2,$$

where $\bar{u}$, s^2: mean and variance of network degree. The network is a random network, if $L \leq 1$. It is a complex network, if $L>1$, and network complexity increases with L. It is obvious that at certain extent the entropy E, defined above, is equivalent to L.

References

Allesina S, Bodini A, Pascual M. 2005. Functional links and robustness in food webs. Philosophical Transactions of the Royal Society B, 364(1524): 1701-1709

Arai HP, Mudry DR. 1983. Protozoan and metazoan parasites of fishes from the headwaters of the Parsnip and McGregor Rivers, British Columbia: a study of possible parasite transfaunations. Canadian Journal of Fisheries and Aquatic Sciences, 40: 1676-1684

Arthur JR, Margolis L, Arai HP. 1976. Parasites of fishes of Aishihik and Stevens Lakes, Yukon Territory, and potential consequences of their interlake transfer through a proposed water diversion for hydroelectrical purposes. Journal of the Fisheries Research Board of Canada, 33: 2489-2499

Barabasi AL. 2009. Scale-free networks: a decade and beyond, Science, 325: 412-413

Barabasi AL, Albert R. 1999. Emergence of scaling in random networks. Science, 286: 509-512

Beehler B. 1983. Frugivory and polygamy in birds of paradise. The Auk, 100(1): 1-12

Bluthgen N, Stork NE, Fiedler K. 2004. Bottom-up control and co-occurrence in complex communities: honeydew and nectar determine a rainforest ant mosaic. Oikos, 106: 344-358

Dunne JA, Williams RJ, Martinez ND. 2002. Food-web structure and network theory: the role of connectance and size. Ecology, 99(20): 12917-12922

Fautin DG, Allen GR. 1997. Field Guides to Anemone Fishes and Their Host Sea Anemones. Western Australian Museum, Australia

Gonenc IE, Koutitonsky VG, Rashleigh B. 2007. Assessment of the Fate and Effects of Toxic Agents on Water Resources. Springer

MacArthur R. 1955. Fluctuation of animal populations and a measure of community stability. Ecology, 36(3): 533-536

Joern A. 1979. Feeding patterns in grasshoppers (Orthoptera: Acrididae): factors influencing diet specialization. Oecologia, 38: 325-347

Ibrahim SS, Eldeeb MAR, Rady MAH. 2011. The role of protein interaction domains in the human cancer network. Network Biology, 2011, 1(1): 59-71

Goemann B, Wingender E, Potapov AP. 2011. Topological peculiarities of mammalian networks with different functionalities: transcription, signal transduction and metabolic networks. Network Biology, 1(3-4): 134-148

Kuang WP, Zhang WJ. 2011. Some effects of parasitism on food web structure: a topological analysis. Network Biology, 1(3-4): 171-185

Leong TS, Holmes JC. 1981. Communities of metazoan parasites in open water fishes of Cold Lake, Alberta. Journal of Fish Biology, 18: 693-713

Martínez-Antonio A. 2011. *Escherichia coli* transcriptional regulatory network. Network Biology, 2011, 1(1): 21-33

National Center for Ecological Analysis and Synthesis. 2011. Interaction Web Database [DB/QL]. www.nceas.ucsb.edu/interactionweb/html/datasets.html

Odum EP. 1983. Basic Ecology. Saunders College Publishing, Philadelphia, USA

Ollerton J, Johnson SD, Cranmer L, et al. 2003. The pollination ecology of an assemblage of grassland asclepiads in South Africa. Annals of Botany, 92: 807-834

Ollerton J, McCollin D, Fautin DG, et al. 2007. Finding NEMO: nestedness engendered by mutualistic organisation in anemonefish and their hosts. Proceedings of the Royal Society of London Series B, 274: 591-598

Paris L, Bazzoni G. 2011. The polarity sub-network in the yeast network of protein-protein interactions.

Network Biology, 1(3-4): 149-158

Pimm SL, Lawton JH, Cohen JE. 1991. Food web patterns and their consequences. Nature, 350: 669-674

Poulin B, Wright SJ, Lefebvre G, et al. 1999. Interspecific synchrony and asynchrony in the fruiting phenologies of congeneric bird-dispersed plants in Panama. Journal of Tropical Ecology, 15: 213-227

Rodriguez A, Infante D. 2011. Characterization *in silico* of flavonoids biosynthesis in *Theobroma cacao* L. Network Biology, 2011, 1(1): 34-45

Rossberg AG, Matsuda H, Amemiya T, et al. 2005. An explanatory model for food-web structure and evolution. Ecological Complexity, 2: 312-321

Small E. 1976. Insect pollinators of the Mer Bleue peat bog of Ottawa. Canadian Field Naturalist, 90:22-28

Sokal RR, Rohlf FJ. 1995. Biometry: the Principles and Practice of Statistics in Biological Research (3rd Edition). W. H. Freeman and Company, New York, USA

Tacutu R, Budovsky A, Yanai H, et al. 2011.Immunoregulatory network and cancer-associated genes: molecular links and relevance to aging. Network Biology, 1(2): 112-120

Townsend CR, Thompson RM, McIntosh AR, et al. 1998. Disturbance, resource supply and food-web architecture in streams. Ecology Letters, 1: 200-209

Zhang WJ. 2007. Methodology on Ecology Research. Sun Yat-sen University, Guangzhou, China

Zhang WJ. 2011a. Constructing ecological interaction networks by correlation analysis: hints from community sampling. Network Biology, 1(2): 81-98

Zhang WJ. 2011b. Network Biology: an exciting frontier science. Network Biology, 1(1): 79-80

Zhang WJ. 2012. Computational Ecology: Graphs, Networks and Agent-based Modeling. World Scientific, Singapore

20

Several mathematical methods for identifying crucial nodes in networks

WenJun Zhang

School of Life Sciences, Sun Yat-sen University, Guangzhou 510275, China; International Academy of Ecology and Environmental Sciences, Hong Kong

E-mail: zhwj@mail.sysu.edu.cn, wjzhang@iaees.org

Abstract

Crucial nodes in a network refer to those nodes that their existence is so important in preserving topological structure of the network and they independently determine the network structure. In this study I introduced and proposed several mathematical methods for identifying crucial nodes in networks. They fall into three categories, node perturbation, network analysis, and network dynamics. Node perturbation methods include adjacency matrix index, degree or flow change index, node perturbation index, etc. Network dynamics methods include network evolution modeling, etc. Network analysis methods include node degree, criticality index, branch flourishing index, node importance index, etc. Advantages and advantages of these methods were discussed. Finally, I suggested that some of these methods may also be used to identify crucial links (connections) in networks. In this case, the change of a link refers to presence/absence of a link, or change of flow in the link, etc.

Keywords networks; crucial nodes; identification; node perturbation; network dynamics; network analysis; crucial links (connections); mathematical methods.

1 Introduction

Crucial nodes are a few of nodes that govern the structure of a network (Junker, 2006). Their missing or even small changes will substantially change the network. Identification of crucial nodes is a fundamental problem in network analysis (Pimm et al., 1991; Montoya et al., 2006; Butts, 2009; Ding, 2012). In this study, I introduced or proposed several mathematical methods for identifying crucial nodes in networks based on previous studies. These methods can be further used in a wider area of network science, such as cancer networks, metabolic networks, etc (Krogan et al., 2006; Ibrahim et al., 2011; Tacutu et al., 2011; Budovsky and Fraifeld, 2012).

2 Methods

Firstly I define the crucial nodes in a network as that with the following features:

(1) Their existence is so important in preserving topological structure of the network (Zhang, 2012a);

(2) They independently determine network structure;

(3) These nodes are closely related to other nodes in the network.

Several mathematical methods for identifying crucial nodes in networks are described as follows.

2.1 Node perturbation index

I define node perturbation index (NP) as

$$NP=dN/dn/N$$

or

$$NP=dN/dn$$

where N: measure of network structure; n: state value or proportion of a known node in the network. Theoretically, NP of all nodes in the network are normally distributed, i.e., $NP{\approx}0$ for most nodes. Crucial nodes have NP much larger or less than 0.

There are many measures of network structure, i.e., total links, total number of nodes, network flow (Latham, 2006), degree distribution (Zhang, 2011; Zhang and Zhan, 2011), aggregation index, coefficient of variation, entropy (Zhang and Zhan, 2011; Zhang, 2012a), and other measures (Paine, 1992; Power et al., 1996; Dunne et al., 2002; Montoya and Sole, 2003; Allesina et al., 2005; Barabasi, 2009).

Another definition of NP is

$$NP=(N_0\text{-}N_t)/N_0/n_0$$

or

$$NP=(N_0\text{-}N_t)/N_0$$

where N_t, N_0: measure of network structure after and before a node is completely removed from the network, respectively; n_0: state value or proportion of the node in the network before the node is removed from the network. $NP{\approx}1$, if the functionality of the node is positively proportional to its state value or proportion in the network; $NP{\approx}\text{-}1$, if the functionality of the node is negatively proportional to its state value or proportion in the network; $NP{>>}1$, if the node is a crucial node.

Node perturbation index, NP, is a general index that can be further materialized in various ways.

2.2 Criticality index

Criticality index is defined as

$$CI_i=\sum_{c=1}^{n}(1+C_{bc})/d_c+\sum_{e=1}^{m}(1+C_{fe})/f_e$$

where CI_i: value of criticality index of node i; n: the number of source nodes directing to target node i; d_c: the number of target nodes of the c-th source node, and C_{bc}: the backward-oriented criticality index of the c-th source node. Similarly, m: the number of target nodes of source node i; f_e: the number of source nodes of the e-th target node, and C_{fe}: the forward-oriented criticality index of the e-th target node.

The nodes with larger CI tend to be crucial nodes. This index is characterized by the following features: (1) considering both forward- and backward-oriented between-node relations; (2) only nodes within the same network can be compared for their relative importance.

I define this index based on the keystone index, etc (Jordán et al., 1999, 2006; Jordán, 2001; Zhang, 2012a).

2.3 Degree change index

I define degree change index as

$$DC_i=\sum_{j=1}^{v}[|(O_{tj}\text{-}O_{0j})/O_{0j}|+|(I_{tj}\text{-}I_{0j})/I_j|]$$

or

$$DC_i=\sum_{j=1}^{v}(|O_{tj}-O_{0j}|+|I_{tj}-I_{0j}|)$$

where DC_i: value of degree change index of node i; v: total number of nodes in the network; O_{tj},O_{0j}: out-degree of node j after and before node i is changed respectively; I_{tj}, I_{0j}: in-degree of node j after and before node i is changed respectively.

The nodes with larger DC tend to be crucial nodes.

2.4 Flow change index

I define flow change index as

$$FC_i=\sum_{j=1}^{v}[|(FO_{tj}-FO_{0j})/FO_{0j}|+|(FI_{tj}-FI_{0j})/FI_j|]$$

or

$$FC_i=\sum_{j=1}^{v}(|FO_{tj}-FO_{0j}|+|FI_{tj}-FI_{0j}|)$$

where FC_i: value of flow change index of node i; v: total number of nodes in the network; FO_{tj},FO_{0j}: outflow of node j after and before node i is changed respectively; FI_{tj}, FI_{0j}: influx of node j after and before node i is changed respectively. The nodes with larger FC tend to be crucial nodes.

Another flow change index is defined as

$$FC_k=\sum_{i}\sum_{j<i}|f_{ijt}-f_{ij0}|$$

where f_{ijt}, f_{ij0}: flow between node i and j after and before node k is changed. The nodes with larger FC tend to be crucial nodes.

2.5 Adjacency matrix index

Following the definition of Zhang (2012a), suppose the adjacency matrix of a network with v nodes is $D=(d_{ij})_{v\times v}$. If $d_{ij}=d_{ji}=0$, then there is not connection from v_i to v_j; if $d_{ij}=-d_{ji}$, and $|d_{ij}|=1$, then there is only a directed connection from v_i to v_j; if $d_{ij}=d_{ji}=1$, then there is only an undirected connection from v_i to v_j; if $d_{ij}=d_{ji}=2$, then there are two parallel connections from v_i to v_j; if $d_{ii}=3$, then v_i has a loop; if $d_{ii}=4$, then v_i is a isolated node; if $d_{ii}=5$, then v_i is a isolated node and it has a loop. $i,j=1,2,\ldots, v$.

I define adjacency matrix index as

$$AD_k=\sum_{i}\sum_{j}|d_{ijt}-d_{ij0}|$$

where d_{ijt}, d_{ij0}: value of the element d_{ij} after and before node k is changed. The nodes with larger AD tend to be crucial nodes.

2.6 Centrality index

Centrality indices are widely used (Scardoni and Laudanna, 2012). The first centrality index is betweenness centrality (Navia et al., 2010). It measures how central a given node is in terms of being adjacent to many shortest paths in the network. It is based on quantifying how often node i is on the shortest path between each pair of nodes j and k. The standardized centrality index for node i is

$$C_i = 2\sum_{j \le k} g_{jk}(i)/g_{jk}/[(v-1)(v-2)]$$

where $i \ne j$ and k, g_{jk} is the number of equally shortest paths between nodes j and k, and $g_{jk}(i)$ is the number of these shortest paths to which node i is adjacent, v is the total number of nodes. The denominator is twice the number of pairs of nodes without node i. If C_i is large for trophic group i, the loss of this node will have many rapidly spreading effects in the network.

The second centrality index is closeness centrality. It measures how close a node is to the rest of nodes. It is based on the proximity principle and quantifies how short the minimal paths from a given node to all other nodes are (Wassermann and Faust, 1994). The standardized form is

$$CC_i = (v-1)/\sum_{j=1}^{v} d_{ij}$$

where $i \ne j$, and d_{ij} is the length of the shortest path between nodes i and j in the network. The smallest value of CC_i will be for that trophic group that upon being removed will affect the majority of other groups.

2.7 Branch flourishing index

I define the branch flourishing index of a node as

$$BF_i = \sum_{j \ne i} (n_{ij} \times ml_{ij})$$

where BF_i: branch flourishing index of the node i; n_{ij}: the total number of paths (chains) between nodes i and j. ml_{ij}: the mean path (chain) length of all paths (chains) between nodes i and j, $j \ne i$; v: the total number of nodes in the network.

The nodes with larger NS tend to be crucial nodes.

2.8 Node degree

Node degree (number of connections of node) is always treated as the simplest index for measuring node importance. The nodes with more links tend to be crucial nodes.

2.9 Connections and between-node connection strength

Various measures on strength (e.g., correlation such as linear correlation, partial correlation, Spearman correlation) and number of connections (interactions) can be used to determine crucial nodes (Paine, 1980; Zhang, 2007, 2011, 2012b; Ding, 2012). For the statistic networks (Zhang, 2012b), a node with more connections (d_i) and larger mean correlation (mc_i) tends to be a crucial node. For example, we may judge the nodes with both connections and mean correlation larger than that of 95% of other nodes as crucial nodes in the network. A simple index for this criterion is

$$CS_i = d_i \times mc_i$$

Here I propose another index, node importance index, for identifying crucial nodes in statistic networks

$$SC_i = \sum_{j \ne i} d_{ij}$$

where SC_i: node importance index of the node i; d_{ij}: the path (chain) strength between nodes i and j in the network, $j \neq i$; v: the total number of nodes in the network. d_{ij} can be defined in different ways. For example,

$$d_{ij}=\max_{n_{ij}} \prod_t |r_{kl}|$$

where r_{kl}: the correlation between nodes k and l in the path (chain) t between nodes i and j, t=1, 2, ... , n_{ij}; n_{ij}: the total number of paths (chains) between nodes i and j. The nodes with larger SC tend to be crucial nodes.

2.10 Network evolution method

Network evolution modeling (Zhang, 2012c) can be used to find crucial nodes. The nodes that cause greater changes of network structure during network evolution are crucial nodes. Sensitivity analysis can be conducted in network evolution modeling to find crucial nodes. For example, we may change the sequence and time of a node joining the network to investigate its impact on the network.

Other evolution (or succession) methods can also be used (Bond, 1989; Rossberg et al., 2005).

3 Discussion

Above methods fall into three categories, node perturbation, network analysis, and network dynamics. Node perturbation methods, such as adjacency matrix index, degree or flow change index, node perturbation index, etc., identify crucial nodes by comparing structural changes of the network resulted from changes of each node. Therefore these methods need a large amount of experiments. From the view of definition of crucial node, however, they are highly reliable methods. Network dynamics methods include network evolution modeling (e.g., community assembly modeling), etc. These methods need to have a deep insight into mechanism of network dynamics and need to build an ideal model for network evolution. They are also high reliable but a lot of works should be done before they can normally function. Network analysis methods, like node degree, criticality index, centrality index, branch flourishing index, etc., need the information of network itself only, and thus cost much less than other methods. Nevertheless, they identify crucial nodes only by analyzing static connection structure of nodes and are thus less reliable than other methods. Connection strength-connection number method (e.g., node importance index) above is mainly a network analysis method. However, if the connection strength is measured by between-node correlation in the process of network evolution, it is then a network dynamics method.

Some of these methods may also be used to identify crucial links (connections) in networks. In this case, the change of a link refers to presence/absence of a link, or change of flow in the link, etc.

References

Allesina S, Bodini A, Pascual M. 2005. Functional links and robustness in food webs. Philosophical Transactions of the Royal Society B, 364(1524): 1701-1709

Barabasi AL. 2009. Scale-free networks: a decade and beyond. Science, 325: 412-413

Bond WJ. 1989. The tortoise and the hare: ecology of angiosperm dominance and gymnosperm persistence. Biological Journal of the Linnean Society, 36: 227-249

Budovsky A, Fraifeld VE. 2012. Medicinal plants growing in the Judea region: network approach for searching potential therapeutic targets. Network Biology, 2(3): 84-94

Butts CT. 2009. Revisiting the foundations of network analysis. Science, 325: 414-416

Ding DW. 2012. Identification of crucial nodes in biological networks. Network Biology, 2(3): 118-120

Dunne JA, Williams RJ, Martinez ND. 2002. Food-web structure and network theory: the role of connectance and size. Ecology, 99(20): 12917-12922

Huang JQ, Zhang WJ. 2012. Analysis on degree distribution of tumor signaling networks. Network Biology, 2(3): 95-109

Ibrahim SS, Eldeeb MAR, Rady MAH. 2011. The role of protein interaction domains in the human cancer network. Network Biology, 1(1): 59-71

Jordán F, Takacs-Santa A, Molnar I. 1999. Are liability theoretical quest for key stones. Oikos, 86: 453-462

Jordán F. 2001. Trophic fields. Community Ecology, 2: 181-185

Jordán F, LiuW, Davis AJ. 2006. Topological keystone species: Measures of positional importance in food webs. Oikos, 112: 535-546

Junker BH, Koschutzki D, Schreiber F. 2006. Exploration of biological network centralities with CentiBiN. BMC Bioinformatics, 7: 219

Krogan NJ, Cagney G, Yu HY, et al. 2006. Global landscape of protein complexes in the yeast *Saccharomyces cerevisiae*. Nature, 440: 637-643

Latham LG. 2006. Network flow analysis algorithms. Ecological Modelling, 192: 586-600

Montoya JM, Pimm SL, Sole RV. 2006. Ecological networks and their fragility. Nature, 442: 259-264

Montoya JM, Sole RV. 2003. Topological properties of food webs: from real data to community assembly models. Oikos, 102: 614-622

Navia AF, Cortés E, Mejía-Falla PA. 2010. Topological analysis of the ecological importance of elasmobranch fishes: A food web study on the Gulf of Tortugas, Colombia. Ecological Modelling, 221: 2918-2926

Pimm SL, Lawton JH, Cohen JE. 1991. Food web patterns and their consequences. Nature, 350: 669-674

Paine RT. 1980. Food webs: linkage, interaction strength and community infrastructure. Journal of Animal Ecology, 49: 667-686

Paine RT. 1992. Food-web analysis through field measurement of per capita interaction strength. Nature, 355: 73-75

Power ME, Tilman D, Estes JA, et al. 1996. Challenges in the quest for keys. Bioscience, 46: 609-620

Rossberg AG, Matsuda H, Amemiya T, et al. 2005. An explanatory model for food-web structure and evolution. Ecological Complexity, 2: 312-321

Scardoni G, Laudanna C. 2012. Centralities based analysis of complex networks. In: New Frontiers in Graph Theory (Zhang YG, ed). 323-348, InTech, Crotia

Tacutu R, Budovsky A, Yanai H, et al. 2011.Immunoregulatory network and cancer-associated genes: molecular links and relevance to aging. Network Biology, 1(2): 112-120

Wasserman S, Faust K. 1994. Social Network Analysis: Methods and Applications. Cambridge University Press, Cambridge, UK

Zhang WJ. 2007. Computer inference of network of ecological interactions from sampling data. Environmental Monitoring and Assessment, 124: 253-261

Zhang WJ. 2011. Constructing ecological interaction networks by correlation analysis: hints from community sampling. Network Biology, 1(2): 81-98

Zhang WJ, Zhan CY. 2011. An algorithm for calculation of degree distribution and detection of network type: with application in food webs. Network Biology, 1(3-4): 159-170

Zhang WJ. 2012a. Computational Ecology: Graphs, Networks and Agent-based Modeling. World Scientific, Singapore

Zhang WJ. 2012b. How to construct the statistic network? An association network of herbaceous plants constructed from field sampling. Network Biology, 2(2): 57-68

Zhang WJ. 2012c. Modeling community succession and assembly: A novel method for network evolution. Network Biology, 2(2): 69-78

PERMISSIONS

LIST OF CONTRIBUTORS

Mohamed Ragab Abdel Gawwad, Jasmin Šutković, Emina Zahirović, Faruk Berat Akcesme and Betul Akcesme
Genetics and Bioengineering department, International University of Sarajevo, Ilidza, 71220 Bosnia and Herzegovina

Lizhi Zhang
Department of Molecular Genetics, The Ohio State University, 484 West 12th Avenue, Columbus, OH 43210, USA

Alessandro Ferrarini
Department of Evolutionary and Functional Biology, University of Parma, Via G. Saragat 4, I-43100 Parma, Italy

Carrie J. Byron
Department of Marine Sciences, University of New England, 11 Hills Beach Road, Biddeford, ME 04005, USA

Craig Tennenhouse
Department of Mathematical Sciences, University of New England, 11 Hills Beach Road, Biddeford, ME 04005, USA

Amar Ćemanović, Jasmin Šutković and Mohamed Ragab Abdel Gawwad
Genetics and Bioengineering department, International University of Sarajevo, Ilidza, 71220 Bosnia and Herzegovina

LiQin Jiang and Ping Liang
School of Life Sciences, Sun Yat-sen University, Guangzhou 510275, China

WenJun Zhang
School of Life Sciences, Sun Yat-sen University, Guangzhou 510275, China
International Academy of Ecology and Environmental Sciences, Hong Kong

Jyotsna Dogra, Navdeep Prashar, Shruti Jain and Meenakshi Sood
Department of Electronic and Communication Engineering, Jaypee Institute of Information Technology, Solan-173234, India

Muhammad Shakil, Muhammad Shahzad and H. A. Wahab
Department of Mathematics, Hazara University, Manshera, Pakistan

Muhammad Naeem
Department of Information Technology, Hazara University, Manshera, Pakistan

Saira Bhatti
Department of Mathematics, COMSATS Institute of Information Technology, Abbottabad, Pakistan

Alessandro Ferrarini
Department of Evolutionary and Functional Biology, University of Parma, Via G. Saragat 4, I-43100 Parma, Italy

Muhammad Shakil and H. A. Wahab
Department of Mathematics, Hazara University, Manshera, Pakistan

Ruth S. Pérez-Alfaro, Edgardo Galán-Vásquez and Agustino Martínez-Antonio
Departamento de Ingeniería Genética, Centro de Investigación y de Estudios Avanzados del IPN, Unidad Irapuato, Km. 9.6 Libramiento Norte Carr. Irapuato-León 36821 Irapuato, Guanajuato, México

Moisés Santillán
Centro de Investigación y de Estudios Avanzados del IPN, Unidad Monterrey, Vía del Conocimiento 201, 66600 Apodaca NL, México

YanHong Qi
Sun Yat-sen University Libraries, Sun Yat-sen University, Guangzhou 510275, China

GuangHua Liu
Guangdong AIB Polytech College, Guangzhou 510507, China

ChenYuan Zhan
School of Life Sciences, Sun Yat-sen University, Guangzhou 510275, China

Index